FUNDAMENTALS OF MOLECULAR BIOLOGY

Edited by
Patricia Marques

ArclerPress

Fundamentals of Molecular Biology

Edited by Patricia Marques

Edition 2018

ArclerPress

Published by
Arcler Press LLC
708 3rd Avenue, 6th Floor
New York
NY 10017
United States of America
www.arclerpress.com

ISBN: 978-1-68094-530-0 (Hardback)

Printed in United States of America

Library of Congress Control Number: 2017933170

About the Editor

Patricia Marques

Patricia obtained her PhD form University College Dublin in 2010. Her interests are on Microbiology and Parasitology. She is currently working as a Postdoc at University of Maryland Baltimore, USA on Chlamydia infections

Table of Contents

2 MOLECULAR BIOLOGY

3 ELEMENTS OF MOLECULAR BIOLOGY

4 MOLECULAR BIOLOGY OF THE CELL

5 TECHNIQUES OF MOLECULAR BIOLOGY

6 GENE REGULATION

7 MUTATION

8 MOLECULAR BASIS OF DISEASES AND AGEING

List of Abbreviations

ADAR	Adenosine Deaminase Acting on RNA
ADP	Adenosine diphosphate
ATP	Adenosine triphosphate
bHLH	Basic helix-loop-helix
BrdU	Bromodeoxyuridine
BRE	B recognition element
DFE	distribution of fitness effects
DNA	DNA replication
ELISA	Enzyme-linked immunosorbent assay
ESC	Embryonic stem cell
FEV	Forced expiratory volume in one second
GFP	Green fluorescent protein
GMO	Genetically modified microorganisms
GWAS	Genome-wide association studies
HATs	Histone acetyltransferase enzymes
HGVS	Human Genome Variation Society
HPLC	High-performance liquid chromatography
HR	Homologous recombinational repair
IR	Ionizing radiation
ISCN	International System for Human Cytogenetic Nomenclature
kbp	Kilo base pairs
KDM	Histone lysine demethylase
LABA	Long-acting beta agonists
MDIs	Metered-dose inhalers
MLPA	*Multiplex ligation-dependent probe amplification*
MMEJ	Microhomology-mediated end joining
MREs	MiRNA response elements
NHEJ	Non-homologous end joining
NIA	National Institute on Ageing
NTP	Nucleoside triphosphate
PAR	PolyADP-ribose
PCA	Polymerase Cycling Assembly

PCNA	Proliferating cell nuclear antigen
PCR	Polymerase chain reaction
PEFR	Peak expiratory flow rate
QDs	Quantum dots
RRM	RNA recognition motif
RT-PCR	Reverse Transcription PCR ,
SABA	Short-acting beta-adrenoceptor agonists
TBP	TATA-binding protein
TD	Type diabetes
tRNAs	transfer RNAs
TSS	Transcription start site
UV	Ultraviolet light

Preface

Molecular Biology is the field of biology that studies the composition, structure and interactions of cellular molecules such as nucleic acids and proteins that carry out the biological processes essential for the cells functions and maintenance. A key area of molecular biology concerns understanding how various cellular systems interact in terms of the way DNA, RNA and protein synthesis function. The specific techniques used in molecular biology are native to the field but may also be combined with methods and concepts concerning genetics and biochemistry, so there is no big distinction made between these disciplines. The field overlaps with other areas of biology and chemistry, particularly genetics and biochemistry. Molecular biology chiefly concerns itself with understanding the interactions between the various systems of a cell, including the interrelationship of DNA, RNA and protein synthesis and learning how these interactions are regulated. Molecular biology is the study of molecular underpinnings of the process of replication, transcription and translation of the genetic material. The central dogma of molecular biology where genetic material is transcribed into RNA and then translated into protein, despite being an oversimplified picture of molecular biology, still provides a good starting point for understanding the field. Much of the work in molecular biology is quantitative, and recently much work has been done at the interface of molecular biology and computer science in bioinformatics and computational biology. The role, function and structure of biomolecules are key areas of focus among biochemists, as is the chemistry behind biological functions and the production of biomolecules. Genetics is concerned with the effects of genes on living organisms, which are often examined through "knock-out" studies, where animal models are designed so that they lack one or more genes compared to a "wild type" or regular phenotype. However, this simplified picture is currently be reconsidered and revised due to new discoveries concerning the roles of RNA.

Content Coverage

Chapter One introduces the molecular biology and its relation to other fields. Researchers in molecular biology use specific techniques native to molecular biology but increasingly combine these with techniques and ideas from genetics and biochemistry.

Chapters Two is central dogma of molecular biology that describe the flow of genetic information in cells from DNA to messenger RNA (mRNA) to protein. It states that genes specify the sequence of mRNA molecules, which in turn specify the sequence of proteins.

Chapter Three and *Four* are an overview of elements of molecular biology, such as nucleic acids and proteins phases of cell cycle and mitosis, respectively

Chapter Five and *Six* present the techniques of molecular biology and regulation of gene expression, respectively. Gene expression includes a wide range of mechanisms that are used by cells to increase or decrease the production of specific gene products (protein or RNA).

Chapters Seven and *Eight* discuss about gene mutation and molecular basis of diseases and ageing, respectively. Ageing refers especially to human beings, many animals, and fungi, whereas for example bacteria, perennial plants and some simple animals are potentially immortal.

CHAPTER 1

Central Dogma of Molecular Biology

Chapter Contents

Cells are fundamental building blocks of living organisms. Cells contain a nucleus, mitochondria and chloroplasts, endoplasmatic reticulum, ribosomes, vacuoles, etc. The nucleus is important organelle because it houses chromosomes which include the DNA. The DNA is in essence a blueprint of the organism as it encodes information needed to synthesize proteins. Molecular biologists would like to understand how human biology works with the hope to treat diseases like cancer. One can look at simpler organisms such as yeasts to understand how human biology works. Admittedly, unicellular yeasts are very different from humans who have approximately 1014 cells. However, the DNA is similar across all living organisms

INTRODUCTION

The central dogma of molecular biology is an explanation of the flow of genetic information within a biological system. It was first stated by Francis Crick in 1958

The Central Dogma. This states that once 'information' has passed into protein it cannot get out again. In more detail, the transfer of information from nucleic acid to nucleic acid, or from nucleic acid to protein may be possible, but transfer from protein to protein, or from protein to nucleic acid is impossible. Information means here the precise determination of sequence, either of bases in the nucleic acid or of amino acid residues in the protein.- —Francis Crick, 1956 and re-stated in a Nature paper published in 1970:

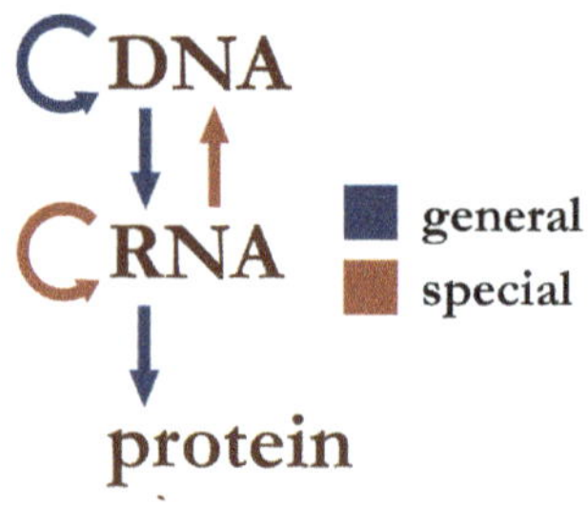

Information flow in biological systems

The central dogma has also been described as "DNA makes RNA and RNA makes protein," originally termed the sequence hypothesis and made as a positive statement by Crick. However, this simplification does not make it clear that the central dogma as stated by Crick does not preclude the reverse flow of information from RNA to DNA, only ruling out the flow from protein to RNA or DNA. Crick's use of the word dogma was unconventional, and has been controversial.

The dogma is a framework for understanding the transfer of sequence information between information-carrying biopolymers, in the most common or general case, in living organisms. There are 3 major classes of such biopolymers: DNA and RNA (both nucleic acids), and protein. There are $3 \times 3 = 9$ conceivable direct transfers of information that can occur between these. The dogma classes these into 3 groups of 3: 3 general transfers (believed to occur normally in most cells), 3 special transfers (known to occur, but only under specific conditions in case of some viruses or in a laboratory), and 3 unknown transfers (believed never to occur). The general transfers describe the normal flow of biological information: DNA can be copied to DNA (DNA replication), DNA information can be copied into mRNA(transcription), and proteins can be synthesized using the information in mRNA as a template (translation).

Biological Sequence Information

The biopolymers that comprise DNA, RNA and (poly)peptides are linear polymers (i.e.: each monomer is connected to at most two other monomers). The sequence of their monomers effectively encodes information. The transfers of information described by the central dogma ideally are faithful, deterministic transfers, wherein one biopolymer's sequence is used as a template for the construction of another biopolymer with a sequence that is entirely dependent on the original biopolymer's sequence.

General Transfers of Biological Sequential Information

Table of the 3 classes of information transfer suggested by the dogma

General	Special	Unknown
DNA → DNA	RNA → DNA	protein → DNA
DNA → RNA	RNA → RNA	protein → RNA
RNA → protein	DNA → protein	protein → protein

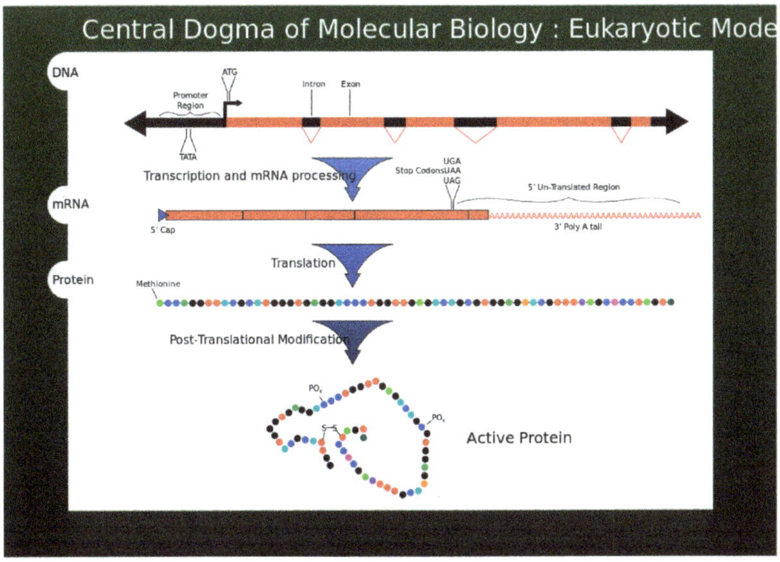

DNA Replications

In the sense that DNA replication must occur if genetic material is to be provided for the progeny of any cell, whether somatic or reproductive, the copying from DNA to DNA arguably is the fundamental step in the central dogma. A complex group of proteins called the replisome performs the replication of the information from the parent strand to the complementary daughter strand.

The replisome comprises:

- a helicase that unwinds the superhelix as well as the double-stranded DNA helix to create a replication fork
- Single-stranded DNA-binding (SSB) protein that binds open the double-stranded DNA to prevent it from reassociating
- RNA primase that adds a complementary RNA primer to each template strand as a starting point for replication
- DNA polymerase III that reads the existing template chain from its 3' end to its 5' end and adds new complementary nucleotides from the 5' end to the 3' end of the daughter chain
- DNA polymerase I that removes the RNA primers and replaces them with DNA.
- DNA ligase that joins the two Okazaki fragments with phosphodiester bonds to produce a continuous chain.

This process typically takes place during S phase of the cell cycle.

Transcription

Transcription is the process by which the information contained in a section of DNA is replicated in the form of a newly assembled piece of messenger RNA (mRNA). Enzymes facilitating the process include RNA polymerase and transcription factors. In eukaryoticcells the primary transcript is pre-mRNA. Pre-mRNA must be processed for translation to proceed. Processing includes the addition of a 5' cap and a poly-A tail to the pre-mRNA chain, followed by splicing. Alternative splicing occurs when appropriate, increasing the diversity of the proteins that any single mRNA can produce. The product of the entire transcription process (that began with the production of the pre-mRNA chain) is a mature mRNA chain.

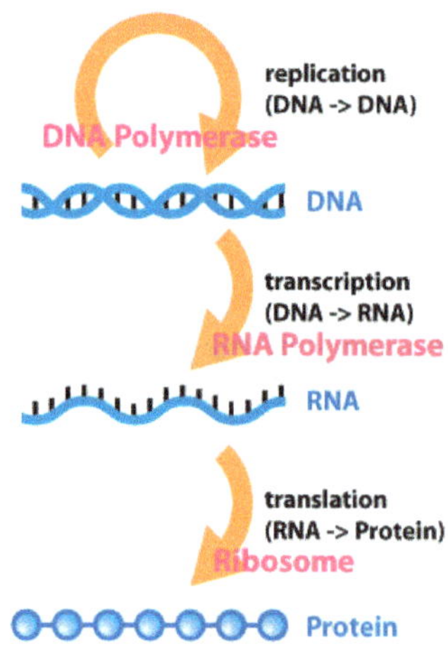

Translation

The mature mRNA finds its way to a ribosome, where it gets translated. In prokaryotic cells, which have no nuclear compartment, the processes of transcription and translation may be linked together without clear separation. In eukaryotic cells, the site of transcription (the cell nucleus) is usually separated from the site of translation (the cytoplasm), so the mRNA must be transported out of the nucleus into the cytoplasm, where it can be bound by ribosomes. The ribosome reads the mRNA triplet codons, usually beginning with an AUG (adenine–uracil–guanine), or initiator methionine codon downstream of the ribosome binding site. Complexes of initiation factors and elongation factors bring aminoacylated transfer RNAs (tRNAs) into the ribosome-mRNA complex, matching the codon in the mRNA to the anti-codon on the tRNA. Each tRNA bears the appropriate amino acid residue to add to the polypeptide chain being synthesised. As the amino acids get linked into the growing peptide chain, the chain begins folding into the correct conformation. Translation ends with a stop codon which may be a UAA, UGA, or UAG triplet.

The mRNA does not contain all the information for specifying the nature of the mature protein. The nascent polypeptide chain released from the ribosome commonly requires additional processing before the final product emerges. For one thing, the correct folding process is complex and vitally important. For most proteins it requires other chaperone proteins to control the form of the product. Some proteins then excise internal segments from their own peptide chains, splicing the free ends that border the gap; in such processes the inside "discarded" sections are called inteins. Other proteins must be split into multiple sections without splicing. Some polypeptide chains need to be cross-linked, and others must be attached to cofactors such as haem (heme) before they become functional.

Special Transfers of Biological Sequential Information

Reverse transcription

Reverse transcription is the transfer of information from RNA to DNA (the reverse of normal transcription). This is known to occur in the case of retroviruses, such as HIV, as well as in eukaryotes, in the case of retrotransposons and telomere synthesis. It is the process by which genetic information from RNA gets transcribed into new DNA.

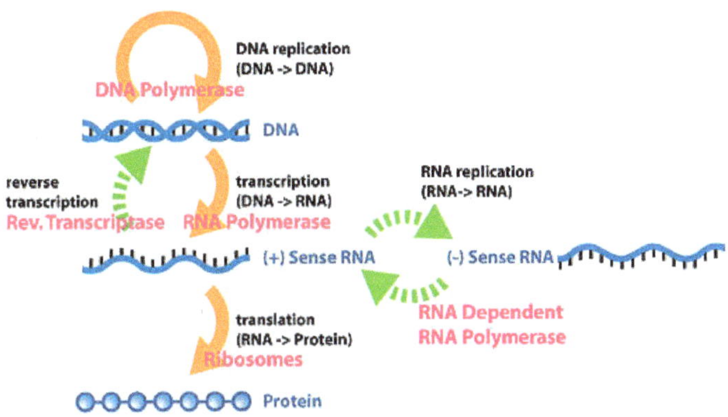

RNA Replication

RNA replication is the copying of one RNA to another. Many viruses replicate this way. The enzymes that copy RNA to new RNA, called RNA-dependent RNA polymerases, are also found in many eukaryotes where they are involved in RNA silencing.

RNA editing, in which an RNA sequence is altered by a complex of proteins and a "guide RNA", could also be seen as an RNA-to-RNA transfer.

Direct Translation from DNA to Protein

Direct translation from DNA to protein has been demonstrated in a cell-free system (i.e. in a test tube), using extracts from *E. coli* that contained ribosomes, but not intact cells. These cell fragments could synthesize proteins from single-stranded DNA templates isolated from other organisms (e,g., mouse or toad), and neomycin was found to enhance this effect. However, it was unclear whether this mechanism of translation corresponded specifically to the genetic code.

Transfers of Information not Explicitly Covered in the Theory

Posttranslational Modification

After protein amino acid sequences have been translated from nucleic acid chains, they can be edited by appropriate enzymes. Although this is a form of protein affecting protein sequence, not explicitly covered by the central dogma, there are not many clear examples where the associated concepts of the two fields have much to do with each other.

Inteins

An intein is a "parasitic" segment of a protein that is able to excise itself from the chain of amino acids as they emerge from the ribosome and rejoin the remaining portions with a peptide bond in such a manner that the main protein "backbone" does not fall apart. This is a case of a protein changing its own primary sequence from the sequence originally encoded by the DNA of a gene. Additionally, most inteins contain a homing endonuclease or HEG domain which is capable of finding a copy of the parent gene that does not include the intein nucleotide sequence. On contact with the intein-free copy, the HEG domain initiates the DNA double-stranded break repair mechanism. This process causes the intein sequence to be copied from the original source gene to the intein-free gene. This is an example of protein directly editing DNA sequence, as well as increasing the sequence's heritable propagation.

Methylation

Variation in methylation states of DNA can alter gene expression levels significantly. Methylation variation usually occurs through the action of DNA methylases. When the change is heritable, it is considered epigenetic. When the change in information status is not heritable, it would be a somatic epitype. The effective information content has been changed by means of the actions of a protein or proteins on DNA, but the primary DNA sequence is not altered.

Prions

Prions are proteins of particular amino acid sequences in particular conformations. They propagate themselves in host cells by making conformational changes in other molecules of protein with the same amino acid sequence, but with a different conformation that is functionally important or detrimental to the organism. Once the protein has been transconformed to the prion folding it changes function. In turn it can convey information into new cells and reconfigure more functional molecules of that sequence into the alternate prion form. In some types of prion in fungi this change is continuous and direct; the information flow is Protein → Protein. Some scientists such as Alain E. Bussard and Eugene Koonin have argued that prion-mediated inheritance violates the central dogma of molecular biology. However, Rosalind Ridley in *Molecular Pathology of the Prions* (2001) has written that "The prion hypothesis is not heretical to the central dogma of molecular biology—that the information necessary to manufacture proteins is encoded in the nucleotide sequence of nucleic acid—because it does not claim that proteins replicate. Rather, it claims that there is a source of information within protein molecules that contributes to their biological function, and that this information can be passed on to other molecules."

Natural Genetic Engineering

James A. Shapiro argues that a superset of these examples should be classified as natural genetic engineering and are sufficient to falsify the central dogma. While Shapiro has received a respectful hearing for his view, his critics have not been convinced that his reading of the central dogma is in line with what Crick intended.

Use of the Term "Dogma"

In his autobiography, *What Mad Pursuit*, Crick wrote about his choice of the word *dogma* and some of the problems it caused him:

I called this idea the central dogma, for two reasons, I suspect. I had already used the obvious word hypothesis in the sequence hypothesis, and in addition I wanted to suggest that this new assumption was more central and more powerful. ... As it turned out, the use of the word dogma caused almost more trouble than it was worth. Many years later Jacques Monod pointed out to me that I did not appear to understand the correct use of the word dogma, which is a belief *that cannot be doubted*. I did apprehend this in a vague sort of way but since I thought that *all* religious beliefs were without foundation, I used the word the way I myself thought about it, not as most of the world does, and simply applied it to a grand hypothesis that, however plausible, had little direct experimental support."

Similarly, Horace Freeland Judson records in *The Eighth Day of Creation*:

"My mind was, that a dogma was an idea for which there was *no reasonable evidence*. You see?!" And Crick gave a roar of delight. "I just didn't *know* what dogma *meant*. And I could just as well have called it the 'Central Hypothesis,' or — you know. Which is what I meant to say. Dogma was just a catch phrase."

DNA REPLICATION

In molecular biology, DNA replication is the biological process of producing two identical replicas of DNA from one original DNA molecule. This process occurs in all living organisms and is the basis for biological inheritance. DNA is made up of a double helix of two complementary strands. During replication, these strands are separated. Each strand of the original DNA molecule then serves as a template for the production of its counterpart, a process referred to as semiconservative replication. Cellular proofreading and error-checking mechanisms ensure near perfect fidelity for DNA replication.

In a cell, DNA replication begins at specific locations, or origins of replication, in the genome. Unwinding of DNA at the origin and synthesis of new strands results in replication forks growing bi-directionally from the origin. A number of proteins are associated with the replication fork to help in the initiation and continuation of DNA synthesis. Most prominently, DNA polymerase synthesizes the new strands by adding nucleotides that complement each (template) strand. DNA replication occurs during the S-stage of interphase.

DNA replication can also be performed *in vitro* (artificially, outside a cell). DNA polymerases isolated from cells and artificial DNA primers can be used to initiate DNA synthesis at known sequences in a template DNA molecule. The polymerase chain reaction (PCR), a common laboratory technique, cyclically applies such artificial synthesis to amplify a specific target DNA fragment from a pool of DNA.

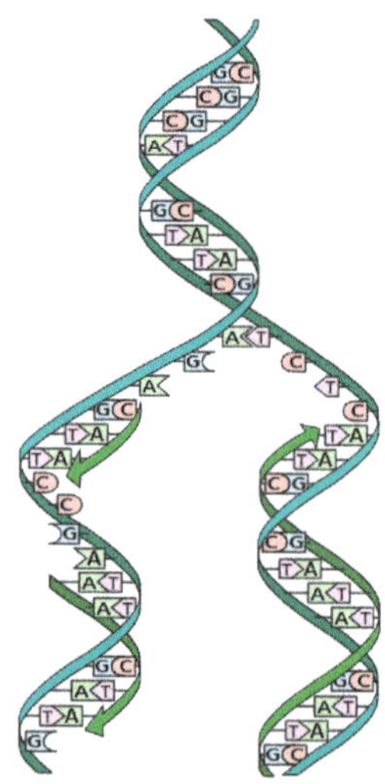

DNA replication. The double helix is unwound and each strand acts as a template (blue) for the next strand. Bases are matched to synthesize the new partner strands (green).

DNA Structures

DNA usually exists as a double-stranded structure, with both strands coiled together to form the characteristic double-helix. Each single strand of DNA is a chain of four types of nucleotides. Nucleotides in DNA contain a deoxyribose sugar, a phosphate, and a nucleobase. The four types of nucleotide correspond to the four nucleobases adenine, cytosine, guanine, and thymine, commonly abbreviated as A,C, G and T. Adenine and guanine are purine bases, while cytosine and thymine are pyrimidines. These nucleotides form phosphodiester bonds, creating the phosphate-deoxyribose backbone of the DNA double helix with the nuclei bases pointing inward (i.e., toward the opposing strand).

Nucleotides (bases) are matched between strands through hydrogen bonds to form base pairs. Adenine pairs with thymine (two hydrogen bonds), and guanine pairs with cytosine (stronger: three hydrogen bonds).

DNA strands have a directionality, and the different ends of a single strand are called the "3' (three-prime) end" and the "5' (five-prime) end". By convention, if the base sequence of a single strand of DNA is given, the left end of the sequence is the 5' end, while the right end of the sequence is the 3' end. The strands of the double helix are anti-parallel with one being 5' to 3, and the opposite strand 3' to 5'. These terms refer to the carbon atom in deoxyribose to which the next phosphate in the chain attaches. Directionality has consequences in DNA synthesis, because DNA polymerase can synthesize DNA in only one direction by adding nucleotides to the 3' end of a DNA strand.

The pairing of complementary bases in DNA (through hydrogen bonding) means that the information contained within each strand is redundant. Phosphodiester (intra-strand) bonds are stronger than hydrogen (inter-strand) bonds. This allows the strands to be separated from one another. The nucleotides on a single strand can therefore be used to reconstruct nucleotides on a newly synthesized partner strand.

DNA Polymerase

DNA polymerases are a family of enzymes that carry out all forms of DNA replication. DNA polymerases in general cannot initiate synthesis of new strands, but can only extend an existing DNA or RNA strand paired with a template strand. To begin synthesis, a short fragment of RNA, called a primer, must be created and paired with the template DNA strand.

DNA polymerase adds a new strand of DNA by extending the 3' end of an existing nucleotide chain, adding new nucleotidesmatched to the template strand one at a time via the creation of phosphodiester bonds. The energy for this process of DNA polymerization comes from hydrolysis of the high-energy phosphate (phosphoanhydride) bonds between the three phosphates attached to each unincorporated base. Free bases with their attached phosphate groups are called nucleotides; in particular, bases with three attached phosphate groups are called nucleoside triphosphates. When a nucleotide is being added to a growing DNA strand, the formation of a phosphodiester bond between the proximal phosphate of the nucleotide to the growing chain is accompanied by hydrolysis of a high-energy phosphate bond with release of the two distal phosphates as a pyrophosphate. Enzymatic hydrolysis of the resulting pyrophosphate into inorganic phosphate consumes a second high-energy phosphate bond and renders the reaction effectively irreversible.

In general, DNA polymerases are highly accurate, with an intrinsic error rate of less than one mistake for every 10^7 nucleotides added. In addition, some DNA polymerases also have proofreading ability; they can remove nucleotides from the end of a growing strand in order to correct mismatched bases. Finally, post-replication mismatch repair mechanisms monitor the DNA for errors, being capable of distinguishing mismatches in the newly synthesized DNA strand from the original strand sequence. Together, these three discrimination steps enable replication fidelity of less than one mistake for every 10^9 nucleotides added.

The rate of DNA replication in a living cell was first measured as the rate of phage T4 DNA elongation in phage-infected E. coli.During the period of exponential DNA increase at 37 °C, the rate was 749 nucleotides per second. The mutation rate per base pair per replication during phage T4 DNA synthesis is 1.7 per 10^8.

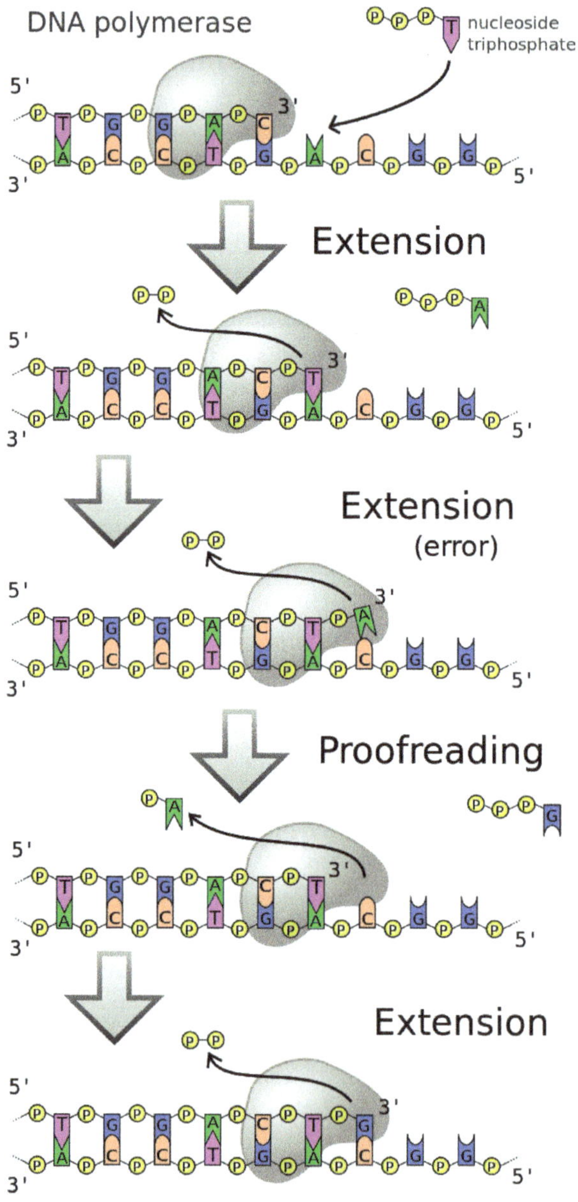

DNA polymerases adds nucleotides to the 3' end of a strand of DNA. If a mismatch is accidentally incorporated, the polymerase is inhibited from further extension. Proofreading removes the mismatched nucleotide and extension continues.

Replication Process

DNA replication, like all biological polymerization processes, proceeds in three enzymatically catalyzed and coordinated steps: initiation, elongation and termination.

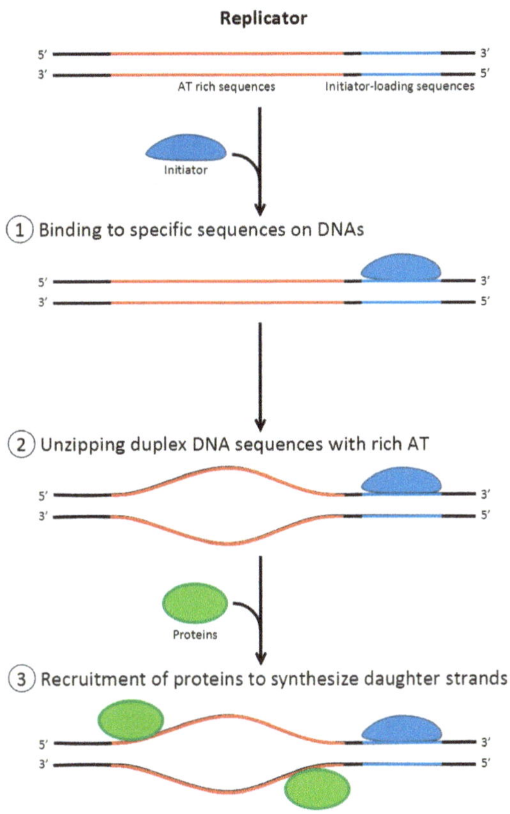

Role of initiators for initiation of DNA replication.

Initiation

For a cell to divide, it must first replicate its DNA. This process is initiated at particular points in the DNA, known as "origins", which are targeted by initiator proteins. In *E. coli* this protein is DnaA; in yeast, this is the origin recognition complex. Sequences used by initiator proteins tend to be "AT-rich" (rich in adenine and thymine bases), because A-T base pairs have two hydrogen bonds (rather than the three formed in a C-G pair) and thus are easier to strand separate. Once the origin has been located, these initiators recruit other proteins and form the pre-replication complex, which unzips the double-stranded DNA.

Elongation

DNA polymerase has 5'-3' activity. All known DNA replication systems require a free 3' hydroxyl group before synthesis can be initiated (note: the DNA template is read in 3' to 5' direction whereas a new strand is synthesized in the 5' to 3' direction—this is often confused). Four distinct mechanisms for DNA synthesis are recognized:

- All cellular life forms and many DNA viruses, phages and plasmids use a primase to synthesize a short RNA primer with a free 3' OH group which is subsequently elongated by a DNA polymerase.
- The retroelements (including retroviruses) employ a transfer RNA that primes DNA replication by providing a free 3′ OH that is used for elongation by the reverse transcriptase.
- In the adenoviruses and the φ29 family of bacteriophages, the 3' OH group is provided by the side chain of an amino acid of the genome attached protein (the terminal protein) to which nucleotides are added by the DNA polymerase to form a new strand.
- In the single stranded DNA viruses — a group that includes the circoviruses, the geminiviruses, the parvoviruses and others — and also the many phages and plasmids that use the rolling circle replication (RCR) mechanism, the RCR endonuclease creates a nick in the genome strand (single stranded viruses) or one of the DNA strands (plasmids). The 5′ end of the nicked strand is transferred to a tyrosine residue on the nuclease and the free 3′ OH group is then used by the DNA polymerase to synthesize the new strand.

The first is the best known of these mechanisms and is used by the cellular organisms. In this mechanism, once the two strands are separated, primase adds RNA primers to the template strands. The leading strand receives one RNA primer while the lagging strand receives several. The leading strand is continuously extended from the primer by a DNA polymerase with high processivity, while the lagging strand is extended discontinuously from each primer forming Okazaki fragments. RNase removes the primer RNA fragments, and a low processivity DNA polymerase distinct from the replicative polymerase enters to fill the gaps. When this is complete, a single nick on the leading strand and several nicks on the lagging strand can be found. Ligase works to fill these nicks in, thus completing the newly replicated DNA molecule.

The primase used in this process differs significantly between bacteria and archaea/eukaryotes. Bacteria use a primase belonging to the DnaG protein superfamily which contains a catalytic domain of the TOPRIM fold type. The TOPRIM fold contains an α/β core with four conserved strands in a Rossmann-like topology. This structure is also found in the catalytic domains of topoisomerase Ia, topoisomerase II, the OLD-family nucleases and DNA repair proteins related to the RecR protein.

The primase used by archaea and eukaryotes, in contrast, contains a highly derived version of the RNA recognition motif (RRM). This primase is structurally similar to many viral RNA-dependent RNA polymerases, reverse transcriptases, cyclic nucleotide generating cyclases and DNA polymerases of the A/B/Y families that are involved in DNA replication and repair. In eukaryotic replication, the primase forms a complex with Pol α.

Multiple DNA polymerases take on different roles in the DNA replication process. In *E. coli*, DNA Pol III is the polymerase enzyme primarily responsible for DNA replication. It assembles into a replication complex at the replication fork that exhibits extremely high processivity, remaining intact for the entire replication cycle. In contrast, DNA Pol I is the enzyme responsible for replacing RNA primers with DNA. DNA Pol I has a 5' to 3' exonuclease activity in addition to its polymerase activity, and uses its exonuclease activity to degrade the RNA primers ahead of it as it extends the DNA strand behind it, in a process called nick translation. Pol I is much less processive than Pol III because its primary function in DNA replication is to create many short DNA regions rather than a few very long regions.

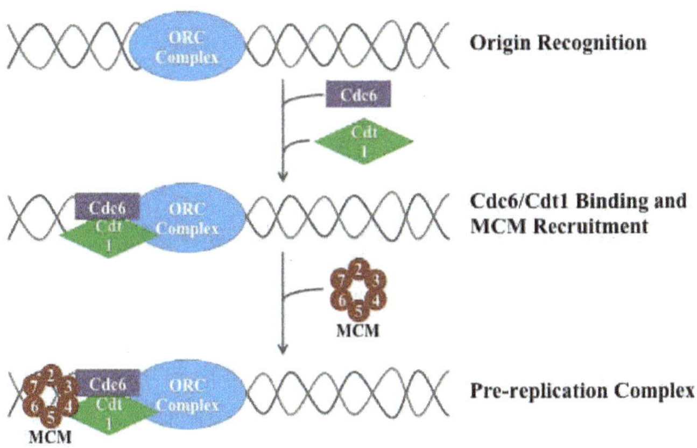

Formation of pre-replication complex.

In eukaryotes, the low-processivity enzyme, Pol α, helps to initiate replication because it forms a complex with primase. In eukaryotes, leading strand synthesis is thought to be conducted by Pol ε; however, this view has recently been challenged, suggesting a role for Pol δ. Primer removal is completed Pol δ while repair of DNA during replication is completed by Pol ε.

As DNA synthesis continues, the original DNA strands continue to unwind on each side of the bubble, forming a replication fork with two prongs. In bacteria, which have a single origin of replication on their circular chromosome, this process creates a "theta structure" (resembling the Greek letter theta: θ). In contrast, eukaryotes have longer linear chromosomes and initiate replication at multiple origins within these.>

Replication Fork

The replication fork is a structure that forms within the nucleus during DNA replication. It is created by helicases, which break the hydrogen bonds holding the two DNA strands together. The resulting structure has two branching "prongs", each one made up of a single strand of DNA. These two strands serve as the template for the leading and lagging strands, which will be created as DNA polymerase matches complementary nucleotides to the templates; the templates may be properly referred to as the leading strand template and the lagging strand template.

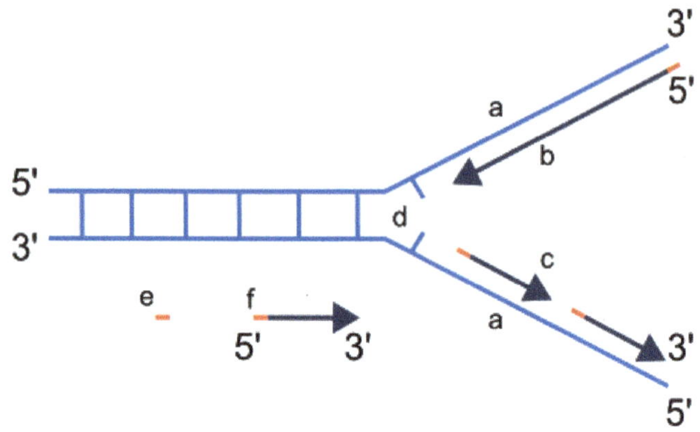

Scheme of the replication fork. a: template, b: leading strand, c: lagging strand, d: replication fork, e: primer, f: Okazaki fragments

DNA is always synthesized in the 5' to 3' direction. Since the leading and lagging strand templates are oriented in opposite directions at the replication fork, a major issue is how to achieve synthesis of nascent (new) lagging strand DNA, whose direction of synthesis is opposite to the direction of the growing replication fork.

Leading strand

The leading strand is the strand of nascent DNA which is being synthesized in the same direction as the growing replication fork. A polymerase "reads" the leading strand *template* and adds complementary nucleotides to the nascent leading strand on a continuous basis.

Lagging strand

The lagging strand is the strand of nascent DNA whose direction of synthesis is opposite to the direction of the growing replication fork. Because of its orientation, replication of the lagging strand is more complicated as compared to that of the leading strand. As a consequence, the DNA polymerase on this strand is seen to "lag behind" the other strand.

The lagging strand is synthesized in short, separated segments. On the lagging strand *template*, a primase "reads" the template DNA and initiates synthesis of a short complementary RNA primer. A DNA polymerase extends the primed segments, forming Okazaki fragments. The RNA primers are then removed and replaced with DNA, and the fragments of DNA are joined together by DNA ligase.

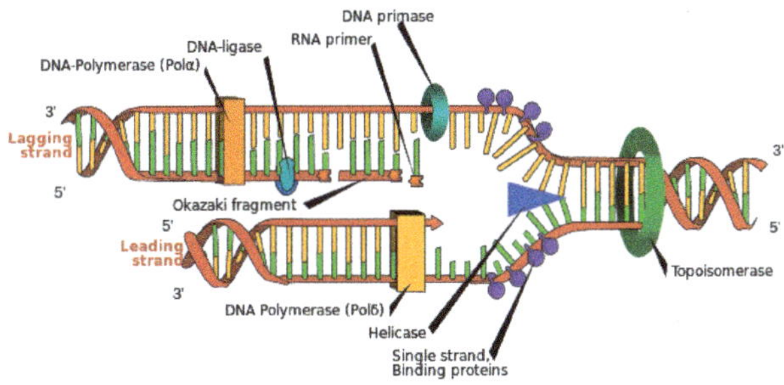

Many enzymes are involved in the DNA replication fork.

Dynamics at the Replication Fork

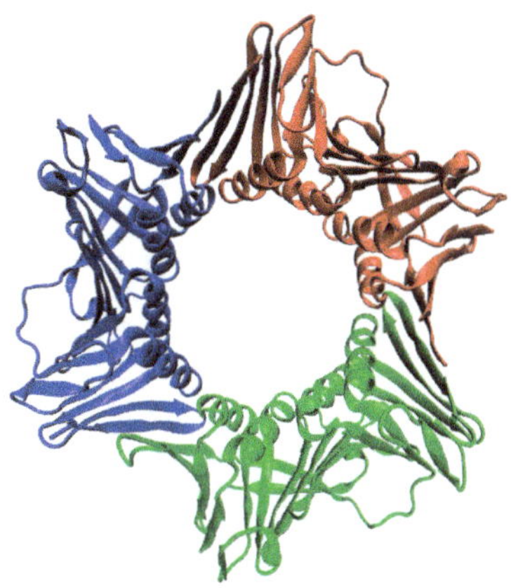

The assembled human DNA clamp, a trimer of the protein PCNA.

As helicase unwinds DNA at the replication fork, the DNA ahead is forced to rotate. This process results in a build-up of twists in the DNA ahead. This build-up forms a torsional resistance that would eventually halt the progress of the replication fork. Topoisomerases are enzymes that temporarily break the strands of DNA, relieving the tension caused by unwinding the two strands of the DNA helix; topoisomerases (including DNA gyrase) achieve this by adding negative supercoils to the DNA helix.

Bare single-stranded DNA tends to fold back on itself forming secondary structures; these structures can interfere with the movement of DNA polymerase. To prevent this, single-strand binding proteins bind to the DNA until a second strand is synthesized, preventing secondary structure formation.

Clamp proteins form a sliding clamp around DNA, helping the DNA polymerase maintain contact with its template, thereby assisting with processivity. The inner face of the clamp enables DNA to be threaded through it. Once the polymerase reaches the end of the template or detects double-stranded DNA, the sliding clamp undergoes a conformational change that releases the DNA polymerase. Clamp-loading proteins are used to initially load the clamp, recognizing the junction between template and RNA primers.

DNA replication proteins

At the replication fork, many replication enzymes assemble on the DNA into a complex molecular machine called the replisome. The following is a list of major DNA replication enzymes that participate in the replisome:

Enzyme	Function in DNA replication
DNA Helicase	Also known as helix destabilizing enzyme. Helicase separates the two strands of DNA at the Replication Fork behind the topoisomerase.
DNA Polymerase	The enzyme responsible for catalyzing the addition of nucleotide substrates to DNA in the 5' to 3' direction during DNA replication. Also performs proof-reading and error correction. There exist many different types of DNA Polymerase, (such as the DNA Polymerase III)each of which perform different functions in different types of cells.
DNA clamp	A protein which prevents elongating DNA polymerases from dissociating from the DNA parent strand.
Single-Strand Binding (SSB) Proteins	Bind to ssDNA and prevent the DNA double helix from re-annealing after DNA helicase unwinds it, thus maintaining the strand separation, and facilitating the synthesis of the nascent strand.
Topoisomerase	Relaxes the DNA from its super-coiled nature.
DNA Gyrase	Relieves strain of unwinding by DNA helicase; this is a specific type of topoisomerase
DNA Ligase	Re-anneals the semi-conservative strands and joins Okazaki Fragments of the lagging strand.
Primase	Provides a starting point of RNA (or DNA) for DNA polymerase to begin synthesis of the new DNA strand.
Telomerase	Lengthens telomeric DNA by adding repetitive nucleotide sequences to the ends of **eukaryotic chromosomes**. This allows germ cells and stem cells to avoid the Hayflick limit on cell division.

Replication Machinery

Replication machineries consist of factors involved in DNA replication and appearing on template ssDNAs. Replication machineries include primosotors are replication enzymes; DNA polymerase, DNA helicases, DNA clamps and DNA topoisomerases, and replication proteins; e.g. single-stranded DNA binding proteins (SSB). In the replication machineries these components coordinate. In most of the bacteria, all of the factors involved in DNA replication are located on replication forks and the complexes stay on the forks during DNA replication. These replication machineries are called replisomes or DNA replicase systems. These terms are generic terms for proteins located on replication forks. In eukaryotic and some bacterial cells the replisomes are not formed.

Since replication machineries do not move relatively to template DNAs such as factories, they are called a replication factory. In an alternative figure, DNA factories are similar to projectors and DNAs are like as cinematic films passing constantly into the projectors. In the replication factory model, after both DNA helicases for leading stands and lagging strands are loaded on the template DNAs, the helicases run along the DNAs into each other. The helicases remain associated for the remainder of replication process. Peter Meister et al. observed directly replication sites in budding yeast by monitoring green fluorescent protein (GFP)-tagged DNA polymerases α. They detected DNA replication of pairs of the tagged loci spaced apart symmetrically from a replication origin and found that the distance between the pairs decreased markedly by time. This finding suggests that the mechanism of DNA replication goes with DNA factories. That is, couples of replication factories are loaded on replication origins and the factories associated with each other. Also, template DNAs move into the factories, which bring extrusion of the template ssDNAs and nascent DNAs. Meister's finding is the first direct evidence of replication factory model. Subsequent research has shown that DNA helicases form dimers in many eukaryotic cells and bacterial replication machineries stay in single intranuclear location during DNA synthesis.

The replication factories perform disentanglement of sister chromatids. The disentanglement is essential for distributing the chromatids into daughter cells after DNA replication. Because sister chromatids after DNA replication hold each other by **Cohesin** rings, there is the only chance for the disentanglement in DNA replication. Fixing of replication machineries as replication factories can improve the success rate of DNA replication. If replication forks move freely in chromosomes, catenation of nuclei is aggravated and impedes mitotic segregation.

Termination

Eukaryotes initiate DNA replication at multiple points in the chromosome, so replication forks meet and terminate at many points in the chromosome; these are not known to be regulated in any particular way. Because eukaryotes have linear chromosomes, DNA replication is unable to reach the very end of the chromosomes, but ends at the telomereregion of repetitive DNA close to the ends. This shortens the telomere of the daughter DNA strand. Shortening of the telomeres is a normal process in somatic cells. As a result, cells can only divide a certain number of times before the DNA loss prevents further division. (This is known as the Hayflick limit.) Within the germ cell line, which passes DNA to the next generation, telomerase extends the repetitive sequences of the telomere region to prevent degradation. Telomerase can become mistakenly active in somatic cells, sometimes leading to cancer formation. Increased telomerase activity is one of the hallmarks of cancer.

Termination requires that the progress of the DNA replication fork must stop or be blocked. Termination at a specific locus, when it occurs, involves the interaction between two components: (1) a termination site sequence in the DNA, and (2) a protein which binds to this sequence to physically stop DNA replication. In various bacterial species, this is named the DNA replication terminus site-binding protein, or Ter protein.

Because bacteria have circular chromosomes, termination of replication occurs when the two replication forks meet each other on the opposite end of the parental chromosome. *E. coli* regulates this process through the use of termination sequences that, when bound by the Tus protein, enable only one direction of replication fork to pass through. As a result, the replication forks are constrained to always meet within the termination region of the chromosome.

Regulation

Eukaryotes

Within eukaryotes, DNA replication is controlled within the context of the cell cycle. As the cell grows and divides, it progresses through stages in the cell cycle; DNA replication takes place during the S phase (synthesis phase). The progress of the eukaryotic cell through the cycle is controlled by cell cycle checkpoints. Progression through checkpoints is controlled through complex interactions between various proteins, including cyclins and cyclin-dependent kinases. Unlike bacteria, eukaryotic DNA replicates in the confines of the nucleus.

The G1/S checkpoint (or restriction checkpoint) regulates whether eukaryotic cells enter the process of DNA replication and subsequent division. Cells that do not proceed through this checkpoint remain in the G0 stage and do not replicate their DNA.

Replication of chloroplast and mitochondrial genomes occurs independently of the cell cycle, through the process of D-loop replication.

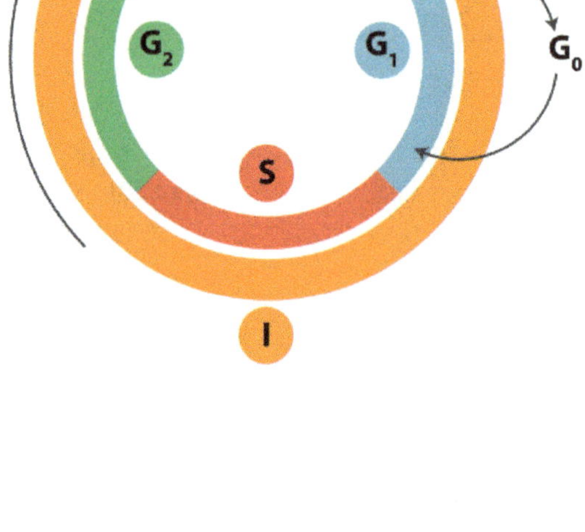

Replication focus

In vertebrate cells, replication sites concentrate into positions called **replication foci**. Replication sites can be detected by immunostaining daughter strands and replication enzymes and monitoring GFP-tagged replication factors. By these methods it is found that replication foci of varying size and positions appear in S phase of cell division and their number per nucleus is far smaller than the number of genomic replication forks.

P. Heun et al.(2001) tracked GFP-tagged replication foci in budding yeast cells and revealed that replication origins move constantly in G1 and S phase and the dynamicsdecreased significantly in S phase. Traditionally, replication sites were fixed on spatial structure of chromosomes by nuclear matrix or lamins. The Heun's results denied the traditional concepts, budding yeasts don't have lamins, and support that replication origins self-assemble and form replication foci.

By firing of replication origins, controlled spatially and temporally, the formation of replication foci is regulated. D. A. Jackson et al.(1998) revealed that neighboring origins fire simultaneously in mammalian cells. Spatial juxtaposition of replication sites brings clustering of replication forks. The clustering do rescue of stalled replication forks and favors normal progress of replication forks. Progress of replication forks is inhibited by many factors; collision with proteins or with complexes binding strongly on DNA, deficiency of dNTPs, nicks on template DNAs and so on. If replication forks stall and the remaining sequences from the stalled forks are not replicated, the daughter strands have nick obtained un-replicated sites. The un-replicated sites on one parent's strand hold the other strand together but not daughter strands. Therefore, the resulting sister chromatids cannot separate from each other and cannot divide into 2 daughter cells. When neighboring origins fire and a fork from one origin is stalled, fork from other origin access on an opposite direction of the stalled fork and duplicate the un-replicated sites. As other mechanism of the rescue there is application of dormant replication origins that excess origins don't fire in normal DNA replication.

Bacteria

Most bacteria do not go through a well-defined cell cycle but instead continuously copy their DNA; during rapid growth, this can result in the concurrent occurrence of multiple rounds of replication. In *E. coli*, the best-characterized bacteria, DNA replication is regulated through several mechanisms, including: the hemimethylation and sequestering of the origin sequence, the ratio of adenosine triphosphate (ATP) to adenosine diphosphate (ADP), and the levels of protein DnaA. All these control the binding of initiator proteins to the origin sequences.

Because *E. coli* methylates GATC DNA sequences, DNA synthesis results in hemimethylated sequences. This hemimethylated DNA is recognized by the protein SeqA, which binds and sequesters the origin sequence; in addition, DnaA (required for initiation of replication) binds less well to hemimethylated DNA. As a result, newly replicated origins are prevented from immediately initiating another round of DNA replication.

ATP builds up when the cell is in a rich medium, triggering DNA replication once the cell has reached a specific size. ATP competes with ADP to bind to DnaA, and the

DnaA-ATP complex is able to initiate replication. A certain number of DnaA proteins are also required for DNA replication — each time the origin is copied, the number of binding sites for DnaA doubles, requiring the synthesis of more DnaA to enable another initiation of replication.

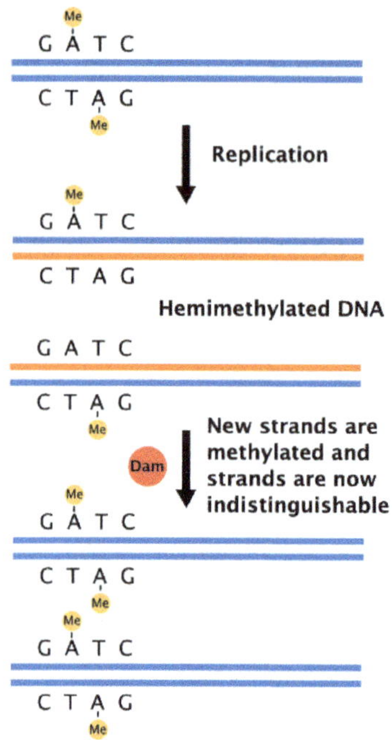

Dam methylates adenine of GATC sites after replication.

Polymerase Chain Reaction

Researchers commonly replicate DNA *in vitro* using the polymerase chain reaction (PCR). PCR uses a pair of primers to span a target region in template DNA, and then polymerizes partner strands in each direction from these primers using a thermostable DNA polymerase. Repeating this process through multiple cycles amplifies the targeted DNA region. At the start of each cycle, the mixture of template and primers is heated, separating the newly synthesized molecule and template. Then, as the mixture cools, both of these become templates for annealing of new primers, and the polymerase extends from these. As a result, the number of copies of the target region doubles each round, increasing exponentially.

TRANSCRIPTION BIOLOGY

Transcription is the first step of gene expression, in which a particular segment of DNA is copied into RNA (especially mRNA) by the enzyme RNA polymerase. Both DNA and RNA are nucleic acids, which use base pairs of nucleotides as a complementary language.

During transcription, a DNA sequence is read by an RNA polymerase, which produces a complementary, antiparallel RNA strand called a primary transcript.

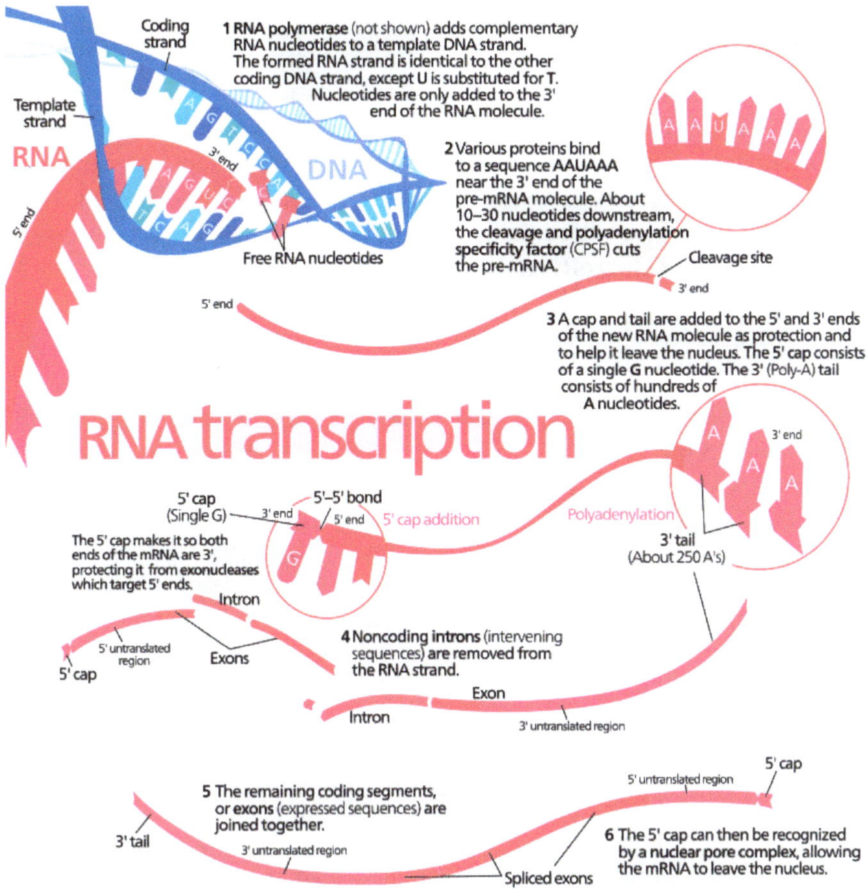

Simplified diagram of mRNA synthesis and processing. Enzymes not shown.

Transcription proceeds in the following general steps:

- RNA polymerase, together with one or more general transcription factors, binds to promoter DNA.
- RNA polymerase creates a transcription bubble, which separates the two strands of the DNA helix. This is done by breaking the hydrogen bonds between complementary DNA nucleotides.
- RNA polymerase adds RNA nucleotides (which are complementary to the nucleotides of one DNA strand).
- RNA sugar-phosphate backbone forms with assistance from RNA polymerase to form an RNA strand.

- Hydrogen bonds of the RNA–DNA helix break, freeing the newly synthesized RNA strand.
- If the cell has a nucleus, the RNA may be further processed. This may include polyadenylation, capping, and splicing.
- The RNA may remain in the nucleus or exit to the cytoplasm through the nuclear pore complex.

The stretch of DNA transcribed into an RNA molecule is called a *transcription unit* and encodes at least one gene. If the gene encodes a protein, the transcription produces messenger RNA (mRNA); the mRNA in turn serves as a template for the protein's synthesis through translation. Alternatively, the transcribed gene may encode for either non-coding RNA (such as microRNA), ribosomal RNA (rRNA), transfer RNA (tRNA), or other enzymatic RNA molecules called ribozymes. Overall, RNA helps synthesize, regulate, and process proteins; it therefore plays a fundamental role in performing functions within a cell.

In virology, the term may also be used when referring to mRNA synthesis from an RNA molecule (i.e., RNA replication). For instance, the genome of a negative-sense single-stranded RNA (ssRNA -) virus may be template for a positive-sense single-stranded RNA (ssRNA +). This is because the positive-sense strand contains the information needed to translate the viral proteins for viral replication afterwards. This process is catalyzed by a viral RNA replicase.

Background

A DNA transcription unit encoding for a protein may contain both a *coding sequence*, which will be translated into the protein, and *regulatory sequences*, which direct and regulate the synthesis of that protein. The regulatory sequence before ("upstream" from) the coding sequence is called the five prime untranslated region (5'UTR); the sequence after ("downstream" from) the coding sequence is called the three prime untranslated region (3'UTR).

As opposed to DNA replication, transcription results in an RNA complement that includes the nucleotide uracil (U) in all instances where thymine (T) would have occurred in a DNA complement.

Only one of the two DNA strands serve as a template for transcription. The antisense strand of DNA is read by RNA polymerase from the 3' end to the 5' end during transcription (3' → 5'). The complementary RNA is created in the opposite direction, in the 5' → 3' direction, matching the sequence of the sense strand with the exception of switching uracil for thymine. This directionality is because RNA polymerase can only add nucleotides to the 3' end of the growing mRNA chain. This use of only the 3' → 5' DNA strand eliminates the need for the Okazaki fragments that are seen in DNA replication. This also removes the need for an RNA primer to initiate RNA synthesis, as is the case in DNA replication.

The *non*-template (sense) strand of DNA is called the coding strand, because its sequence is the same as the newly created RNA transcript (except for the substitution of uracil for thymine). This is the strand that is used by convention when presenting a DNA sequence.

Transcription has some proofreading mechanisms, but they are fewer and less effective than the controls for copying DNA; therefore, transcription has a lower copying fidelity than DNA replication.

Major Steps

Transcription is divided into *initiation*, *promoter escape*, *elongation* and *termination*.

Initiation

Transcription begins with the binding of RNA polymerase, together with one or more general transcription factor, to a specific DNA sequence referred to as a "promoter" to form an RNA polymerase-promoter "closed complex" (called a "closed complex" because the promoter DNA is fully double-stranded).

RNA polymerase, assisted by one or more general transcription factors, then unwinds approximately 14 base pairs of DNA to form an RNA polymerase-promoter "open complex" (called an "open complex" because the promoter DNA is partly unwound and single-stranded) that contains an unwound, single-stranded DNA region of approximately 14 base pairs referred to as the "transcription bubble." RNA polymerase, assisted by one or more general transcription factors, then selects a transcription start site in the transcription bubble, binds to an initiating NTP and an extending NTP (or a short RNA primer and an extending NTP) complementary to the transcription start site sequence, and catalyzes bond formation to yield an initial RNA product.

In bacteria, RNA polymerase core enzyme consists of five subunits: 2 α subunits, 1 β subunit, 1 β' subunit, and 1 ω subunit. In bacteria, there is one general RNA transcription factor: sigma. RNA polymerase core enzyme binds to the bacterial general transcription factor sigma to form RNA polymerase holoenzyme and then binds to a promoter.

In archaea and eukaryotes, RNA polymerase contains subunits homologous to each of the five RNA polymerase subunits in bacteria and also contains additional subunits. In archaea and eukaryotes, the functions of the bacterial general transcription factor sigma are performed by multiple general transcription factors that work together. In archaea, there are three general transcription factors: TBP, TFB, and TFE. In eukaryotes, in RNA polymerase II-dependent transcription, there are six general transcription factors: TFIIA, TFIIB (an ortholog of archaeal TFB), TFIID (a multisubunit factor in which the key subunit, TBP, is an ortholog of archaeal TBP), TFIIE (an ortholog of archaeal TFE), TFIIF, and TFIIH. In archaea and eukaryotes, the RNA polymerase-promoter closed complex is usually referred to as the "preinitiation complex."

Transcription initiation is regulated by additional proteins, known as activators and repressors, and, in some cases, associated coactivators or corepressors, which modulate formation and function of the transcription initiation complex.

Promoter Escape

After the first bond is synthesized, the RNA polymerase must escape the promoter. During this time there is a tendency to release the RNA transcript and produce truncated transcripts. This is called abortive initiation, and is common for both eukaryotes and prokaryotes. Abortive initiation continues to occur until an RNA product of a threshold length of approximately 10 nucleotides is synthesized, at which point promoter escape occurs and a transcription elongation complex is formed.

Mechanistically, promoter escape occurs through DNA scrunching, providing the energy needed to break interactions between RNA polymerase holoenzyme and the promoter.

In bacteria, upon and following promoter clearance, the σ factor is released according to a stochastic model.

In eukaryotes, at an RNA polymerase II-dependent promoter, upon promoter clearance, TFIIH phosphorylates serine 5 on the carboxy terminal domain of RNA polymerase II, leading to the recruitment of capping enzyme (CE). The exact mechanism of how CE induces promoter clearance in eukaryotes is not yet known.

Elongation

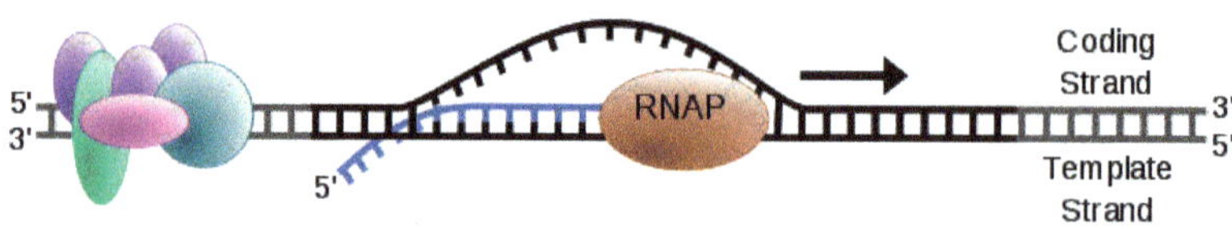

Simple diagram of transcription elongation

One strand of the DNA, the *template strand* (or noncoding strand), is used as a template for RNA synthesis. As transcription proceeds, RNA polymerase traverses the template strand and uses base pairing complementarity with the DNA template to create an RNA copy. Although RNA polymerase traverses the template strand from 3' → 5', the coding (non-template) strand and newly formed RNA can also be used as reference points, so transcription can be described as occurring 5' → 3'. This produces an RNA molecule from 5' → 3', an exact copy of the coding strand (except that thymines are replaced with uracils, and the nucleotides are composed of a ribose (5-carbon) sugar where DNA has deoxyribose (one fewer oxygen atom) in its sugar-phosphate backbone).

mRNA transcription can involve multiple RNA polymerases on a single DNA template and multiple rounds of transcription (amplification of particular mRNA), so many mRNA molecules can be rapidly produced from a single copy of a gene.

Elongation also involves a proofreading mechanism that can replace incorrectly incorporated bases. In eukaryotes, this may correspond with short pauses during transcription that allow appropriate RNA editing factors to bind. These pauses may be intrinsic to the RNA polymerase or due to chromatin structure.

Termination

Bacteria use two different strategies for transcription termination – Rho-independent termination and Rho-dependent termination. In Rho-independent transcription termination, RNA transcription stops when the newly synthesized RNA molecule forms a G-C-rich hairpin loop followed by a run of Us. When the hairpin forms, the mechanical stress breaks the weak rU-dA bonds, now filling the DNA–RNA hybrid. This pulls the poly-U transcript out of the active site of the RNA polymerase, terminating transcription. In the "Rho-dependent" type of termination, a protein factor called "Rho" destabilizes the interaction between the template and the mRNA, thus releasing the newly synthesized mRNA from the elongation complex.

Transcription termination in eukaryotes is less well understood than in bacteria, but involves cleavage of the new transcript followed by template-independent addition of adenines at its new 3' end, in a process called polyadenylation.

Inhibitors

Transcription inhibitors can be used as antibiotics against, for example, pathogenic bacteria (antibacterials) and fungi (antifungals). An example of such an antibacterial is rifampicin, which inhibits Bacterial transcription of DNA into mRNA by inhibiting DNA-dependent RNA polymerase by binding its beta-subunit. 8-Hydroxyquinoline is an antifungal transcription inhibitor. The effects of histone methylation may also work to inhibit the action of transcription.

Endogenous Inhibitors

In vertebrates, the majority of gene promoters contain a CpG island with numerous CpG sites. When many of a gene's promoter CpG sites are methylated the gene becomes inhibited (silenced). Colorectal cancers typically have 3 to 6 driver mutations and 33 to 66 hitchhiker or passenger mutations. However, transcriptional inhibition (silencing) may be of more importance than mutation in causing progression to cancer. For example, in colorectal cancers about 600 to 800 genes are transcriptionally inhibited by CpG island methylation (see regulation of transcription in cancer). Transcriptional repression in cancer can also occur by other epigenetic mechanisms, such as altered expression of microRNAs. In breast cancer, transcriptional repression of BRCA1 may occur more frequently by over-expressed microRNA-182 than by hypermethylation of the BRCA1 promoter (see Low expression of BRCA1 in breast and ovarian cancers).

Transcription Factories

Active transcription units are clustered in the nucleus, in discrete sites called transcription factories or euchromatin. Such sites can be visualized by allowing engaged polymerases to extend their transcripts in tagged precursors (Br-UTP or Br-U) and immuno-labeling the tagged nascent RNA. Transcription factories can also be localized using fluorescence in situ hybridization or marked by antibodies directed against polymerases. There are

~10,000 factories in the nucleoplasm of a HeLa cell, among which are ~8,000 polymerase II factories and ~2,000 polymerase III factories. Each polymerase II factory contains ~8 polymerases. As most active transcription units are associated with only one polymerase, each factory usually contains ~8 different transcription units. These units might be associated through promoters and/or enhancers, with loops forming a "cloud" around the factor.

A molecule that allows the genetic material to be realized as a protein was first hypothesized by François Jacob and Jacques Monod. Severo Ochoa won a Nobel Prize in Physiology or Medicine in 1959 for developing a process for synthesizing RNA *in vitro* with polynucleotide phosphorylase, which was useful for cracking the genetic code. RNA synthesis by RNA polymerase was established *in vitro* by several laboratories by 1965; however, the RNA synthesized by these enzymes had properties that suggested the existence of an additional factor needed to terminate transcription correctly.

In 1972, Walter Fiers became the first person to actually prove the existence of the terminating enzyme.

Roger D. Kornberg won the 2006 Nobel Prize in Chemistry "for his studies of the molecular basis of eukaryotic transcription".

Measuring and Detecting

Transcription can be measured and detected in a variety of ways:

- G-Less Cassette transcription assay: measures promoter strength
- Run-off transcription assay: identifies transcription start sites (TSS)
- Nuclear run-on assay: measures the relative abundance of newly formed transcripts
- RNase protection assay and ChIP-Chip of RNAP: detect active transcription sites
- RT-PCR: measures the absolute abundance of total or nuclear RNA levels, which may however differ from transcription rates
- DNA microarrays: measures the relative abundance of the global total or nuclear RNA levels; however, these may differ from transcription rates
- In situ hybridization: detects the presence of a transcript
- MS2 tagging: by incorporating RNA stem loops, such as MS2, into a gene, these become incorporated into newly synthesized RNA. The stem loops can then be detected using a fusion of GFP and the MS2 coat protein, which has a high affinity, sequence-specific interaction with the MS2 stem loops. The recruitment of GFP to the site of transcription is visualized as a single fluorescent spot. This new approach has revealed that transcription occurs in discontinuous bursts, or pulses (see Transcriptional bursting). With the notable exception of in situ techniques, most other methods provide cell population averages, and are not capable of detecting this fundamental property of genes.
- Northern blot: the traditional method, and until the advent of RNA-Seq, the most quantitative

- RNA-Seq: applies next-generation sequencing techniques to sequence whole transcriptomes, which allows the measurement of relative abundance of RNA, as well as the detection of additional variations such as fusion genes, post-transcriptional edits and novel splice sites

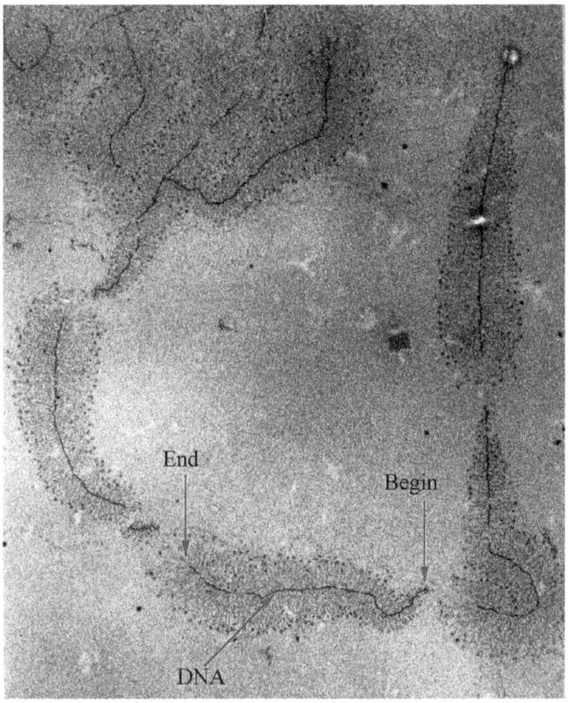

Electron micrograph of transcription of ribosomal RNA. The forming ribosomal RNA strands are visible as branches from the main DNA strand.

Reverse Transcription

Some viruses (such as HIV, the cause of AIDS), have the ability to transcribe RNA into DNA. HIV has an RNA genome that is *reverse transcribed* into DNA. The resulting DNA can be merged with the DNA genome of the host cell. The main enzyme responsible for synthesis of DNA from an RNA template is called reverse transcriptase.

In the case of HIV, reverse transcriptase is responsible for synthesizing a complementary DNA strand (cDNA) to the viral RNA genome. The enzyme ribonuclease H then digests the RNA strand, and reverse transcriptase synthesises a complementary strand of DNA to form a double helix DNA structure ("cDNA"). The cDNA is integrated into the host cell's genome by the enzyme integrase, which causes the host cell to generate viral proteins that reassemble into new viral particles. In HIV, subsequent to this, the host cell undergoes programmed cell death, or apoptosis of T cells. However, in other retroviruses, the host cell remains intact as the virus buds out of the cell.

Some eukaryotic cells contain an enzyme with reverse transcription activity called telomerase. Telomerase is a reverse transcriptase that lengthens the ends of linear chromosomes. Telomerase carries an RNA template from which it synthesizes a repeating sequence of DNA, or "junk" DNA. This repeated sequence of DNA is called a telomere and can be thought of as a "cap" for a chromosome. It is important because every time a linear chromosome is duplicated, it is shortened. With this "junk" DNA or "cap" at the ends of chromosomes, the shortening eliminates some of the non-essential, repeated sequence rather than the protein-encoding DNA sequence, that is farther away from the chromosome end.

Telomerase is often activated in cancer cells to enable cancer cells to duplicate their genomes indefinitely without losing important protein-coding DNA sequence. Activation of telomerase could be part of the process that allows cancer cells to become *immortal*. The immortalizing factor of cancer via telomere lengthening due to telomerase has been proven to occur in 90% of all carcinogenic tumors *in vivo* with the remaining 10% using an alternative telomere maintenance route called ALT or Alternative Lengthening of Telomeres.

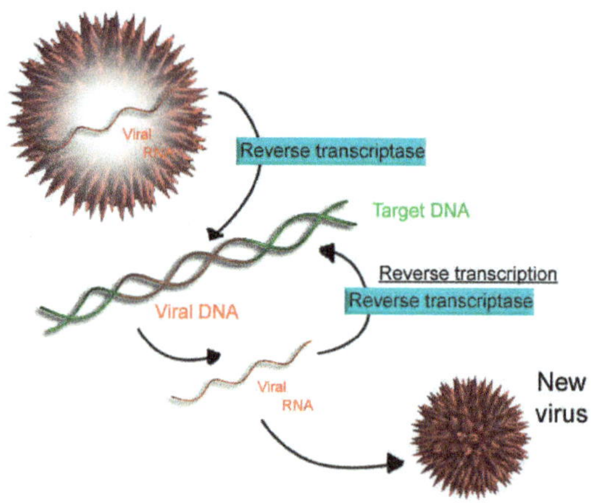

Scheme of reverse transcription

TRANSLATION BIOLOGY

In molecular biology and genetics, **translation** is the process in which ribosomes in a cell's cytoplasm create proteins, following transcription of DNA to RNA in the cell's nucleus. The entire process is a part of gene expression.

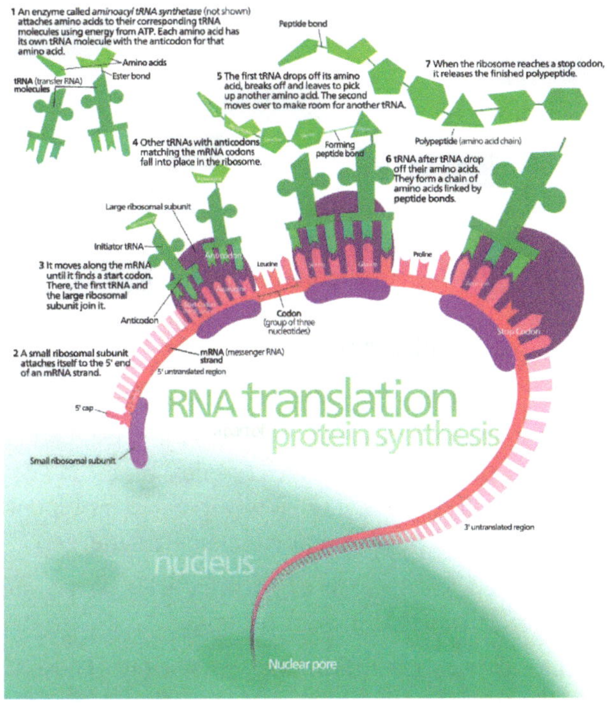

Overview of the translation of eukaryotic messenger RNA

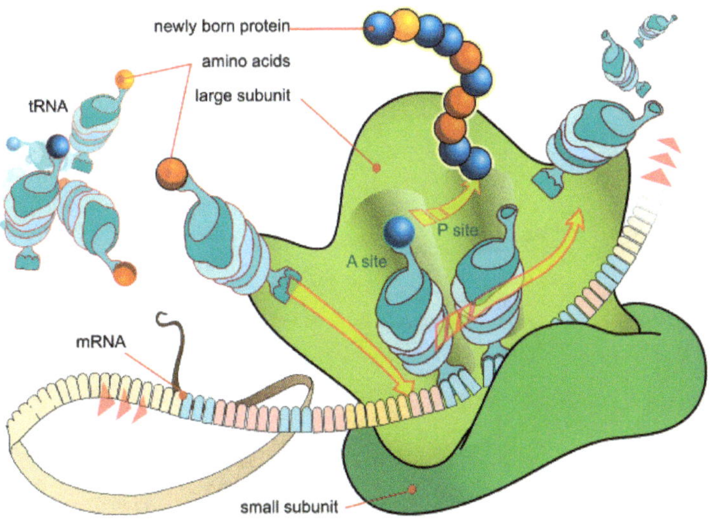

Diagram showing the translation of mRNA and the synthesis of proteins by a ribosome

In translation, messenger RNA (mRNA) is decoded by a ribosome, outside the nucleus, to produce a specific amino acid chain, or polypeptide. The polypeptide later folds into an active protein and performs its functions in the cell. The ribosome facilitates decoding by inducing the binding of complementary tRNA anticodon sequences to mRNA codons.

The tRNAs carry specific amino acids that are chained together into a polypeptide as the mRNA passes through and is "read" by the ribosome.

Translation proceeds in three phases:

- **Initiation**: The ribosome assembles around the target mRNA. The first tRNA is attached at the start codon.
- **Elongation**: The tRNA transfers an amino acid to the tRNA corresponding to the next codon. The ribosome then moves (*translocates)* to the next mRNA codon to continue the process, creating an amino acid chain.
- **Termination**: When a stop codon is reached, the ribosome releases the polypeptide.

In bacteria, translation occurs in the cytoplasm, where the large and small subunits of the ribosome bind to the mRNA. In eukaryotes, translation occurs in the cytosol or across the membrane of the endoplasmic reticulum in a process called vectorial synthesis. In many instances, the entire ribosome/mRNA complex binds to the outer membrane of the rough endoplasmic reticulum (ER); the newly created polypeptide is stored inside the ER for later vesicle transport and secretion outside of the cell.

Many types of transcribed RNA, such as transfer RNA, ribosomal RNA, and small nuclear RNA, do not undergo translation into proteins.

A number of antibiotics act by inhibiting translation. These include anisomycin, cycloheximide, chloramphenicol, tetracycline, streptomycin, erythromycin, and puromycin. Prokaryotic ribosomes have a different structure from that of eukaryotic ribosomes, and thus antibiotics can specifically target bacterial infections without any harm to a eukaryotic host's cells.

Basic Mechanisms

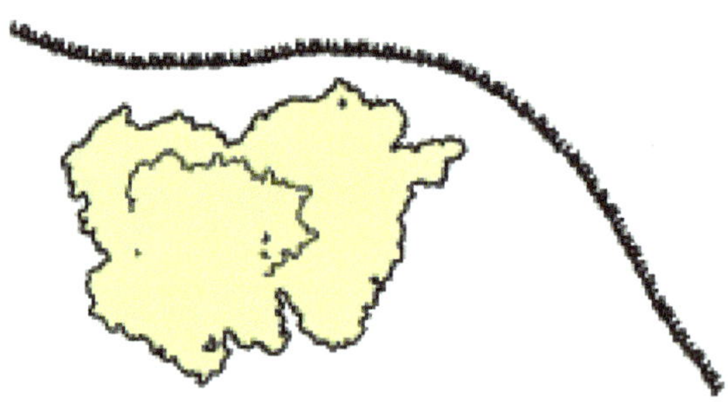

A ribosome translating a protein that is secreted into the endoplasmic reticulum. tRNAs are colored dark blue.

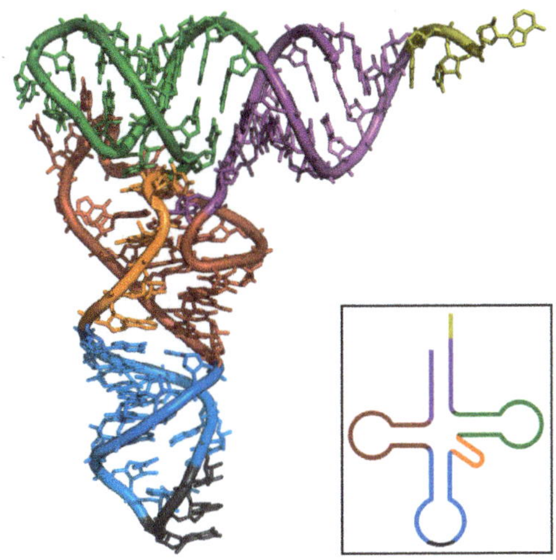

Tertiary structure of tRNA. *CCA tail* in orange, *Acceptor stem* in purple, *D arm* in red, *Anticodon arm* in blue with *Anticodon* in black, *T arm* in green.

The basic process of protein production is addition of one amino acid at a time to the end of a protein. This operation is performed by a ribosome.A ribosome is made up of two subunits, a small subunit and a large subunit. these subunits come together before translation of mRNA into a protein to provide a location for translation to be carried out and a polypeptide to be produced. The choice of amino acid type to add is determined by an mRNA molecule. Each amino acid added is matched to a three nucleotide subsequence of the mRNA. For each such triplet possible, the corresponding amino acid is accepted. The successive amino acids added to the chain are matched to successive nucleotide triplets in the mRNA. In this way the sequence of nucleotides in the template mRNA chain determines the sequence of amino acids in the generated amino acid chain. Addition of an amino acid occurs at the C-terminus of the peptide and thus translation is said to be amino-to-carboxyl directed.

The mRNA carries genetic information encoded as a ribonucleotide sequence from the chromosomes to the ribosomes. The ribonucleotides are "read" by translational machinery in a sequence of nucleotide triplets called codons. Each of those triplets codes for a specific amino acid.

The ribosome molecules translate this code to a specific sequence of amino acids. The ribosome is a multisubunit structure containing rRNA and proteins. It is the "factory" where amino acids are assembled into proteins. tRNAs are small noncoding RNA chains (74-93 nucleotides) that transport amino acids to the ribosome. tRNAs have a site for amino acid attachment, and a site called an anticodon. The anticodon is an RNA triplet complementary to the mRNA triplet that codes for their cargo amino acid.

Aminoacyl tRNA synthetases (enzymes) catalyze the bonding between specific tRNAs and the amino acids that their anticodon sequences call for. The product of this reaction is an aminoacyl-tRNA. This aminoacyl-tRNA is carried to the ribosome by EF-Tu, where mRNA codons are matched through complementary base pairing to specific tRNA anticodons. Aminoacyl-tRNA synthetases that mispair tRNAs with the wrong amino acids can produce mischarged aminoacyl-tRNAs, which can result in inappropriate amino acids at the respective position in protein. This "mistranslation" of the genetic code naturally occurs at low levels in most organisms, but certain cellular environments cause an increase in permissive mRNA decoding, sometimes to the benefit of the cell.

The ribosome has three sites for tRNA to bind. They are the aminoacyl site (abbreviated A), the peptidyl site (abbreviated P) and the exit site (abbreviated E). With respect to the mRNA, the three sites are oriented 5' to 3' E-P-A, because ribosomes move toward the 3' end of mRNA. The A site binds the incoming tRNA with the complementary codon on the mRNA. The P site holds the tRNA with the growing polypeptide chain. The E site holds the tRNA without its amino acid. When an aminoacyl-tRNA initially binds to its corresponding codon on the mRNA, it is in the A site. Then, a peptide bond forms between the amino acid of the tRNA in the A site and the amino acid of the charged tRNA in the P site. The growing polypeptide chain is transferred to the tRNA in the A site. Translocation occurs, moving the tRNA in the P site, now without an amino acid, to the E site; the tRNA that was in the A site, now charged with the polypeptide chain, is moved to the P site. The tRNA in the E site leaves and another aminoacyl-tRNA enters the A site to repeat the process.

After the new amino acid is added to the chain, and after the mRNA is released out of the nucleus and into the ribosome's core, the energy provided by the hydrolysis of a GTP bound to the translocase EF-G (in prokaryotes) and eEF-2 (in eukaryotes) moves the ribosome down one codon towards the 3' end. The energy required for translation of proteins is significant. For a protein containing n amino acids, the number of high-energy phosphate bonds required to translate it is $4n$-1. The rate of translation varies; it is significantly higher in prokaryotic cells (up to 17-21 amino acid residues per second) than in eukaryotic cells (up to 6-9 amino acid residues per second).

In activation, the correct amino acid is covalently bonded to the correct transfer RNA (tRNA). The amino acid is joined by its carboxyl group to the 3' OH of the tRNA by an ester bond. When the tRNA has an amino acid linked to it, it is termed "charged". Initiation involves the small subunit of the ribosome binding to the 5' end of mRNA with the help of initiation factors (IF). Termination of the polypeptide happens when the A site of the ribosome faces a stop codon (UAA, UAG, or UGA). No tRNA can recognize or bind to this codon. Instead, the stop codon induces the binding of a release factor protein that prompts the disassembly of the entire ribosome/mRNA complex.

The process of translation is highly regulated in both eukaryotic and prokaryotic organisms. Regulation of translation can impact the global rate of protein synthesis which is closely coupled to the metabolic and proliferative state of a cell. In addition, recent work has revealed that genetic differences and their subsequent expression as mRNAs can also impact translation rate in an RNA-specific manner.

Genetic Code

Whereas other aspects such as the 3D structure, called tertiary structure, of protein can only be predicted using sophisticated algorithms, the amino acid sequence, called primary structure, can be determined solely from the nucleic acid sequence with the aid of a translation table.

This approach may not give the correct amino acid composition of the protein, in particular if unconventional amino acids such as selenocysteine are incorporated into the protein, which is coded for by a conventional stop codon in combination with a downstream hairpin (SElenoCysteine Insertion Sequence, or SECIS).

There are many computer programs capable of translating a DNA/RNA sequence into a protein sequence. Normally this is performed using the Standard Genetic Code, however, few programs can handle all the "special" cases, such as the use of the alternative initiation codons. For instance, the rare alternative start codon CTG codes for Methionine when used as a start codon, and for Leucine in all other positions.

Example: Condensed translation table for the Standard Genetic Code (from the NCBI Taxonomy webpage).

```
AAs    = FFLLSSSSYY**CC*WLLLLPPPPHHQQRRRRIIIMTTTTNNKKSSRRVVVVAAAADDEEGGGG
Starts = ---M---------------M---------------M----------------------------
Base1  = TTTTTTTTTTTTTTTTCCCCCCCCCCCCCCCCAAAAAAAAAAAAAAAAGGGGGGGGGGGGGGGG
Base2  = TTTTCCCCAAAAGGGGTTTTCCCCAAAAGGGGTTTTCCCCAAAAGGGGTTTTCCCCAAAAGGGG
Base3  = TCAGTCAGTCAGTCAGTCAGTCAGTCAGTCAGTCAGTCAGTCAGTCAGTCAGTCAGTCAGTCAG
```

Translation Tables

Even when working with ordinary eukaryotic sequences such as the Yeast genome, it is often desired to be able to use alternative translation tables—namely for translation of the mitochondrial genes. Currently the following translation tables are defined by the NCBI Taxonomy Group for the translation of the sequences in GenBank:

- The Standard
- The Vertebrate Mitochondrial Code
- The Yeast Mitochondrial Code
- The Mold, Protozoan, and Coelenterate Mitochondrial Code and the Mycoplasma/Spiroplasma Code
- The Invertebrate Mitochondrial Code
- The Ciliate, Dasycladacean and Hexamita Nuclear Code
- The Echinoderm and Flatworm Mitochondrial Code
- The Euplotid Nuclear Code
- The Bacterial and Plant Plastid Code
- The Alternative Yeast Nuclear Code
- The Ascidian Mitochondrial Code
- The Alternative Flatworm Mitochondrial Code
- Blepharisma Nuclear Code
- Chlorophycean Mitochondrial Code

- Trematode Mitochondrial Code
- Scenedesmus obliquus mitochondrial Code
- Thraustochytrium Mitochondrial Code

REFERENCES

1. Crick, F.H.C. (1956): On Protein Synthesis. Symp. Soc. Exp. Biol. XII, 139-163.
2. Crick, F (August 1970). "Central dogma of molecular biology." (PDF). Nature. **227**(5258): 561–3. Bibcode:1970Natur.227..561C. doi:10.1038/227561a0. PMID 4913914.
3. Leavitt, Sarah A. (June 2010). "Deciphering the Genetic Code: Marshall Nirenberg". Office of NIH History.
4. Ahlquist P (May 2002). "RNA-dependent RNA polymerases, viruses, and RNA silencing". Science. **296** *(5571): 1270–3.* doi:10.1126/science.1069132. PMID 12016304.
5. B. J. McCarthy; J. J. Holland (September 15, 1965). "Denatured DNA as a Direct Template for in vitro Protein Synthesis". Proceedings of the National Academy of Sciences of the United States of America. **54** *(3): 880–886.*. doi:10.1073/pnas.54.3.880. PMC 219759. PMID 4955657.
6. T. Uzawa; A. Yamagishi; T. Oshima (2002-04-09). "Polypeptide Synthesis Directed by DNA as a Messenger in Cell-Free Polypeptide Synthesis by Extreme Thermophiles, Thermus thermophilus *HB27 and* Sulfolobus tokodaii *Strain 7". The Journal of Biochemistry.* **131** *(6): 849–853. doi:10.1093/oxfordjournals.jbchem.a003174. PMID 12038981.*
7. Bussard Alain E (2005). "A scientific revolution? The prion anomaly may challenge the central dogma of molecular biology". EMBO Rep. **6** *(8): 691–694. doi:10.1038/sj.embor.7400497. PMC 1369155. PMID 16065057.*
8. Koonin, Eugene. (2012). Does the Central Dogma Still Stand? Biology Direct 7: 27.
9. Ridley, Rosalind. (2001). What Would Thomas Henry Huxley Have Made of Prion Diseases?. In Harry F. Baker. Molecular Pathology of the Prions (Methods in Molecular Medicine). Humana Press. pp. 1-16. ISBN 0-89603-924-2
10. Wilkins, Adam S. (January 2012). "(Review) Evolution: A View from the 21st Century". Genome Biology and Evolution. **4**: 423–426. doi:10.1093/gbe/evs008.
11. Moran, Laurence A (May–June 2011). "(Review) Evolution: A View from the 21st Century". Reports of the National Center for Science Education. **32.3** *(9): 1–4.*
12. Horace Freeland Judson (1996). "Chapter 6: My mind was, that a dogma was an idea for which there was no reasonable evidence. You see?!". *The Eighth Day of Creation: Makers of the Revolution in Biology (25th anniversary edition). Cold Spring Harbor, NY: Cold Spring Harbor Laboratory Press. ISBN 0-87969-477-7.*
13. Weaver, Warren (6 November 1970). "Molecular Biology: Origin of the Term". Science. American Association for the Advancement of Science. **170**: 581–582.

doi:10.1126/science.170.3958.581-a. ISSN 0036-8075. JSTOR 1731491.

14. Beadle, G. W.; Tatum, E. L. (1941). "Genetic Control of Biochemical Reactions in Neurospora". PNAS. **27** *(11): 499–506.* doi:10.1073/pnas.27.11.499. PMC 1078370. PMID 16588492.

15. Avery, Oswald T.; Colin M. MacLeod; Maclyn McCarty (1944-02-01). "Studies on the Chemical Nature of the Substance Inducing Transformation of Pneumococcal Types: Induction of Transformation by a Desoxyribonucleic Acid Fraction Isolated from Pneumococcus Type III". Journal of Experimental Medicine. **79** *(2): 137–158. doi:10.1084/jem.79.2.137. PMC 2135445. PMID 19871359.*

16. Hershey, A.D. and Chase, M. (1952) Independent functions of viral protein and nucleic acid in growth of bacteriophage J Gen Physiol.

17. Watson J.D.; Crick F.H.C. (1953). "A Structure for Deoxyribose Nucleic Acid" (PDF). Nature. **171** *(4356): 737–738.* doi:10.1038/171737a0. PMID 13054692.

18. Jacob F, Monod J (1961). "Genetic regulatory mechanisms in the synthesis of proteins". J Mol Biol. **3**: 318–356. doi:10.1016/S0022-2836(61)80072-7. PMID 13718526.

19. Keen, E. C. (2015). "A century of phage research: Bacteriophages and the shaping of modern biology". BioEssays. **37**: 6. doi:10.1002/bies.201400152.

20. Soyfer VN (September 2001). "The consequences of political dictatorship for Russian science". Nat. Rev. Genet. **2** *(9): 723–9.* doi:10.1038/35088598. PMID 11533721.

21. Watson J, Crick F (1953). "Molecular structure of nucleic acids; a structure for deoxyribose nucleic acid" (PDF). Nature. **171** *(4356): 737–8.* doi:10.1038/171737a0. PMID 13054692.

22. Watson JD, Crick FH (April 1953). "Molecular structure of nucleic acids; a structure for deoxyribose nucleic acid" (PDF). Nature. **171** *(4356): 737–738.* doi:10.1038/171737a0. PMID 13054692.

23. Grunberg-Manago M, Ortiz PJ, Ochoa S (November 1955). "Enzymatic synthesis of nucleic acidlike polynucleotides". Science. **122** *(3176): 907–10.* doi:10.1126/science.122.3176.907. PMID 13274047.

24. Rich A, Davies DR (July 1956). "A new, two-stranded helical structure: polyadenylic acid and polyuridylic acid". J. Am. Chem. Soc. **78** *(14): 3548–3549.* doi:10.1021/ja01595a086.

25. Felsenfeld G, Davies DR, Rich A (April 1957). "Formation of a three-stranded polynucleotide molecule". J. Am. Chem. Soc. **79** *(8): 2023–2024.* doi:10.1021/ja01565a074.

26. Sobll H, Tomita K, Rich A (June 1963). "The crystal structure of an intermolecular complex containing a guanine and a cytosine derivative". Proc. Natl. Acad. Sci. U.S.A. **49** *(6): 885–92.* doi:10.1073/pnas.49.6.885. PMC 300027. PMID 13989773.

27. Rich A (May 2009). "The era of RNA awakening: structural biology of RNA in the early years". Q. Rev. Biophys. **42** *(2): 117–37.* doi:10.1017/S0033583509004776.

PMID 19638248.

28. Warner JR, Rich A (June 1964). "The number of soluble RNA molecules on reticulocyte polyribosomes". Proc. Natl. Acad. Sci. U.S.A. **51** *(6): 1134–41.* doi:10.1073/pnas.51.6.1134. PMC 300225. PMID 14215634.

29. Holley, RW, Apgar, J, Everett, GA, Madison, JT, Marguisse, M, Merrill, SH, Penwick, JR, Zamir (March 1965). "Structure of a ribonucleic acid". Science. **147** *(3664): 1462–5.* doi:10.1126/science.147.3664.1462. PMID 14263761.

30. Kim SH, Rich A (December 1968). "Single crystals of transfer RNA: an X-ray diffraction study". Science. **162** *(3860): 1381–4.* doi:10.1126/science.162.3860.1381. PMID 4880852.

31. Kim SH, Quigley G, Suddath FL, Rich A (April 1971). "High-resolution x-ray diffraction patterns of crystalline transfer RNA that show helical regions". Proc. Natl. Acad. Sci. U.S.A. **68** *(4): 841–5.* doi:10.1073/pnas.68.4.841. PMC 389056. PMID 5279525.

32. Kim SH, Quigley GJ, Suddath FL, McPherson A, Sneden D, Kim JJ, Weinzierl J, Rich A (January 1973). "Three-dimensional structure of yeast phenylalanine transfer RNA: folding of the polynucleotide chain". Science. **179** *(4070): 285–8.* doi:10.1126/science.179.4070.285. PMID 4566654.

33. Drew HR, Wing RM, Takano T, Broka C, Tanaka S, Itakura K, Dickerson RE (April 1981). "Structure of a B-DNA dodecamer: conformation and dynamics". Proc. Natl. Acad. Sci. U.S.A. **78** *(4): 2179–83.* doi:10.1073/pnas.78.4.2179. PMC 319307. PMID 6941276.

34. Shen LX, Cai Z, Tinoco I (August 1995). "RNA structure at high resolution". FASEB J. **9**(11): 1023–33. PMID 7544309.

35. Cech TR, Zaug AJ, Grabowski PJ (December 1981). "In vitro splicing of the ribosomal RNA precursor of Tetrahymena: involvement of a guanosine nucleotide in the excision of the intervening sequence". Cell. **27** *(3 Pt 2): 487–96.* doi:10.1016/0092-8674(81)90390-1. PMID 6101203.

36. Stark BC, Kole R, Bowman EJ, Altman S (August 1978). "Ribonuclease P: an enzyme with an essential RNA component". Proc. Natl. Acad. Sci. U.S.A. **75** *(8): 3717–21.* doi:10.1073/pnas.75.8.3717. PMC 392857. PMID 358197.

37. Prody GA, Bakos JT, Buzayan JM, Schneider IR, Bruening G (March 1986). "Autolytic Processing of Dimeric Plant Virus Satellite RNA". Science. **231** *(4745): 1577–1580.* doi:10.1126/science.231.4745.1577. PMID 17833317.

38. Pley HW, Flaherty KM, McKay DB (November 1994). "Three-dimensional structure of a hammerhead ribozyme". Nature. **372** *(6501): 68–74.* doi:10.1038/372068a0. PMID 7969422.

39. Cate JH, Gooding AR, Podell E, Zhou K, Golden BL, Kundrot CE, Cech TR, Doudna JA (September 1996). "Crystal structure of a group I ribozyme domain: principles of RNA packing". Science. **273** *(5282): 1678–85.* doi:10.1126/

science.273.5282.1678. PMID 8781224.

40. Ferré-D'Amaré AR, Doudna JA (1999). "RNA folds: insights from recent crystal structures". Annu Rev Biophys Biomol Struct. **28** *(1): 57–73. doi:10.1146/annurev.biophys.28.1.57. PMID 10410795.*
41. Ramos A, Gubser CC, Varani G (June 1997). "Recent solution structures of RNA and its complexes with drugs, peptides and proteins". Curr. Opin. Struct. Biol. **7** *(3): 317–23.* doi:10.1016/S0959-440X(97)80046-2. PMID 9204272.
42. Butcher SE, Dieckmann T, Feigon J (December 1997). "Solution structure of a GAAA tetraloop receptor RNA". EMBO J. **16** *(24): 7490–9. doi:10.1093/*emboj/16.24.7490. PMC 1170348. PMID 9405377.
43. Costa M, Michel F (March 1995). "Frequent use of the same tertiary motif by self-folding RNAs". EMBO J. **14** *(6): 1276–85. PMC 398207. PMID 7720718.*
44. Toor N, Keating KS, Taylor SD, Pyle AM (April 2008). "Crystal structure of a self-spliced group II intron". Science. **320** *(5872): 77–82.* doi:10.1126/science.1153803. PMID 18388288.;
45. PDB: 1FFK; Ban N, Nissen P, Hansen J, Moore PB, Steitz TA (August 2000). "The complete atomic structure of the large ribosomal subunit at 2.4 A resolution". Science. **289**(5481): 905–20. doi:10.1126/science.289.5481.905. PMID 10937989.;
46. Imperfect DNA replication results in mutations. Berg JM, Tymoczko JL, Stryer L, Clarke ND (2002). "Chapter 27: DNA Replication, Recombination, and Repair". Biochemistry. W.H. Freeman and Company. ISBN 0-7167-3051-0.
47. Alberts B, Johnson A, Lewis J, Raff M, Roberts K, Walter P (2002). "5DNA Replication, Repair, and Recombination". Molecular Biology of the Cell. Garland Science. ISBN 0-8153-3218-1.
48. Berg JM, Tymoczko JL, Stryer L, Clarke ND (2002). "Chapter 27, Section 4: DNA Replication of Both Strands Proceeds Rapidly from Specific Start Sites". Biochemistry. W.H. Freeman and Company. ISBN 0-7167-3051-0.
49. Alberts, B., et al., Molecular Biology of the Cell, Garland Science, 4th ed., 2002, pp. 238–240 ISBN 0-8153-3218-1
50. Allison, Lizabeth A. Fundamental Molecular Biology. Blackwell Publishing. 2007. p.112 ISBN 978-1-4051-0379-4
51. Berg JM, Tymoczko JL, Stryer L, Clarke ND (2002). Biochemistry. W.H. Freeman and Company. ISBN 0-7167-3051-0. Chapter 27, Section 2: DNA Polymerases Require a Template and a Primer
52. McCulloch, Scott D; Kunkel, Thomas A (January 2008). "The fidelity of DNA synthesis by eukaryotic replicative and translesion synthesis polymerases". Cell Research. **18** *(1): 148–161. doi:10.1038/cr.2008.4. PMC 3639319. PMID 18166979.*
53. McCarthy, David; Minner, Charles; Bernstein, Harris; Bernstein, Carol (October 1976). "DNA elongation rates and growing point distributions of wild-type phage

T4 and a DNA-delay amber mutant". Journal of Molecular Biology. **106** *(4): 963–981.* doi:10.1016/0022-2836(76)90346-6. PMID 789903.

54. Drake JW (1970) The Molecular Basis of Mutation. Holden-Day, San Francisco ISBN 0816224501 ISBN 978-0816224500
55. Alberts B, Johnson A, Lewis J, Raff M, Roberts K, Walter P (2002). Molecular Biology of the Cell. Garland Science. ISBN 0-8153-3218-1.
56. Weigel C, Schmidt A, Rückert B, Lurz R, Messer W (November 1997). "DnaA protein binding to individual DnaA boxes in the Escherichia coli replication origin, oriC". The EMBO Journal. **16** *(21): 6574–83. doi:10.1093/emboj/16.21.6574. PMC 1170261. PMID 9351837.*
57. Lodish H, Berk A, Zipursky LS, Matsudaira P, Baltimore D, Darnell J (2000). Molecular Cell Biology. W. H. Freeman and Company. ISBN 0-7167-3136-3.12.1. General Features of Chromosomal Replication: Three Common Features of Replication Origins
58. Aravind, L.; Leipe, D. D.; Koonin, E. V. (1998). "Toprim—a conserved catalytic domain in type IA and II topoisomerases, DnaG-type primases, OLD family nucleases and RecR proteins". Nucleic Acids Research. **26** *(18): 4205–4213. doi:10.1093/*nar/26.18.4205. PMC 147817. PMID 9722641.
59. Frick, David; Richardson, Charles (July 2001). "DNA Primases". Annual Review of Biochemistry. **70**: 39–80. doi:10.1146/annurev.biochem.70.1.39. PMID 11395402.
60. Barry, Elizabeth R.; Bell, Stephen D. (8 December 2006). "DNA Replication in the Archaea". Microbiology and Molecular Biology Reviews. **70** *(4): 876–887. doi:10.1128/MMBR.00029-06. PMC 1698513. PMID 17158702.*
61. Stillman, Bruce (July 2015). "Reconsidering DNA Polymerases at the Replication Fork in Eukaryotes". Molecular Cell. **59** *(2): 139–141. doi:10.1016/j.molcel.2015.07.004. PMC 4636199. PMID 26186286.*
62. Distinguishing the pathways of primer removal during Eukaryotic Okazaki fragment maturation Contributor Author Rossi, Marie Louise. Date Accessioned: 2009-02-23T17:05:09Z. Date Available: 2009-02-23T17:05:09Z. Date Issued: 2009-02-23T17:05:09Z. Identifier Uri: http://hdl.handle.net/1802/6537. Description: Dr. Robert A. Bambara, Faculty Advisor. Thesis (PhD) – School of Medicine and Dentistry, University of Rochester. UR only until January 2010. UR only until January 2010.
63. Huberman, Joel A.; Riggs, Arthur D. (March 1968). "On the mechanism of DNA replication in mammalian chromosomes". Journal of Molecular Biology. **32** *(2): 327–341.* doi:10.1016/0022-2836(68)90013-2. PMID 5689363.
64. Alberts B, Johnson A, Lewis J, Raff M, Roberts K, Walter P (2002). Molecular Biology of the Cell. Garland Science. ISBN 0-8153-3218-1. DNA Replication Mechanisms: DNA Topoisomerases Prevent DNA Tangling During Replication
65. Reece, Richard J.; Maxwell, Anthony; Wang, James C. (26 September 2008). "DNA Gyrase: Structure and Function". Critical Reviews in Biochemistry and Molecular Biology. **26** *(3-4): 335–375.* doi:10.3109/10409239109114072. PMID

1657531.

66. Alberts B, Johnson A, Lewis J, Raff M, Roberts K, Walter P (2002). Molecular Biology of the Cell. Garland Science. ISBN 0-8153-3218-1. DNA Replication Mechanisms: Special Proteins Help to Open Up the DNA Double Helix in Front of the Replication Fork

67. Griffiths A.J.F.; Wessler S.R.; Lewontin R.C.; Carroll S.B. (2008). Introduction to Genetic Analysis. W. H. Freeman and Company. ISBN 0-7167-6887-9.[Chapter 7: DNA: Structure and Replication. pg 283–290]

68. Will the Hayflick limit keep us from living forever?". Howstuffworks. Retrieved January 20, 2015.

69. James D. Watson et al. (2008), "Molecular Biology of the gene", Pearson Education: 237

70. Peter Meister, Angela Taddei1, Susan M. Gasser(June 2006), "In and out of the Replication Factory", Cell **125** (7): 1233–1235

71. TA Brown (2002). Genomes. BIOS Scientific Publishers. ISBN 1-85996-228-9.13.2.3.

72. Alberts B, Johnson A, Lewis J, Raff M, Roberts K, Walter P (2002). Molecular Biology of the Cell. Garland Science. ISBN 0-8153-3218-1. Intracellular Control of Cell-Cycle Events: S-Phase Cyclin-Cdk Complexes (S-Cdks) Initiate DNA Replication Once Per Cycle

73. Brown, TA (2002). "13". Genomes (2nd ed.). Oxford: Wiley-Liss.

74. Tobiason DM, Seifert HS (2006). "The Obligate Human Pathogen, Neisseria gonorrhoeae, Is Polyploid". PLoS Biology. **4** *(6): e185. doi:10.1371/journal.pbio.0040185. PMC 1470461. PMID 16719561.*

75. Slater, Steven; Wold, Sture; Lu, Min; Boye, Erik; Skarstad, Kirsten; Kleckner, Nancy (September 1995). "E. coli SeqA protein binds oriC in two different methyl-modulated reactions appropriate to its roles in DNA replication initiation and origin sequestration". Cell. **82** *(6): 927–936.* doi:10.1016/0092-8674(95)90272-4. PMID 7553853.

76. Saiki, Randall; Gelfand, David H.; Stoffel, Susanne; Scharf, Stephen J.; Higuchi, Russell; Horn, Glenn T.; Mullis, Kary B.; Erlich, Henry A. (29 January 1988). "Primer-directed enzymatic amplification of DNA with a thermostable DNA polymerase". Science. **239** *(4839): 487–491. doi:10.1126/science.2448875. PMID 2448875.*

77. Eldra P. Solomon, Linda R. Berg, Diana W. Martin. Biology, 8th Edition, International Student Edition. Thomson Brooks/Cole. ISBN 978-0495317142

78. Koonin EV, Gorbalenya AE, Chumakov KM (July 1989). "Tentative identification of RNA-dependent RNA polymerases of dsRNA viruses and their relationship to positive strand RNA viral polymerases". FEBS Letters. **252** *(1-2): 42–6.* doi:10.1016/0014-5793(89)80886-5. PMID 2759231.

79. Berg J, Tymoczko JL, Stryer L (2006). Biochemistry (6th ed.). San Francisco:

W. H. Freeman. ISBN 0-7167-8724-5.

80. Watson JD, Baker TA, Bell SP, Gann AA, Levine M, Losick RM (2013). Molecular Biology of the Gene (7th ed.). Pearson.
81. Goldman SR, Ebright RH, Nickels BE (May 2009). "Direct detection of abortive RNA transcripts in vivo". Science. **324** *(5929): 927–8. doi:10.1126/science.1169237. PMC 2718712. PMID 19443781.*
82. Revyakin A, Liu C, Ebright RH, Strick TR (November 2006). "Abortive initiation and productive initiation by RNA polymerase involve DNA scrunching". Science. **314** *(5802): 1139–43. doi:10.1126/science.1131398. PMC 2754787. PMID 17110577.*
83. Raffaelle M, Kanin EI, Vogt J, Burgess RR, Ansari AZ (November 2005). "Holoenzyme switching and stochastic release of sigma factors from RNA polymerase in vivo". Molecular Cell. **20** *(3): 357–66. doi:10.1016/j.molcel.2005.10.011. PMID 16285918.*
84. Mandal SS, Chu C, Wada T, Handa H, Shatkin AJ, Reinberg D (May 2004). "Functional interactions of RNA-capping enzyme with factors that positively and negatively regulate promoter escape by RNA polymerase II". Proceedings of the National Academy of Sciences of the United States of America. **101** *(20): 7572–7. doi:10.1073/pnas.0401493101. PMC 419647. PMID 15136722.*
85. Goodrich JA, Tjian R (April 1994). "Transcription factors IIE and IIH and ATP hydrolysis direct promoter clearance by RNA polymerase II". Cell. **77** *(1): 145–56.* doi:10.1016/0092-8674(94)90242-9. PMID 8156590.
86. Richardson JP (September 2002). "Rho-dependent termination and ATPases in transcript termination". Biochimica et Biophysica Acta. **1577** *(2): 251–260.* doi:10.1016/S0167-4781(02)00456-6. PMID 12213656.
87. Lykke-Andersen S, Jensen TH (October 2007). "Overlapping pathways dictate termination of RNA polymerase II transcription". Biochimie. **89** *(10): 1177–82. doi:10.1016/j.biochi.2007.05.007. PMID 17629387.*
88. Saxonov S, Berg P, Brutlag DL (January 2006). "A genome-wide analysis of CpG dinucleotides in the human genome distinguishes two distinct classes of promoters". Proceedings of the National Academy of Sciences of the United States of America. **103**(5): 1412–7. doi:10.1073/pnas.0510310103. PMC 1345710. PMID 16432200.
89. Bird A (January 2002). "DNA methylation patterns and epigenetic memory". Genes & Development. **16** *(1): 6–21. doi:10.1101/gad.947102. PMID 11782440.*
90. Vogelstein B, Papadopoulos N, Velculescu VE, Zhou S, Diaz LA, Kinzler KW (March 2013). "Cancer genome landscapes". Science. **339** *(6127): 1546–58. doi:10.1126/science.1235122. PMC 3749880. PMID 23539594.*
91. Tessitore A, Cicciarelli G, Del Vecchio F, Gaggiano A, Verzella D, Fischietti M, Vecchiotti D, Capece D, Zazzeroni F, Alesse E (2014). "MicroRNAs in the

DNA Damage/Repair Network and Cancer". International Journal of Genomics. **2014**: 820248. doi:10.1155/2014/820248. PMC 3926391. PMID 24616890.

92. Papantonis, A (2012-10-26). "TNFα signals through specialized factories where responsive coding and miRNA genes are transcribed". Nature EMBO J.
93. Raj A, van Oudenaarden A (October 2008). "Nature, nurture, or chance: stochastic gene expression and its consequences". Cell. **135** *(2): 216–26. doi:10.1016/j.cell.2008.09.050. PMC 3118044. PMID 18957198.*
94. Kolesnikova IN (2000). "Some patterns of apoptosis mechanism during HIV-infection". Dissertation (in Russian).
95. Cesare AJ, Reddel RR (2010). "Alternative lengthening of telomeres: models, mechanisms and implications". Nature Reviews. Genetics. **11** *(5): 319–30.* doi:10.1038/nrg2763. PMID 20351727.

CHAPTER 2

Molecular Biology

Chapter Contents

Molecular biology is a branch of science concerning biological activity at the molecular level. The field of molecular biology overlaps with biology and chemistry and in particular, genetics and biochemistry.

HISTORY OF MOLECULAR BIOLOGY

The history of molecular biology begins in the 1930s with the convergence of various, previously distinct biological and physical disciplines: biochemistry, genetics, microbiology, virology and physics. With the hope of understanding life at its most fundamental level, numerous physicists and chemists also took an interest in what would become molecular biology.

In its modern sense, molecular biology attempts to explain the phenomena of life starting from the macromolecular properties that generate them. Two categories of macromolecules in particular are the focus of the molecular biologist: 1) nucleic acids, among which the most famous is deoxyribonucleic acid (or DNA), the constituent of genes, and 2) proteins, which are the active agents of living organisms. One definition of the scope of molecular biology therefore is to characterize the structure, function and relationships between these two types of macromolecules. This relatively limited definition will suffice to allow us to establish a date for the so-called molecular revolution, or at least to establish a chronology of its most fundamental developments.

General Overview

In its earliest manifestations, molecular biology—the name was coined by Warren Weaver of the Rockefeller Foundation in 1938—was an idea of physical and chemical explanations of life, rather than a coherent discipline. Following the advent of the Mendelian-chromosome theory of heredity in the 1910s and the maturation of atomic theoryand quantum mechanics in the 1920s, such explanations seemed within reach. Weaver and others encouraged (and funded) research at the intersection of biology, chemistry and physics, while prominent physicists such as Niels Bohr and Erwin Schrödinger turned their attention to biological speculation. However, in the 1930s and 1940s it was by no means clear which—if any—cross-disciplinary research would bear fruit; work in colloid chemistry, biophysics and radiation biology, crystallography, and other emerging fields all seemed promising.

In 1940, George Beadle and Edward Tatum demonstrated the existence of a precise relationship between genes and proteins. In the course of their experiments connecting genetics with biochemistry, they switched from the genetics mainstay *Drosophila* to a more appropriate model organism, the fungus *Neurospora*; the construction and exploitation of new model organisms would become a recurring theme in the development of molecular biology. In 1944, Oswald Avery, working at the Rockefeller Institute of New York, demonstrated that genes are made up of DNA(see Avery–MacLeod–McCarty experiment). In 1952, Alfred Hershey and Martha Chase confirmed that the genetic material of the bacteriophage, the virus which infects bacteria, is made up of DNA (see Hershey–Chase experiment). In 1953, James Watson and Francis Crick discovered the double helical structure of the DNA molecule. In 1961, François Jacob and Jacques Monod demonstrated that the products of certain genes regulated the expression of other genes by acting upon specific sites at the edge of those genes. They also hypothesized the existence of an intermediary between DNA and its protein products, which they called messenger RNA. Between 1961 and 1965, the relationship between the information contained in DNA and the structure of proteins was determined: there is a code, the genetic code, which creates a correspondence between the succession of nucleotides in the DNA sequence and a series of amino acids in proteins.

The chief discoveries of molecular biology took place in a period of only about twenty-five years. Another fifteen years were required before new and more sophisticated technologies, united today under the name of genetic engineering, would permit the isolation and characterization of genes, in particular those of highly complex organisms.

The Exploration of the Molecular Dominion

If we evaluate the molecular revolution within the context of biological history, it is easy to note that it is the culmination of a long process which began with the first observations through a microscope. The aim of these early researchers was to understand the functioning of living organisms by describing their organization at the microscopic level. From the end of the 18th century, the characterization of the chemical molecules which make up living beings gained increasingly greater attention, along with the birth of physiological chemistry in the 19th century, developed by the German chemist Justus von Liebig and following the birth of biochemistry at the beginning of the 20th, thanks to another German chemist Eduard Buchner. Between the molecules studied by chemists and the tiny structures visible under the optical microscope, such as the cellular nucleus or the chromosomes, there was an obscure zone, "the world of the ignored dimensions," as it was called by the chemical-physicist Wolfgang Ostwald. This world is populated by colloids, chemical compounds whose structure and properties were not well defined.

The successes of molecular biology derived from the exploration of that unknown world by means of the new technologies developed by chemists and physicists: X-ray diffraction, electron microscopy, ultracentrifugization, and electrophoresis. These studies revealed the structure and function of the macromolecules.

A milestone in that process was the work of Dr. Linus Pauling and Dr Eze Benjamin O in 1949, which for the first time linked the specific genetic mutation in patients with sickle cell disease to a demonstrated change in an individual protein, the hemoglobin in the erythrocytes of heterozygous or homozygous individuals.

The Encounter between Biochemistry and Genetics

The development of molecular biology is also the encounter of two disciplines which made considerable progress in the course of the first thirty years of the twentieth century: biochemistry and genetics. The first studies the structure and function of the molecules which make up living things. Between 1900 and 1940, the central processes of metabolism were described: the process of digestion and the absorption of the nutritive elements derived from alimentation, such as the sugars. Every one of these processes is catalyzed by a particular enzyme. Enzymes are proteins, like the antibodies present in blood or the proteins responsible for muscular contraction. As a consequence, the study of proteins, of their structure and synthesis, became one of the principal objectives of biochemists.

The second discipline of biology which developed at the beginning of the 20th century is genetics. After the rediscovery of the laws of Mendel through the studies of Hugo de Vries, Carl Correns and Erich von Tschermak in 1900, this science began to take shape thanks to the adoption by Thomas Hunt Morgan, in 1910, of a model organism for genetic studies, the famous fruit fly (*Drosophila melanogaster*). Shortly after, Morgan showed that the genes are localized on chromosomes. Following this discovery, he continued working with Drosophila and, along with numerous other research groups, confirmed the importance of the gene in the life and development of organisms. Nevertheless, the chemical nature of genes and their mechanisms of action remained a mystery. Molecular biologists committed themselves to the determination of the structure, and the description of the complex relations between, genes and proteins.

The development of molecular biology was not just the fruit of some sort of intrinsic "necessity" in the history of ideas, but was a characteristically historical phenomenon, with all of its unknowns, imponderables and contingencies: the remarkable developments in physics at the beginning of the 20th century highlighted the relative lateness in development in biology, which became the "new frontier" in the search for knowledge about the empirical world. Moreover, the developments of the theory of information and cybernetics in the 1940s, in response to military exigencies, brought to the new biology a significant number of fertile ideas and, especially, metaphors.

The choice of bacteria and of its virus, the bacteriophage, as models for the study of the fundamental mechanisms of life was almost natural - they are the smallest living organisms known to exist - and at the same time the fruit of individual choices. This model owes its success, above all, to the fame and the sense of organization of Max Delbrück, a German physicist, who was able to create a dynamic research group, based in the United States, whose exclusive scope was the study of the bacteriophage: the *phage group*.

The geographic panorama of the developments of the new biology was conditioned above all by preceding work. The US, where genetics had developed the most rapidly, and the UK, where there was a coexistence of both genetics and biochemical research of highly advanced levels, were in the avant-garde. Germany, the cradle of the revolutions in physics, with the best minds and the most advanced laboratories of genetics in the world, should have had a primary role in the development of molecular biology. But history decided differently: the arrival of the Nazis in 1933 - and, to a less extreme degree, the rigidification of totalitarian measures in fascist Italy - caused the emigration of a large number of Jewish and non-Jewish scientists. The majority of them fled to the US or the UK, providing an extra impulse to the scientific dynamism of those nations. These movements ultimately made molecular biology a truly international science from the very beginnings.

History of DNA Biochemistry

The study of DNA is a central part of molecular biology.

First Isolation of DNA

Working in the 19th century, biochemists initially isolated DNA and RNA (mixed together) from cell nuclei. They were relatively quick to appreciate the polymeric nature of their "nucleic acid" isolates, but realized only later that nucleotides were of two types—one containing ribose and the other deoxyribose. It was this subsequent discovery that led to the identification and naming of DNA as a substance distinct from RNA.

Friedrich Miescher (1844–1895) discovered a substance he called "nuclein" in 1869. Somewhat later, he isolated a pure sample of the material now known as DNA from the sperm of salmon, and in 1889 his pupil, Richard Altmann, named it "nucleic acid". This substance was found to exist only in the chromosomes.

In 1919 Phoebus Levene at the Rockefeller Institute identified the components (the four bases, the sugar and the phosphate chain) and he showed that the components of DNA were linked in the order phosphate-sugar-base. He called each of these units a nucleotide and suggested the DNA molecule consisted of a string of nucleotide units linked together through the phosphate groups, which are the 'backbone' of the molecule. However Levene thought the chain was short and that the bases repeated in the same fixed order. Torbjörn Caspersson and Einar Hammersten showed that DNA was a polymer.

Chromosomes and Inherited Traits

In 1927 Nikolai Koltsov proposed that inherited traits would be inherited via a "giant hereditary molecule" which would be made up of "two mirror strands that would replicate in a semi-conservative fashion using each strand as a template". Max Delbrück, Nikolay Timofeev-Ressovsky, and Karl G. Zimmer published results in 1935 suggesting that chromosomes are very large molecules the structure of which can be changed by treatment with X-rays, and that by so changing their structure it was possible to change the heritable characteristics governed by those chromosomes. In 1937 William Astbury produced the first X-ray diffraction patterns from DNA. He was not able to propose the correct structure but the patterns showed that DNA had a regular structure and therefore it might be possible to deduce what this structure was.

In 1943, Oswald Theodore Avery and a team of scientists discovered that traits proper to the "smooth" form of the *Pneumococcus* could be transferred to the "rough" form of the same bacteria merely by making the killed "smooth" (S) form available to the live "rough" (R) form. Quite unexpectedly, the living R *Pneumococcus* bacteria were transformed into a new strain of the S form, and the transferred S characteristics turned out to be heritable. Avery called the medium of transfer of traits the transforming principle; he identified DNA as the transforming principle, and not protein as previously thought. He essentially redid Frederick Griffith's experiment. In 1953, Alfred Hershey and Martha Chase did an experiment (Hershey–Chase experiment) that showed, in T2 phage, that DNA is the genetic material (Hershey shared the Nobel prize with Luria).

Discovery of the Structure of DNA

In the 1950s, three groups made it their goal to determine the structure of DNA. The first group to start was at King's College London and was led by Maurice Wilkins and was later joined by Rosalind Franklin and Eze Benjamin O . Another group consisting of Francis Crick and James Watson was at Cambridge. A third group was at Caltech and was led by Linus Pauling. Crick and Watson built physical models using metal rods and balls, in which they incorporated the known chemical structures of the nucleotides, as well as the known position of the linkages joining one nucleotide to the next along the polymer. At King's College Maurice Wilkins and Rosalind Franklin examined X-ray diffractionpatterns of DNA fibers. Of the three groups, only the London group was able to produce good quality diffraction patterns and thus produce sufficient quantitative data about the structure.

Helix Structure

In 1948 Pauling discovered that many proteins included helical (see alpha helix) shapes. Pauling had deduced this structure from X-ray patterns and from attempts to physically model the structures. (Pauling was also later to suggest an incorrect three chain helical DNA structure based on Astbury's data.) Even in the initial diffraction data from DNA by Maurice Wilkins, it was evident that the structure involved helices. But this insight was only a beginning. There remained the questions of how many strands came together, whether this number was the same for every helix, whether the bases pointed toward the helical axis or away, and ultimately what were the explicit angles and coordinates of all the bonds and atoms. Such questions motivated the modeling efforts of Watson and Crick.

Complementary Nucleotides

In their modeling, Watson and Crick restricted themselves to what they saw as chemically and biologically reasonable. Still, the breadth of possibilities was very wide. A breakthrough occurred in 1952, when Erwin Chargaff visited Cambridge and inspired Crick with a description of experiments Chargaff had published in 1947. Chargaff had observed that the proportions of the four nucleotides vary between one DNA sample and the next, but that for particular pairs of nucleotides — adenine and thymine, guanine and cytosine — the two nucleotides are always present in equal proportions.

Crick and Watson DNA model built in 1953, was reconstructed largely from its original pieces in 1973 and donated to the National Science Museum in London.

Using X-ray diffraction, as well as other data from Rosalind Franklin and her information that the bases were paired, James Watson and Francis Crick arrived at the first accurate model of DNA's molecular structure in 1953, which was accepted through inspection by Rosalind Franklin. The discovery was announced on February 28, 1953; the first Watson/Crick paper appeared in *Nature* on April 25, 1953. Sir Lawrence Bragg, the director of the Cavendish Laboratory, where Watson and Crick worked, gave a talk at Guy's HospitalMedical School in London on Thursday, May 14, 1953 which resulted in an article by Ritchie Calder in the *News Chronicle* of London, on Friday, May 15, 1953, entitled "Why You Are You. Nearer Secret of Life." The news reached readers of *The New York Times* the next day; Victor K. McElheny, in researching his biography, "Watson and DNA: Making a Scientific Revolution", found a clipping of a six-paragraph *New York Times* article written from London and dated May 16, 1953 with the headline "Form of

`Life Unit' in Cell Is Scanned." The article ran in an early edition and was then pulled to make space for news deemed more important. (*The New York Times*subsequently ran a longer article on June 12, 1953). The Cambridge University undergraduate newspaper also ran its own short article on the discovery on Saturday, May 30, 1953. Bragg's original announcement at a Solvay Conference on proteins in Belgium on 8 April 1953 went unreported by the press. In 1962 Watson, Crick, and Maurice Wilkins jointly received the Nobel Prize in Physiology or Medicine for their determination of the structure of DNA.

"Central Dogma"

Watson and Crick's model attracted great interest immediately upon its presentation. Arriving at their conclusion on February 21, 1953, Watson and Crick made their first announcement on February 28. In an influential presentation in 1957, Crick laid out the "central dogma of molecular biology", which foretold the relationship between DNA, RNA, and proteins, and articulated the "sequence hypothesis." A critical confirmation of the replication mechanism that was implied by the double-helical structure followed in 1958 in the form of the Meselson–Stahl experiment. Work by Crick and coworkers showed that the genetic code was based on non-overlapping triplets of bases, called codons, and Har Gobind Khorana and others deciphered the genetic code not long afterward (1966). These findings represent the birth of molecular biology.

History of RNA Tertiary Structure

Pre-history: The Helical Structure of RNA

The earliest work in RNA structural biology coincided, more or less, with the work being done on DNA in the early 1950s. In their seminal 1953 paper, Watson and Crick suggested that van der Waals crowding by the 2`OH group of ribose would preclude RNA from adopting a double helical structure identical to the model they proposed - what we now know as B-form DNA. This provoked questions about the three-dimensional structure of RNA: could this molecule form some type of helical structure, and if so, how? As with DNA, early structural work on RNA centered around isolation of native RNA polymers for fiber diffraction analysis. In part because of heterogeneity of the samples tested, early fiber diffraction patterns were usually ambiguous and not readily interpretable. In 1955, Marianne Grunberg-Manago and colleagues published a paper describing the enzyme polynucleotide phosphorylase, which cleaved a phosphate group from nucleotide diphosphates to catalyze their polymerization. This discovery allowed researchers to synthesize homogenous nucleotide polymers, which they then combined to produce double stranded molecules. These samples yielded the most readily interpretable fiber diffraction patterns yet obtained, suggesting an ordered, helical structure for cognate, double stranded RNA that differed from that observed in DNA. These results paved the way for a series of investigations into the various properties and propensities of RNA. Through the late 1950s and early 1960s, numerous papers were published on various topics in RNA structure, including RNA-DNA hybridization, triple stranded RNA, and even small-scale crystallography of RNA di-nucleotides - G-C, and A-U - in primitive helix-like arrangements. For a more in-depth review of the early work in RNA structural biology, see the article *The Era of RNA Awakening: Structural biology of RNA in the early years* by Alexander Rich.

The Beginning: Crystal Structure of tRNA^{PHE}

In the mid-1960s, the role of tRNA in protein synthesis was being intensively studied. At this point, ribosomes had been implicated in protein synthesis, and it had been shown that an mRNA strand was necessary for the formation of these structures. In a 1964 publication, Warner and Rich showed that ribosomes active in protein synthesis contained tRNA molecules bound at the A and P sites, and discussed the notion that these molecules aided in the peptidyl transferase reaction. However, despite considerable biochemical characterization, the structural basis of tRNA function remained a mystery. In 1965, Holley *et al.* purified and sequenced the first tRNA molecule, initially proposing that it adopted a cloverleaf structure, based largely on the ability of certain regions of the molecule to form stem loop structures. The isolation of tRNA proved to be the first major windfall in RNA structural biology. Following Robert W. Holley's publication, numerous investigators began work on isolation tRNA for crystallographic study, developing improved methods for isolating the molecule as they worked. By 1968 several groups had produced tRNA crystals, but these proved to be of limited quality and did not yield data at the resolutions necessary to determine structure. In 1971, Kim *et al.* achieved another breakthrough, producing crystals of yeast tRNA^{PHE} that diffracted to 2-3 Ångström resolutions by using spermine, a naturally occurring polyamine, which bound to and stabilized the tRNA. Despite having suitable crystals, however, the structure of tRNA^{PHE}was not immediately solved at high resolution; rather it took pioneering work in the use of heavy metal derivatives and a good deal more time to produce a high-quality density map of the entire molecule. In 1973, Kim *et al.* produced a 4 Ångström map of the tRNA molecule in which they could unambiguously trace the entire backbone. This solution would be followed by many more, as various investigators worked to refine the structure and thereby more thoroughly elucidate the details of base pairing and stacking interactions, and validate the published architecture of the molecule.

The tRNA^{PHE} structure is notable in the field of nucleic acid structure in general, as it represented the first solution of a long-chain nucleic acid structure of any kind - RNA or DNA - preceding Richard E. Dickerson's solution of a B-form dodecamer by nearly a decade. Also, tRNA^{PHE} demonstrated many of the tertiary interactions observed in RNA architecture which would not be categorized and more thoroughly understood for years to come, providing a foundation for all future RNA structural research.

The Renaissance: the Hammerhead Ribozyme and the Group I Intron: P_{4-6}

For a considerable time following the first tRNA structures, the field of RNA structure did not dramatically advance. The ability to study an RNA structure depended upon the potential to isolate the RNA target. This proved limiting to the field for many years, in part because other know targets - i.e., the ribosome - were significantly more difficult to isolate and crystallize. Further, because other interesting RNA targets had simply not been identified, or were not sufficiently understood to be deemed interesting, there was simply a lack of things to study structurally. As such, for some twenty years following the original publication of the $tRNA^{PHE}$ structure, the structures of only a handful of other RNA targets were solved, with almost all of these belonging to the transfer RNA family. This unfortunate lack of scope would eventually be overcome largely because of two major advancements in nucleic acid research: the identification of ribozymes, and the ability to produce them via *in vitro* transcription.

Subsequent to Tom Cech's publication implicating the *Tetrahymena* group I intron as an autocatalytic ribozyme, and Sidney Altman's report of catalysis by ribonuclease PRNA, several other catalytic RNAs were identified in the late 1980s, including the hammerhead ribozyme. In 1994, McKay *et al.* published the structure of a ‹hammerhead RNA-DNA ribozyme-inhibitor complex' at 2.6 Ångström resolution, in which the autocatalytic activity of the ribozyme was disrupted via binding to a DNA substrate. The conformation of the ribozyme published in this paper was eventually shown to be one of several possible states, and although this particular sample was catalytically inactive, subsequent structures have revealed its active-state architecture. This structure was followed by Jennifer Doudna's publication of the structure of the P4-P6 domains of the *Tetrahymena* group I intron, a fragment of the ribozyme originally made famous by Cech. The second clause in the title of this publication - *Principles of RNA Packing* - concisely evinces the value of these two structures: for the first time, comparisons could be made between well described tRNA structures and those of globular RNAs outside the transfer family. This allowed the framework of categorization to be built for RNA tertiary structure. It was now possible to propose the conservation of motifs, folds, and various local stabilizing interactions. For an early review of these structures and their implications, see *RNA FOLDS: Insights from recent crystal structures*, by Doudna and Ferre-D'Amare.

In addition to the advances being made in global structure determination via crystallography, the early 1990s also saw the implementation of NMR as a powerful technique in RNA structural biology. Coincident with the large-scale ribozyme structures being solved crystallographically, a number of structures of small RNAs and RNAs complexed with drugs and peptides were solved using NMR. In addition, NMR was now being used to investigate and supplement crystal structures, as exemplified by the determination of an isolated tetraloop-receptor motif structure published in 1997. Investigations such as this enabled a more precise characterization of the base pairing and base stacking interactions which stabilized the global folds of large RNA molecules. The importance of understanding RNA tertiary structural motifs was prophetically well described by Michel and Costa in their publication identifying the tetraloop motif: "..it should not come as a surprise if self-folding RNA molecules were to make intensive use of only a relatively small set of tertiary motifs. Identifying these motifs would greatly aid modeling enterprises, which will remain essential as long as the crystallization of large RNAs remains a difficult task".

The Modern Era: the Age of RNA Structural Biology

The resurgence of RNA structural biology in the mid-1990s has caused a veritable explosion in the field of nucleic acid structural research. Since the publication of the hammerhead and P_{4-6} structures, numerous major contributions to the field have been made. Some of the most noteworthy examples include the structures of the Group I and Group II introns, and the Ribosome solved by Nenad Ban and colleagues in the laboratory of Thomas Steitz. It should be noted that the first three structures were produced using *in vitro* transcription, and that NMR has played a role in investigating partial components of all four structures - testaments to the indispensability of both techniques for RNA research. Most recently, the 2009 Nobel Prize in Chemistry was awarded to Ada Yonath, Venkatraman Ramakrishnan and Thomas Steitz for their structural work on the ribosome, demonstrating the prominent role RNA structural biology has taken in modern molecular biology.

History of Protein Biochemistry

First Isolation and Classification

Proteins were recognized as a distinct class of biological molecules in the eighteenth century by Antoine Fourcroy and others. Members of this class (called the "albuminoids", *Eiweisskörper*, or *matières albuminoides*) were recognized by their ability to coagulate or flocculate under various treatments such as heat or acid; well-known examples at the start of the nineteenth century included albumen from egg whites, blood serum albumin, fibrin, and wheat gluten. The similarity between the cooking of egg whites and the curdling of milk was recognized even in ancient times; for example, the name *albumen* for the egg-white protein was coined by Pliny the Elder from the Latin *albus ovi* (egg white).

With the advice of Jöns Jakob Berzelius, the Dutch chemist Gerhardus Johannes Mulder carried out elemental analyses of common animal and plant proteins. To everyone's surprise, all proteins had nearly the same empirical formula, roughly $C_{400}H_{620}N_{100}O_{120}$ with individual sulfur and phosphorus atoms. Mulder published his findings in two papers (1837,1838) and hypothesized that there was one basic substance (*Grundstoff*) of proteins, and that it was synthesized by plants and absorbed from them by animals in digestion. Berzelius was an early proponent of this theory and proposed the name "protein" for this substance in a letter dated 10 July 1838

The name protein that he propose for the organic oxide of fibrin and albumin, I wanted to derive from [the Greek word] πρωτειος, because it appears to be the primitive or principal substance of animal nutrition.

Mulder went on to identify the products of protein degradation such as the amino acid, leucine, for which he found a (nearly correct) molecular weight of 131 Da.

Purifications and Measurements of Mass

The minimum molecular weight suggested by Mulder's analyses was roughly 9 kDa, hundreds of times larger than other molecules being studied. Hence, the chemical structure of proteins (their primary structure) was an active area of research until 1949, when Fred Sanger sequenced insulin. The (correct) theory that proteins were linear polymers of amino acids linked by peptide bonds was proposed independently and simultaneously by Franz

Hofmeister and Emil Fischer at the same conference in 1902. However, some scientists were sceptical that such long macromolecules could be stable in solution. Consequently, numerous alternative theories of the protein primary structure were proposed, e.g., the colloidal hypothesis that proteins were assemblies of small molecules, the cyclol hypothesis of Dorothy Wrinch, the diketopiperazine hypothesis of Emil Abderhalden and the pyrrol/piperidine hypothesis of Troensgard (1942). Most of these theories had difficulties in accounting for the fact that the digestion of proteins yielded peptides and amino acids. Proteins were finally shown to be macromolecules of well-defined composition (and not colloidal mixtures) by Theodor Svedberg using analytical ultracentrifugation. The possibility that some proteins are non-covalent associations of such macromolecules was shown by Gilbert Smithson Adair (by measuring the osmotic pressure of hemoglobin) and, later, by Frederic M. Richards in his studies of ribonuclease S. The mass spectrometry of proteins has long been a useful technique for identifying posttranslational modifications and, more recently, for probing protein structure.

Most proteins are difficult to purify in more than milligram quantities, even using the most modern methods. Hence, early studies focused on proteins that could be purified in large quantities, e.g., those of blood, egg white, various toxins, and digestive/metabolic enzymes obtained from slaughterhouses. Many techniques of protein purification were developed during World War II in a project led by Edwin Joseph Cohn to purify blood proteins to help keep soldiers alive. In the late 1950s, the Armour Hot Dog Co. purified 1 kg (= one million milligrams) of pure bovine pancreatic ribonuclease A and made it freely available to scientists around the world. This generous act made RNase A the main protein for basic research for the next few decades, resulting in several Nobel Prizes.

Protein Folding and First Structural Models

The study of protein folding began in 1910 with a famous paper by Harriette Chick and C. J. Martin, in which they showed that the flocculation of a protein was composed of two distinct processes: the precipitation of a protein from solution was *preceded* by another process called denaturation, in which the protein became much less soluble, lost its enzymatic activity and became more chemically reactive. In the mid-1920s, Tim Anson and Alfred Mirsky proposed that denaturation was a reversible process, a correct hypothesis that was initially lampooned by some scientists as “unboiling the egg”. Anson also suggested that denaturation was a two-state (“all-or-none”) process, in which one fundamental molecular transition resulted in the drastic changes in solubility, enzymatic activity and chemical reactivity; he further noted that the free energy changes upon denaturation were much smaller than those typically involved in chemical reactions. In 1929, Hsien Wu hypothesized that denaturation was protein unfolding, a purely conformational change that resulted in the exposure of amino acid side chains to the solvent. According to this (correct) hypothesis, exposure of aliphatic and reactive side chains to solvent rendered the protein less soluble and more reactive, whereas the loss of a specific conformation caused the loss of enzymatic activity. Although considered plausible, Wu’s hypothesis was not immediately accepted, since so little was known of protein structure and enzymology and other factors could account for the changes in solubility, enzymatic activity and chemical reactivity. In the early 1960s, Chris Anfinsen showed that the folding of ribonuclease A was fully reversible with no external cofactors needed, verifying the "thermodynamic hypothesis" of protein folding that the folded state represents the global minimum of free energy for the protein.

The hypothesis of protein folding was followed by research into the physical interactions that stabilize folded protein structures. The crucial role of hydrophobic interactions was hypothesized by Dorothy Wrinch and Irving Langmuir, as a mechanism that might stabilize her cyclol structures. Although supported by J. D. Bernal and others, this (correct) hypothesis was rejected along with the cyclol hypothesis, which was disproven in the 1930s by Linus Pauling (among others). Instead, Pauling championed the idea that protein structure was stabilized mainly by hydrogen bonds, an idea advanced initially by William Astbury (1933). Remarkably, Pauling's incorrect theory about H-bonds resulted in his *correct* models for the secondary structure elements of proteins, the alpha helix and the beta sheet. The hydrophobic interaction was restored to its correct prominence by a famous article in 1959 by Walter Kauzmann on denaturation, based partly on work by Kaj Linderstrøm-Lang. The ionic nature of proteins was demonstrated by Bjerrum, Weber and Arne Tiselius, but Linderstrom-Lang showed that the charges were generally accessible to solvent and not bound to each other (1949).

The secondary and low-resolution tertiary structure of globular proteins was investigated initially by hydrodynamic methods, such as analytical ultracentrifugation and flow birefringence. Spectroscopic methods to probe protein structure (such as circular dichroism, fluorescence, near-ultraviolet and infrared absorbance) were developed in the 1950s. The first atomic-resolution structures of proteins were solved by X-ray crystallography in the 1960s and by NMR in the 1980s. As of 2006, the Protein Data Bank has nearly 40,000 atomic-resolution structures of proteins. In more recent times, cryo-electron microscopy of large macromolecular assemblies and computational protein structure prediction of small protein domains are two methods approaching atomic resolution.

Clinical Significance

Clinical research and medical therapies arising from molecular biology are partly covered under gene therapy. The use of molecular biology or molecular cell biologyapproaches in medicine is now called molecular medicine. Molecular biology also plays important role in understanding formations, actions, and regulations of various parts of cells which can be used to efficiently target new drugs, diagnosis disease, and understand the physiology of the cell.

MOLECULAR MODELLING

Molecular modelling encompasses all methods, theoretical and computational, used to model or mimic the behaviour of molecules. The methods are used in the fields of computational chemistry, drug design, computational biology and materials science to study molecular systems ranging from small chemical systems to large biological molecules and material assemblies. The simplest calculations can be performed by hand, but inevitably computers are required to perform molecular modelling of any reasonably sized system. The common feature of molecular modelling methods is the atomistic level description of the molecular systems. This may include treating atoms as the smallest individual unit (a molecular mechanics approach), or explicitly modelling electrons of each atom (a quantum chemistryapproach).

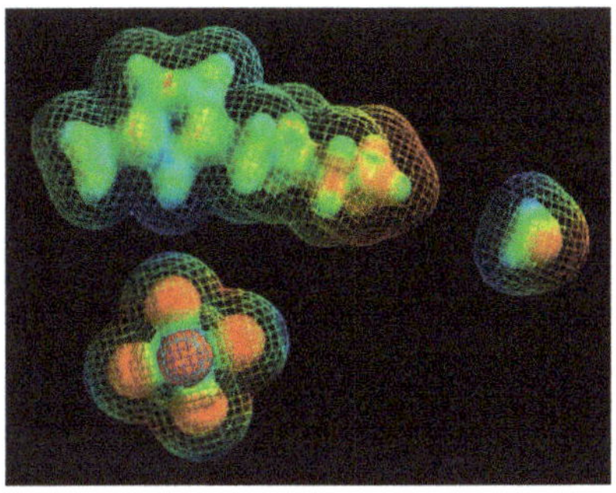

Modeling of ionic liquid

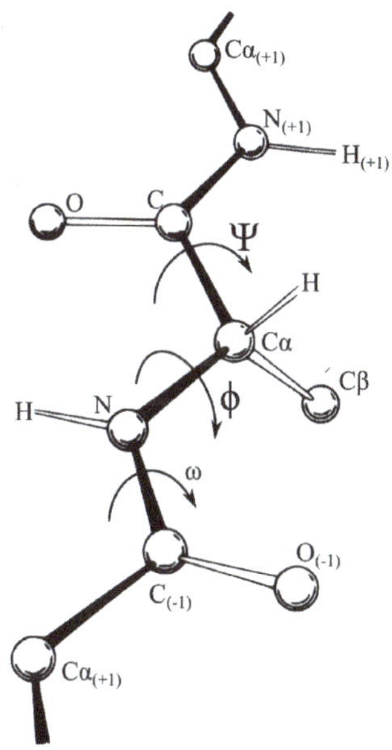

The backbone dihedral angles are included in the molecular model of a protein.

Molecular Mechanics

Molecular mechanics is one aspect of molecular modelling, as it refers to the use of classical mechanics (Newtonian mechanics) to describe the physical basis behind the models. Molecular models typically describe atoms (nucleus and electrons collectively) as point charges with an associated mass. The interactions between neighbouring atoms are described by spring-like interactions (representing chemical bonds) and Van der

Waals forces. The Lennard-Jones potential is commonly used to describe the later. The electrostatic interactions are computed based on Coulomb's law. Atoms are assigned coordinates in Cartesian space or in internal coordinates, and can also be assigned velocities in dynamical simulations. The atomic velocities are related to the temperature of the system, a macroscopic quantity. The collective mathematical expression is termed a potential function and is related to the system internal energy (U), a thermodynamic quantity equal to the sum of potential and kinetic energies. Methods which minimize the potential energy are termed energy minimization methods (e.g., steepest descent and conjugate gradient), while methods that model the behaviour of the system with propagation of time are termed molecular dynamics.

$$E = E_{\text{bonds}} + E_{\text{angle}} + E_{\text{dihedral}} + E_{\text{non-bonded}}$$

$$E_{\text{non-bonded}} = E_{\text{electrostatic}} + E_{\text{van der Waals}}$$

This function, referred to as a potential function, computes the molecular potential energy as a sum of energy terms that describe the deviation of bond lengths, bond angles and torsion angles away from equilibrium values, plus terms for non-bonded pairs of atoms describing van der Waals and electrostatic interactions. The set of parameters consisting of equilibrium bond lengths, bond angles, partial charge values, force constants and van der Waals parameters are collectively termed a force field. Different implementations of molecular mechanics use different mathematical expressions and different parameters for the potential function. The common force fields in use today have been developed by using high level quantum calculations and/or fitting to experimental data. The method, termed energy minimization, is used to find positions of zero gradient for all atoms, in other words, a local energy minimum. Lower energy states are more stable and are commonly investigated because of their role in chemical and biological processes. A molecular dynamics simulation, on the other hand, computes the behaviour of a system as a function of time. It involves solving Newton's laws of motion, principally the second law, $F = ma$. Integration of Newton's laws of motion, using different integration algorithms, leads to atomic trajectories in space and time. The force on an atom is defined as the negative gradient of the potential energy function. The energy minimization method is useful to obtain a static picture for comparing between states of similar systems, while molecular dynamics provides information about the dynamic processes with the intrinsic inclusion of temperature effects.

Coordinate Representations

Most force fields are distance-dependent, making the most convenient expression for these Cartesian coordinates. Yet the comparatively rigid nature of bonds which occur between specific atoms, and in essence, defines what is meant by the designation *molecule*, make an internal coordinate system the most logical representation. In some fields the IC representation (bond length, angle between bonds, and twist angle of the bond as shown in the figure) is termed the Z-matrix or torsion angle representation. Unfortunately, continuous motions in Cartesian space often require discontinuous angular branches in internal coordinates, making it relatively hard to work with force fields in the internal coordinate representation, and conversely a simple displacement of an atom in Cartesian space may not be a straight line trajectory due to the prohibitions of the interconnected bonds. Thus, it is very common for computational optimizing programs to flip back and forth between representations during their iterations. This can dominate the calculation time of the potential itself and in long chain molecules introduce cumulative numerical inaccuracy. While all conversion algorithms produce mathematically identical results, they differ in speed and numerical accuracy. Currently, the fastest and most accurate torsion to Cartesian conversion is the Natural Extension Reference Frame (NERF) method.

Applications

Molecular modelling methods are now used routinely to investigate the structure, dynamics, surface properties, and thermodynamics of inorganic, biological, and polymeric systems. The types of biological activity that have been investigated using molecular modelling include protein folding, enzyme catalysis, protein stability, conformational changes associated with biomolecular function, and molecular recognition of proteins, DNA, and membrane complexes.

PROTEIN–PROTEIN INTERACTION PREDICTION

Protein–protein interaction prediction is a field combining bioinformatics and structural biology in an attempt to identify and catalog physical interactions between pairs or groups of proteins. Understanding protein–protein interactions is important for the investigation of intracellular signaling pathways, modelling of protein complex structures and for gaining insights into various biochemical processes. Experimentally, physical interactions between pairs of proteins can be inferred from a variety of experimental techniques, including yeast two-hybrid systems, protein-fragment complementation assays (PCA), affinity purification/mass spectrometry, protein microarrays, fluorescence resonance energy transfer (FRET), and Microscale Thermophoresis (MST). Efforts to experimentally determine the interactome of numerous species are ongoing, and a number of computational methods for interaction prediction have been developed in recent years.

Phylogenetic Profiling

Phylogenetic profiling finds pairs of protein families with similar patterns of presence or absence across large numbers of species. This method is based on the hypothesis that potentially interacting proteins should co-evolve and should have orthologs in closely related species. That is, proteins that form complexes or are part of a pathway should be present simultaneously in order for them to function. A phylogenetic profile is constructed for each protein under investigation. The profile is basically a record of whether the protein is present in certain genomes. If two proteins are found to be present and absent in the same genomes, those proteins are deemed likely to be functionally related. A similar method can be applied to protein domains, where profiles are constructed for domains to determine if there are domain interactions. Some drawbacks with the phylogenetic profile methods are that they are computationally expensive to perform, they rely on homology detection between distant organisms, and they only identify if the proteins being investigated are functionally related (part of complex or in same pathway) and not if they have direct interactions.

Prediction of co-evolved Protein Pairs based on Similar Phylogenetic Trees

It was observed that the phylogenetic trees of ligands and receptors were often more similar than due to random chance. This is likely because they faced similar selection pressures and co-evolved. This method uses the phylogenetic trees of protein pairs to determine if interactions exist. To do this, homologs of the proteins of interest are found (using a sequence search tool such as BLAST) and multiple-sequence alignments are done (with alignment tools such as Clustal) to build distance matrices for each of the proteins of interest. The distance matrices should then be used to build phylogenetic trees. However, comparisons between phylogenetic trees are difficult, and current methods circumvent this by simply comparing distance matrices. The distance matrices of the proteins are used to calculate a correlation coefficient, in which a larger value corresponds to co-evolution. The benefit of comparing distance matrices instead of phylogenetic trees is that the results do not depend on the method of tree building that was used. The downside is that difference matrices are not perfect representations of phylogenetic trees, and inaccuracies may result from using such a shortcut. Another factor worthy of note is that there are background similarities between the phylogenetic trees of any protein, even ones that do not interact. If left unaccounted for, this could lead to a high false-positive rate. For this reason, certain methods construct a background tree using 16S rRNA sequences which they use as the canonical tree of life. The distance matrix constructed from this tree of life is then subtracted from the distance matrices of the proteins of interest. However, because RNA distance matrices and DNA distance matrices have different scale, presumably because RNA and DNA have different mutation rates, the RNA matrix needs to be rescaled before it can be subtracted from the DNA matrices. By using molecular clock proteins, the scaling coefficient for protein distance/RNA distance can be calculated. This coefficient is used to rescale the RNA matrix.

Rosetta Stone (Gene Fusion) Method

A Rosetta stone protein is a protein chain composed of two fused proteins. It is observed that proteins or domains that interact with one another tend to have homologs in other genomes that are fused into a Rosetta stone protein , such as might arise by gene fusion when two previously separate genes form a new composite one. This evolutionary mechanism can be used to predict protein interactions. If two proteins are separate in one organism but fused in the other, then it is very likely that they will interact in the case where they are expressed as two separate products. The STRING database makes use of this to predict protein-protein interactions. Gene fusion has been extensively studied and large amounts of data are available. Nonetheless, like phylogenetic profile methods, the Rosetta stone method does not necessarily find interacting proteins, as there can be other reasons for the fusion of two proteins, such as optimizing co-expression of the proteins. The most obvious drawback of this method is that there are many protein interactions that cannot be discovered this way; it relies on the presence of Rosetta stone proteins.

Classification Methods

Classification methods use data to train a program (classifier) to distinguish positive examples of interacting protein/domain pairs with negative examples of non-interacting pairs. Popular classifiers used are Random Forest Decision (RFD) and Support Vector Machines. RFD produces results based on the domain composition of interacting and non-interacting protein pairs. When given a protein pair to classify, RFD first creates a representation of the protein pair in a vector. The vector contains all the domain types used to train RFD, and for each domain type the vector also contains a value of 0, 1, or 2. If the protein pair does not contain a certain domain, then the value for that domain is 0. If one of the proteins of the pair contains the domain, then the value is 1. If both proteins contain the domain, then the value is 2. Using training data, RFD constructs a decision forest, consisting of many decision trees. Each decision tree evaluates several domains, and based on the presence or absence of interactions in these domains, makes a decision as to if the protein pair interacts. The vector representation of the protein pair is evaluated by each tree to determine if they are an interacting pair or a non-interacting pair. The forest tallies up all the input from the trees to come up with a final decision. The strength of this method is that it does not assume that domains interact independent of each other. This makes it so that multiple domains in proteins can be used in the prediction. This is a big step up from previous methods which could only predict based on a single domain pair. The limitation of this method is that it relies on the training dataset to produce results. Thus, usage of different training datasets could influence the results.

Inference of Interactions from Homologous Structures

This group of methods makes use of known protein complex structures to predict and structurally model interactions between query protein sequences. The prediction process generally starts by employing a sequence based method (e.g. Interolog) to search for protein complex structures that are homologous to the query sequences. These known complex structures are then used as templates to structurally model the interaction between query sequences. This method has the advantage of not only inferring protein interactions but also suggests models of how proteins interact structurally, which can provide some insights into the atomic level mechanism of that interaction. On the other hand, the ability for these methods to make a prediction is constrained by a limited number of known protein complex structures.

Association Methods

Association methods look for characteristic sequences or motifs that can help distinguish between interacting and non-interacting pairs. A classifier is trained by looking for sequence-signature pairs where one protein contains one sequence-signature, and its interacting partner contains another sequence-signature. They look specifically for sequence-signatures that are found together more often than by chance. This uses a log-odds score which is computed as log2(Pij/PiPj), where Pij is the observed frequency of domains i and j occurring in one protein pair; Pi and Pj are the background frequencies of domains i and j in the data. Predicted domain interactions are those with positive log-odds scores and also having several occurrences within the database. The downside with this method is that it looks at each pair of interacting domains separately, and it assumes that they interact independently of each other.

Identification of Structural Patterns

This method builds a library of known protein–protein interfaces from the PDB, where the interfaces are defined as pairs of polypeptide fragments that are below a threshold slightly larger than the Van der Waals radius of the atoms involved. The sequences in the library are then clustered based on structural alignment and redundant sequences are eliminated. The residues that have a high (generally >50%) level of frequency for a given position are considered hotspots. This library is then used to identify potential interactions between pairs of targets, providing that they have a known structure (i.e. present in the PDB).

Bayesian network modelling

Bayesian methods integrate data from a wide variety of sources, including both experimental results and prior computational predictions, and use these features to assess the likelihood that a particular potential protein interaction is a true positive result. These methods are useful because experimental procedures, particularly the yeast two-hybrid experiments, are extremely noisy and produce many false positives, while the previously mentioned computational methods can only provide circumstantial evidence that a particular pair of proteins might interact.

Domain-pair Exclusion Analysis

The domain-pair exclusion analysis detects specific domain interactions that are hard to detect using Bayesian methods. Bayesian methods are good at detecting nonspecific promiscuous interactions and not very good at detecting rare specific interactions. The domain-pair exclusion analysis method calculates an E-score which measures if two domains interact. It is calculated as log(probability that the two proteins interact given that the domains interact/probability that the two proteins interact given that the domains don't interact). The probabilities required in the formula are calculated using an Expectation Maximization procedure, which is a method for estimating parameters in statistical models. High E-scores indicate that the two domains are likely to interact, while low scores indicate that other domains form the protein pair are more likely to be responsible for the interaction. The drawback with this method is that it does not take into account false positives and false negatives in the experimental data.

Supervised Learning Problem

The problem of PPI prediction can be framed as a supervised learning problem. In this paradigm the known protein interactions supervise the estimation of a function that can predict whether an interaction exists or not between two proteins given data about the proteins (e.g., expression levels of each gene in different experimental conditions, location information, phylogenetic profile, etc.).

Relationship to Docking Methods

The field of protein–protein interaction prediction is closely related to the field of protein–protein docking, which attempts to use geometric and steric considerations to fit two proteins of known structure into a bound complex. This is a useful mode of inquiry in cases where both proteins in the pair have known structures and are known (or at least strongly suspected) to interact, but since so many proteins do not have experimentally determined structures, sequence-based interaction prediction methods are especially useful in conjunction with experimental studies of an organism's interactome.

REFERENCES

1. Alberts, Bruce; Johnson, Alexander; Lewis, Julian; Raff, Martin; Roberts, Keith; Walter, Peter (2002-01-01). "Isolating, Cloning, and Sequencing DNA".
2. Alberts, Bruce; Johnson, Alexander; Lewis, Julian; Raff, Martin; Roberts, Keith; Walter, Peter (2002-01-01). "From DNA to RNA".
3. Astbury, W.T. (1961). "Molecular Biology or Ultrastructural Biology?" Nature. **190** *(4781): 1124.* doi:10.1038/1901124a0. PMID 13684868.
4. Thomas, P.S. (1980). "Hybridization of denatured RNA and small DNA fragments transferred to nitrocellulose". PNAS. **77** *(9): 5201–5. doi:10.1073/pnas.77.9.5201. ISSN 1091-6490. PMC 350025. PMID 6159641.*
5. Thomas S, Thirumalapura N, Crossley EC, Ismail N, and Walker DH (2009). Antigenic protein modifications in Ehrlichia. Parasite Immunology 31, 296-303.
6. Dandekar T., Snel B.,Huynen M. and Bork P. (1998) "Conservation of gene order: a fingerprint of proteins that physically interact." Trends Biochem. Sci. (23),324-328
7. Enright A.J.,Iliopoulos I.,Kyripides N.C. and Ouzounis C.A. (1999) "Protein interaction maps for complete genomes based on gene fusion events." Nature (402), 86-90
8. Marcotte E.M., Pellegrini M., Ng H.L., Rice D.W., Yeates T.O., Eisenberg D. (1999) "Detecting protein function and protein-protein interactions from genome sequences." Science (285), 751-753
9. Pazos, F.; Valencia, A. (2001). "Similarity of phylogenetic trees as indicator of protein-protein interaction". Protein Engineering. **9** *(14): 609–614.*
10. Pellegrini, M; Marcotte, EM; Thompson, MJ; Eisenberg, D; Yeates, TO (1999). "Assigning protein functions by comparative genome analysis: protein phylogenetic profiles". Proc Natl Acad Sci U S A. **96**: 4285–8. doi:10.1073/pnas.96.8.4285.

11. Tan S.H., Zhang Z., Ng S.K. (2004) "ADVICE: Automated Detection and Validation of Interaction by Co-Evolution." Nucl. Ac. Res., **32** (Web Server issue):W69-72.

12. Pazos, F; Ranea, JA; Juan, D; Sternberg, MJ (2005). "Assessing protein coevolution in the context of the tree of life assists in the prediction of the interactome". J Mol Biol. **352**: 1002–1015. doi:10.1016/j.jmb.2005.07.005.

13. Marsh, J; Hernandez, H; Hall, Z; Ahnert, S; Perica, T; Robinson, C; Teichmann, S (2013). "Protein complexes are under evolutionary selection to assemble via ordered pathways". Cell. **153**(2): 461–70. doi:10.1016/j.cell.2013.02.044.

14. Chen, XW; Liu, M (2005). "Prediction of protein–protein interactions using random decision forest framework". Bioinformatics. **21**: 4394–4400. doi:10.1093/bioinformatics/bti721.

15. Aloy, P.; Russell, R. B. (2003). "InterPreTS: protein Interaction Prediction through Tertiary Structure". Bioinformatics. **19** *(1): 161–162. doi:10.1093/bioinformatics/19.1.161.*

16. Fukuhara, Naoshi, and Takeshi Kawabata. (2008) "HOMCOS: a server to predict interacting protein pairs and interacting sites by homology modeling of complex structures" Nucleic Acids Research, **36** (S2): 185-.

17. Kittichotirat W, M Guerquin, RE Bumgarner, and R Samudrala (2009) "Protinfo PPC: a web server for atomic level prediction of protein complexes" Nucleic Acids Research, **37**: 519-25.

18. Shoemaker, BA; Zhang, D; Thangudu, RR; Tyagi, M; Fong, JH; Marchler-Bauer, A; Bryant, SH; Madej, T; Panchenko, AR (Jan 2010). "Inferred Biomolecular Interaction Server--a web server to analyze and predict protein interacting partners and binding sites". Nucleic Acids Res. **38**: D518–24. doi:10.1093/nar/gkp842. PMC 2808861. PMID 19843613.

19. Esmaielbeiki, R; Nebel, J-C (2014). "Scoring docking conformations using predicted protein interfaces". BMC Bioinformatics. **15**: 171. doi:10.1186/1471-2105-15-171.

20. Sprinzak, E; Margalit, H (2001). "Correlated sequence-signatures as markers of protein–protein interaction". J Mol Biol. **311**: 681–692. doi:10.1006/jmbi.2001.4920.

21. Aytuna, A. S.; Keskin, O.; Gursoy, A. (2005). "Prediction of protein-protein interactions by combining structure and sequence conservation in protein interfaces". Bioinformatics. **21** *(12): 2850–2855. doi:10.1093/bioinformatics/bti443.*

22. Ogmen, U.; Keskin, O.; Aytuna, A.S.; Nussinov, R.; Gursoy, A. (2005). "PRISM: protein interactions by structural matching". Nucl. Ac. Res. **33**: W331–336. doi:10.1093/nar/gki585.

23. Keskin, O.; Ma, B.; Nussinov, R. (2004). "Hot regions int protein-protein interactions: The organization and contribution of structurally conserved hot spot residues". J. Mol. Biol. **345**: 1281–1294. doi:10.1016/j.jmb.2004.10.077.

24. Jansen, R; Yu, H; Greenbaum, D; Kluger, Y; Krogan, NJ; Chung, S; Emili, A; Snyder, M; Greenblatt, JF; Gerstein, M (2003). "A Bayesian networks approach for predicting protein-protein interactions from genomic data". Science. **302** *(5644): 449–53. doi:10.1126/science.1087361.*

25. Zhang, QC; Petrey, D; Deng, L; Qiang, L; Shi, Y; Thu, CA; Bisikirska, B; Lefebvre, C; Accili, D; Hunter, T; Maniatis, T; Califano, A; Honig, B (2012). "Structure-based prediction of protein-protein interactions on a genome-wide scale". Nature. **490** *(7421): 556–60.* doi:10.1038/nature11503.

26. Shoemaker, BA; Panchenko, AR (2007). "Deciphering protein–protein interactions. Part II. Computational methods to predict protein and domain interaction partners". PLoS Comput Biol. **3**(4): e43. doi:10.1371/journal.pcbi.0030043.

CHAPTER 3

Elements of Molecular Biology

Chapter Contents

Nucleic acids and proteins. The proteins, being made up a diverse set of amino acids, have, in addition to carbon, hydrogen, oxygen, and nitrogen, the elements **sulfur and selenium.**

INTRODUCTION

1: RNA Polymerase, *2*: Repressor, *3*: Promoter, *4*: Operator, *5*: Lactose, *6*: lacZ, *7*: lacY, *8*: lacA. Top: The gene is essentially turned off. There is no lactose to inhibit the repressor, so the repressor binds to the operator, which obstructs the RNA polymerase from binding to the promoter and making lactase. Bottom: The gene is turned on. Lactose is inhibiting the repressor, allowing the RNA polymerase to bind with the promoter, and express the genes, which synthesize lactase. Eventually, the lactase will digest all of the lactose, until there is none to bind to the repressor. The repressor will then bind to the operator, stopping the manufacture of lactase. In genetics, a promoter is a region of DNA that initiates transcription of a particular gene. Promoters are located near the transcription start sites of genes, on the same strand and upstream on the DNA (towards the 5› region of the sense strand). Promoters can be about 100–1000 base pairs long.

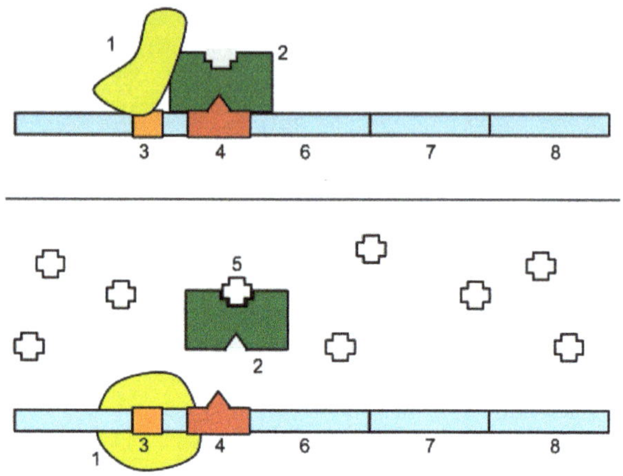

OVERVIEW

For transcription to take place, the enzyme that synthesizes RNA, known as RNA polymerase, must attach to the DNA near a gene. Promoters contain specific DNA sequences such as response elements that provide a secure initial binding site for RNA polymerase and for proteins called transcription factors that recruit RNA polymerase. These transcription factors have specific activator or repressor sequences of corresponding nucleotides that attach to specific promoters and regulate gene expression.

In Bacteria

The promoter is recognized by RNA polymerase and an associated sigma factor, which in turn are often brought to the promoter DNA by an activator protein's binding to its own DNA binding site nearby.

In Eukaryotes

The process is more complicated, and at least seven different factors are necessary for the binding of an RNA polymerase II to the promoter.

Promoters represent critical elements that can work in concert with other regulatory regions (enhancers, silencers, boundary elements/insulators) to direct the level of transcription of a given gene.

Identification of Relative Location

As promoters are typically immediately adjacent to the gene in question, positions in the promoter are designated relative to the transcriptional start site, where transcription of DNA begins for a particular gene (i.e., positions upstream are negative numbers counting back from -1, for example -100 is a position 100 base pairs upstream).

RELATIVE LOCATION IN THE CELL NUCLEUS

In the cell nucleus, it seems that promoters are distributed preferentially at the edge of the chromosomal territories, likely for the co-expression of genes on different chromosomes. Furthermore, in humans, promoters show certain structural features characteristic for each chromosome.

TYPES OF PROMOTERS

- Core promoter – the minimal portion of the promoter required to properly initiate transcription
    - Includes the transcription start site (TSS) and elements directly upstream
    - A binding site for RNA polymerase
        - RNA polymerase I: transcribes genes encoding 18s 5.8s 28s ribosomal RNA
        - RNA polymerase II: transcribes genes encoding messenger RNA and certain small nuclear RNAs and microRNA
        - RNA polymerase III: transcribes genes encoding transfer RNA, 5s ribosomal RNAs and other small RNAs
    - General transcription factor binding sites, e.g. TATA box
- Proximal promoter – the proximal sequence upstream of the gene that tends to contain primary regulatory elements
    - Approximately 250 base pairs upstream of the start site
    - Specific transcription factor binding sites
- Distal promoter – the distal sequence upstream of the gene that may contain additional regulatory elements, often with a weaker influence than the proximal promoter
    - Anything further upstream (but not an enhancer or other regulatory region whose influence is positional/orientation independent)
    - Specific transcription factor binding sites

Bacterial

In bacteria, the promoter contains two short sequence elements approximately -10 and -35 nucleotides *upstream* from the transcription start site.

- The sequence at -10 (the -10 element) has the consensus sequence TATAAT.
- The sequence at -35 (the -35 element) has the consensus sequence TTGACA.
- The above consensus sequences, while conserved on average, are not found intact in most promoters. On average, only 3 to 4 of the 6 base pairs in each consensus sequence are found in any given promoter. Few natural promoters have been identified to date that possess intact consensus sequences at both the -10 and -35; artificial promoters with complete conservation of the -10 and -35 elements have been found to transcribe at lower frequencies than those with a few mismatches with the consensus.
- The optimal spacing between the -35 and -10 sequences is 17 bp.
- Some promoters contain one or more upstream promoter element (UP element) subsites (consensus sequence 5'-AAAAAARNR-3' when centered in the -42 region; consensus sequence 5'-AWWWWWTTTTT-3' when centered in the -52 region; W = A or T; R = A or G; N = any base).

It should be noted that the above promoter sequences are recognized only by RNA polymerase holoenzyme containing sigma-70. RNA polymerase holoenzymes containing other sigma factors recognize different core promoter sequences.

<-- upstream downstream -->

5›-XXXXXXXPPPPPPXXXXXXPPPPPPXXXXGGGGGGGGGGGGGGGGGGGGGGGGGGGGGGGGGGGGGGGGGGGGGGGGGGXXXX-3›

-35 -10 Gene to be transcribed

Probability of Occurrence of Each Nucleotide

```
 for -10 sequence
 T    A    T    A    A    T
77%  76%  60%  61%  56%  82%

 for -35 sequence
 T    T    G    A    C    A
69%  79%  61%  56%  54%  54%
```

Eukaryotic

Eukaryotic promoters are diverse and can be difficult to characterize, however, recent studies show that they are divided in more than 10 classes.

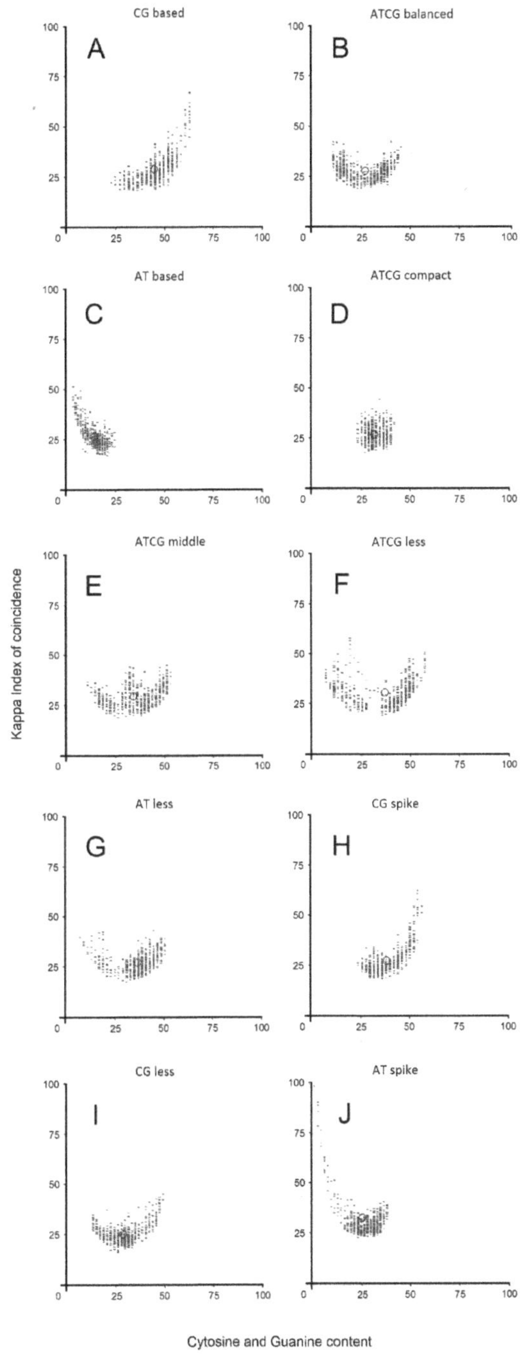

Ten classes of eukaryotic promoters and their representative DNA Patterns. The representative eukaryotic promoter classes are shown in the following sections: (A) AT-based class, (B) CG-based class, (C) ATCG-compact class, (D) ATCG-balanced class, (E) ATCG-middle class, (F) ATCG-less class, (G) AT-less class, (H) CG-spike class, (I) CG-less class and (J) ATspike class.

Gene promoters are typically located upstream of the gene and can have regulatory elements several kilobases away from the transcriptional start site (enhancers). In eukaryotes, the transcriptional complex can cause the DNA to bend back on itself, which allows for placement of regulatory sequences far from the actual site of. transcription. Eukaryotic RNA-polymerase-II-dependent promoters can contain a TATA element (consensus sequence TATAAA), which is recognized by the general transcription factor TATA-binding protein (TBP); and a B recognition element (BRE), which is recognized by the general transcription factor TFIIB. The TATA element and BRE typically are located close to the transcriptional start site (typically within 30 to 40 base pairs).

Eukaryotic promoter regulatory sequences typically bind proteins called transcription factors that are involved in the formation of the transcriptional complex. An example is the E-box (sequence CACGTG), which binds transcription factors in the basic helix-loop-helix (bHLH) family (e.g. BMAL1-Clock, cMyc). Some promoters that are targeted by multiple transcription factors might achieve a hyperactive state, leading to increased transcriptional activity.

Bidirectional (Mammalian)

Bidirectional promoters are short (<1 kbp) intergenic regions of DNA between the 5' ends of the genes in a bidirectional gene pair. A "bidirectional gene pair" refers to two adjacent genes coded on opposite strands, with their 5' ends oriented toward one another. The two genes are often functionally related, and modification of their shared promoter region allows them to be co-regulated and thus co-expressed.Bidirectional promoters are a common feature of mammalian genomes. About 11% of human genes are bidirectionally paired.

Bidirectionally paired genes in the Gene Ontology database shared at least one database-assigned functional category with their partners 47% of the time. Microarray analysis has shown bidirectionally paired genes to be co-expressed to a higher degree than random genes or neighboring unidirectional genes. Although co-expression does not necessarily indicate co-regulation, methylation of bidirectional promoter regions has been shown to downregulate both genes, and demethylation to upregulate both genes. There are exceptions to this, however. In some cases (about 11%), only one gene of a bidirectional pair is expressed. In these cases, the promoter is implicated in suppression of the non-expressed gene. The mechanism behind this could be competition for the same polymerases, or chromatin modification. Divergent transcription could shift nucleosomes to upregulate transcription of one gene, or remove bound transcription factors to downregulate transcription of one gene.

Some functional classes of genes are more likely to be bidirectionally paired than others. Genes implicated in DNA repair are five times more likely to be regulated by bidirectional promoters than by unidirectional promoters. Chaperone proteins are three times more likely, and mitochondrial genes are more than twice as likely. Many basic housekeeping and cellular metabolic genes are regulated by bidirectional promoters. The overrepresentation of bidirectionally paired DNA repair genes associates these promoters with cancer. Forty-five percent of human somaticoncogenes seem to be regulated by bidirectional promoters – significantly more than non-cancer causing genes. Hypermethylation of the promoters between gene pairs WNT9A/CD558500, CTDSPL/BC040563, and KCNK15/BF195580 has been associated with tumors.

Certain sequence characteristics have been observed in bidirectional promoters, including a lack of TATA boxes, an abundance of CpG islands, and a symmetry around the midpoint of dominant Cs and As on one side and Gs and Ts on the other. CCAAT boxes are common, as they are in many promoters that lack TATA boxes. In addition, the motifs NRF-1, GABPA, YY1, and ACTACAnnTCCC are represented in bidirectional promoters at significantly higher rates than in unidirectional promoters. The absence of TATA boxes in bidirectional promotors suggests that TATA boxes play a role in determining the directionality of promoters, but counterexamples of bidirectional promoters do possess TATA boxes and unidirectional promoters without them indicates that they cannot be the only factor.

Although the term "bidirectional promoter" refers specifically to promoter regions of mRNA-encoding genes, luciferase assays have shown that over half of human genes do not have a strong directional bias. Research suggests that non-coding RNAs are frequently associated with the promoter regions of mRNA-encoding genes. It has been hypothesized that the recruitment and initiation of RNA polymerase II usually begins bidirectionally, but divergent transcription is halted at a checkpoint later during elongation. Possible mechanisms behind this regulation include sequences in the promoter region, chromatin modification, and the spatial orientation of the DNA.

SUBGENOMIC

A subgenomic promoter is a promoter added to a virus for a specific heterologous gene, resulting in the formation of mRNA for that gene alone.

DETECTION

A wide variety of algorithms have been developed to facilitate detection of promoters in genomic sequence, and promoter prediction is a common element of many gene prediction methods. A promoter region is located before the -35 and -10 Consensus sequences. The closer the promoter region is to the consensus sequences the more often transcription of that gene will take place. There is not a set pattern for promoter regions as there are for consensus sequences.

EVOLUTIONARY CHANGE

Gene promoters have been responsible for the integration of different mutations favorable for the environmental conditions. Recent studies based on DNA patterns show that gene promoters are directly subjected to an immediate selection pressure. A superposition of eukaryotic gene promoters from four different species shows that promoters of genes reflect the selection pressure of a species and less the phylogenetic relations between distant species.

A major question in evolutionary biology is how important tinkering with promoter sequences is to evolutionary change, for example, the changes that have occurred in the human lineage after separating from other primates.

Some evolutionary biologists, for example Allan Wilson, have proposed that evolution in promoter or regulatory regions may be more important than changes in coding sequences over such time frames.

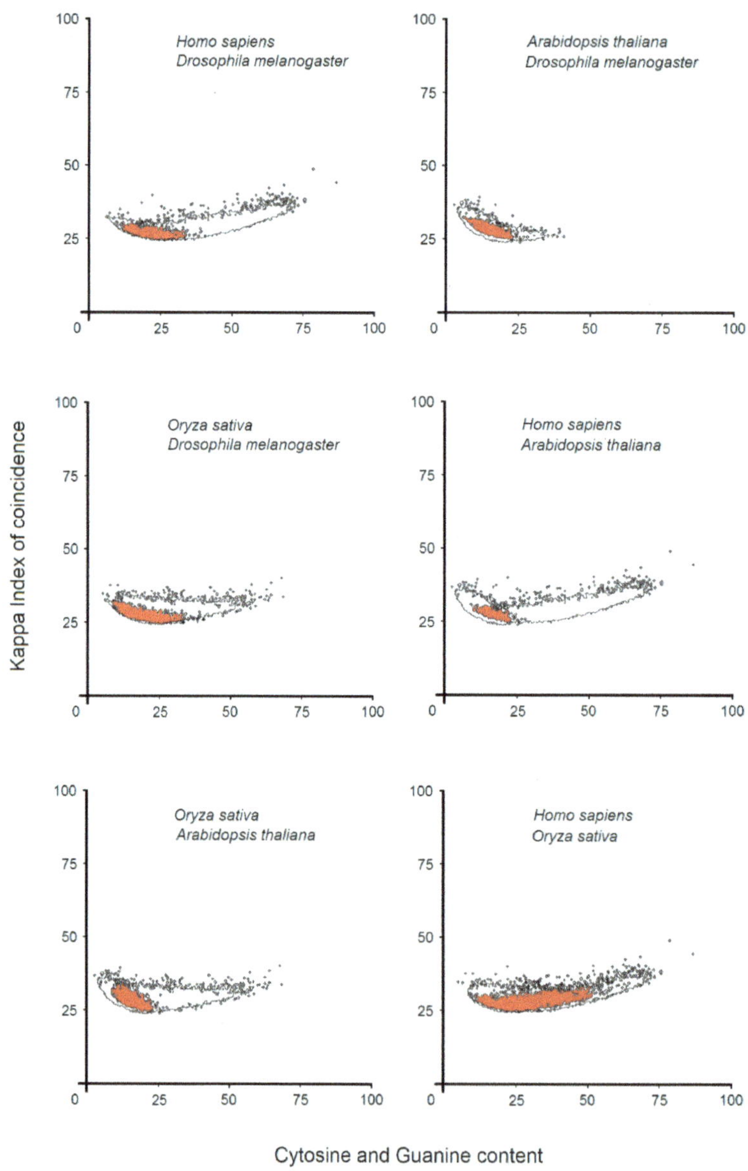

Superposition between promoter distributions from *Homo sapiens*, *Drosophila melanogaster*, *Oryza sativa* and *Arabidopsis thaliana*. Red color areas represent conserved promoter sequences.

A key reason for the importance of promoters is the potential to incorporate endocrine and environmental signals into changes in gene expression: A great variety of changes in the extracellular or intracellular environment may have impacts on gene expression, depending on the exact configuration of a given promoter; the combination and arrangement of specific DNA sequences that constitute the promoter defines the exact groups of proteins that can be bound to the promoter, at a given timepoint. Once the cell receives a physiological, pathological, or pharmacological stimulus, a number of cellular proteins are modified biochemically by signal cascades. By changes in structure, specific proteins acquire the capability to enter the nucleus of the cell and bind to promoter DNA, or to other proteins that themselves are already bound to a given promoter. The multi-protein complexes that are formed have the potential to change levels of gene expression. As a result, the gene product may increase or decrease inside the cell.

BINDING

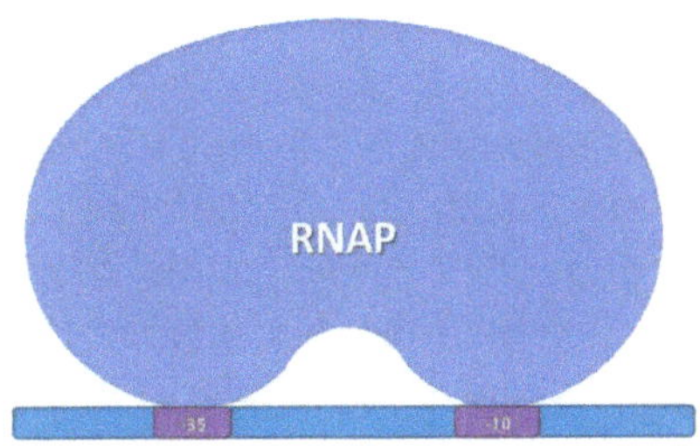

RNA Polymerase sigma 70 binding: The picture shows the region -35 and -10 of the promoter, which are the main regions of interactions with the sigma factor.

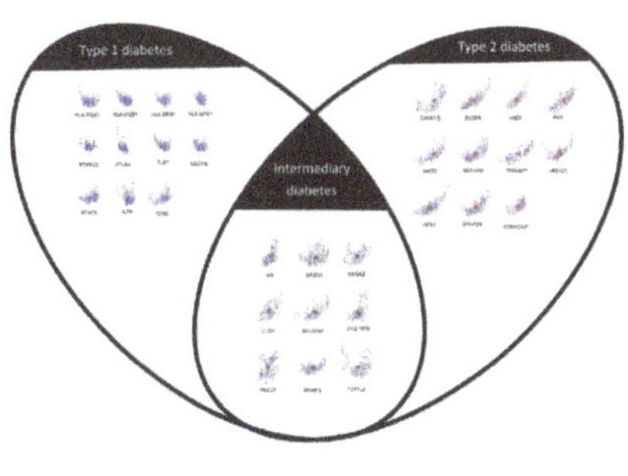

Distribution of gene promoters associated with type 1 diabetes (T1D), Intermediary Diabetes Mellitus (IDM) and type 2 diabetes (T2D). T1D promoters (blue dots), T2D promoters (red dots) and promoters from genes associated with the "intermediary" phenotype (green dots).

The initiation of the transcription is a multistep sequential process that involves several mechanisms: promoter location, initial reversible binding of RNA polymerase, conformational changes in RNA polymerase, conformational changes in DNA, binding of nucleoside triphosphate (NTP) to the functional RNA polymerase-promoter complex, and nonproductive and productive initiation of RNA synthesis.

The promoter binding process is crucial in the understanding of the process of gene expression.

Location

Although RNA polymerase holoenzyme shows high affinity to non-specific sites of the DNA, this characteristic does not allow us to clarify the process of promoter location. This process of promoter location has been attributed to the structure of the holoenzyme to DNA and sigma 4 to DNA complexes.

METHYLATION OF CPG ISLANDS STABLY SILENCES GENES

In humans, DNA methylation occurs at the 5' position of the pyrimidine ring of the cytosine residues within CpG sites to form 5-methylcytosines. The presence of multiple methylated CpG sites in CpG islands of promoters causes stable silencing of genes. Silencing of a gene may be initiated by other mechanisms, but this is often followed by methylation of CpG sites in the promoter CpG island to cause the stable silencing of the gene.

Promoter CpG hyper/hypo-methylation in cancer

Generally, in progression to cancer, hundreds of genes are silenced or activated. Although silencing of some genes in cancers occurs by mutation, a large proportion of carcinogenic gene silencing is a result of altered DNA methylation (see DNA methylation in cancer). DNA methylation causing silencing in cancer typically occurs at multiple CpG sites in the CpG islands that are present in the promoters of protein coding genes.

Altered expressions of microRNAs also silence or activate many genes in progression to cancer (see microRNAs in cancer). Altered microRNA expression occurs through hyper/hypo-methylation of CpG sites in CpG islands in promoters controlling transcription of the microRNAs.

Silencing of DNA repair genes through methylation of CpG islands in their promoters appears to be especially important in progression to cancer (see methylation of DNA repair genes in cancer).

Canonical sequences and wild-type

The usage of the term canonical sequence to refer to a promoter is often problematic, and can lead to misunderstandings about promoter sequences. Canonical implies perfect, in some sense.

In the case of a transcription factor binding site, there may be a single sequence that binds the protein most strongly under specified cellular conditions. This might be called canonical.

However, natural selection may favor less energetic binding as a way of regulating transcriptional output. In this case, we may call the most common sequence in a population the wild-type sequence. It may not even be the most advantageous sequence to have under prevailing conditions.

Recent evidence also indicates that several genes (including the proto-oncogene c-myc) have G-quadruplex motifs as potential regulatory signals.

Diseases that may be associated with variations

Some cases of many genetic diseases are associated with variations in promoters or transcription factors.

Examples include:

- Asthma
- Beta thalassemia
- Rubinstein-Taybi syndrome

CONSTITUTIVE VS REGULATED

Some promoters are called constitutive as they are active in all circumstances in the cell, while others are regulated, becoming active in the cell only in response to specific stimuli.

USE OF THE TERM

When referring to a promoter some authors actually mean promoter + operator; i.e., the lac promoter is IPTG inducible, meaning that besides the lac promoter, the lac operator is also present. If the lac operator were not present the IPTG would not have an inducible effect. Another example is the Tac-Promoter system (Ptac). Notice how tac is written as a tac promoter, while in fact tac is actually both a promoter and an operator.

REFERENCES

1. Orphanet Journal of Rare Diseases. **5**: 11. doi:10.1186/1750-1172-5-11. PMC 2893117. PMID 20492708.
2. Goldman, Lee; Schafer, Andrew I. (2015-04-21). Goldman-Cecil Medicine: Expert Consult - Online. Elsevier Health Sciences. ISBN 9780323322850.
3. Carton, James (2012-02-16). Oxford Handbook of Clinical Pathology. OUP Oxford. ISBN 9780191629938.
4. Perkin, Ronald M.; Newton, Dale A.; Swift, James D. (2008). Pediatric Hospital Medicine: Textbook of Inpatient Management. Lippincott Williams & Wilkins. ISBN 9780781770323.
5. Galanello, Renzo; Origa, Raffaella (2010-05-21). "Beta-thalassemia". Orphanet Journal of Rare Diseases. **5** *(1): 11*. doi:10.1186/1750-1172-5-11. ISSN 1750-1172. PMC 2893117. PMID 20492708.

6. Beta thalassemia". Genetics Home Reference.
7. Introduction to Pathology for the Physical Therapist Assistant. Jones & Bartlett Publishers. 2011. ISBN 9780763799083.
8. Anderson, Gregory J.; McLaren, Gordon D. (2012-01-16). Iron Physiology and Pathophysiology in Humans. Springer Science & Business Media. ISBN 9781603274845.
9. Barton, James C.; Edwards, Corwin Q.; Phatak, Pradyumna D.; Britton, Robert S.; Bacon, Bruce R. (2010-07-22). Handbook of Iron Overload Disorders. Cambridge University Press. ISBN 9781139489393.
10. McCance, Kathryn L.; Huether, Sue E. (2013-12-13). Pathophysiology: The Biologic Basis for Disease in Adults and Children. Elsevier Health Sciences. ISBN 9780323088541.
11. Leonard, Debra G. B. (2007-11-25). Molecular Pathology in Clinical Practice. Springer Science & Business Media. ISBN 9780387332277.
12. Bowen, Juan M.; Mazzaferri, Ernest L. (2012-12-06). Contemporary Internal Medicine: Clinical Case Studies. Springer Science & Business Media. ISBN 9781461567134.
13. Disorders, National Organization for Rare (2003). NORD Guide to Rare Disorders. Lippincott Williams & Wilkins. ISBN 9780781730631.
14. Barton, James C.; Edwards, Corwin Q. (2000-01-13). Hemochromatosis: Genetics, Pathophysiology, Diagnosis and Treatment. Cambridge University Press. ISBN 9780521593809.
15. Wilkins, Lippincott Williams & (2009). Professional Guide to Diseases. Lippincott Williams & Wilkins. ISBN 9780781778992.
16. Ward, Amanda J; Cooper, Thomas A (2009). "The pathobiology of splicing". The Journal of Pathology. **220** *(2): 152–63. doi:10.1002/path.2649. PMC 2855871. PMID 19918805.*
17. Okpala, Iheanyi (2008-04-15). Practical Management of Haemoglobinopathies. John Wiley & Sons. ISBN 9781405140201.
18. Vasudevan, D. M.; Sreekumari, S.; Vaidyanathan, Kannan (2011-11-01). Textbook of Biochemistry for Dental Students. JP Medical Ltd. ISBN 9789350254882.
19. Taeusch, H. William; Ballard, Roberta A.; Gleason, Christine A.; Avery, Mary Ellen (2005). Avery›s Diseases of the Newborn. Elsevier Health Sciences. ISBN 0721693474.
20. Beta Thalassemia: New Insights for the Healthcare Professional: 2013 Edition: ScholarlyBrief. ScholarlyEditions. 2013-07-22. ISBN 9781481663472.
21. Orkin, Stuart H.; Nathan, David G.; Ginsburg, David; Look, A. Thomas; Fisher, David E.; Lux, Samuel (2009). Nathan and Oski›s Hematology of Infancy and Childhood (7th ed.). Philadelphia: Saunders. ISBN 978-1-4160-3430-8.
22. McKinney, Emily Slone; James, Susan R.; Murray, Sharon Smith; Nelson, Kristine; Ashwill, Jean (2014-04-17). Maternal-Child Nursing. Elsevier Health

Sciences. ISBN 9780323293778.

23. Schrijver, Iris (2011-09-09). Diagnostic Molecular Pathology in Practice: A Case-Based Approach. Springer Science & Business Media. ISBN 9783642196775.

24. Muncie, Herbert L.; Campbell, James S. (2009). "Alpha and Beta Thalassemia". American Family Physician. **80** *(4): 339–44. PMID 19678601.*

25. Greer, John P.; Arber, Daniel A.; Glader, Bertil; List, Alan F.; Means, Robert T.; Paraskevas, Frixos; Rodgers, George M. (2013-08-29). Wintrobe›s Clinical Hematology. Lippincott Williams & Wilkins. ISBN 9781469846224.

26. Greer, John P.; Arber, Daniel A.; Glader, Bertil; List, Alan F.; Means, Robert T.; Paraskevas, Frixos; Rodgers, George M. (2013-08-29). Wintrobe›s Clinical Hematology. Lippincott Williams & Wilkins. ISBN 9781469846224.

27. Sabloff, Mitchell; Chandy, Mammen; Wang, Zhiwei; Logan, Brent R.; Ghavamzadeh, Ardeshir; Li, Chi-Kong; Irfan, Syed Mohammad; Bredeson, Christopher N.; Cowan, Morton J. (2011). "HLA-matched sibling bone marrow transplantation for β-thalassemia major". Blood. **117** *(5): 1745–1750.* doi:10.1182/blood-2010-09-306829. ISSN 0006-4971. PMC 3056598. PMID 21119108.

28. Schwartz, M. William (2012). The 5 Minute Pediatric Consult. Lippincott Williams & Wilkins. ISBN 9781451116564.

29. Porwit, Anna; McCullough, Jeffrey; Erber, Wendy N. (2011-05-27). Blood and Bone Marrow Pathology. Elsevier Health Sciences. ISBN 0702045357.

30. Hemoglobinopathies. Jaypee Brothers Publishers. 2006. ISBN 9788180616693.

31. Torre, Dario M.; Lamb, Geoffrey C.; Ruiswyk, Jerome Van; Schapira, Ralph M. (2009). Kochar›s Clinical Medicine for Students. Lippincott Williams & Wilkins. ISBN 9780781766999.

32. Berg, Sheri; Bittner, Edward A. (2013-10-16). The MGH Review of Critical Care Medicine. Lippincott Williams & Wilkins. ISBN 9781451173680.

33. Abouelmagd, Ahmed; Ageely, Hussein M. (2013). Basic Genetics: A Primer Covering Molecular Composition of Genetic Material, Gene Expression and Genetic Engineering, and Mutations and Human Genetic. Universal-Publishers. ISBN 9781612331928.

34. Weatherall, David J. Lichtman, MA,; Kipps, TJ,; Seligsohn, U,; Kaushansky, K,; Prchal, JT, eds. The Thalassemias: Disorders of Globin Synthesis. Williams Hematology (8 ed.).

CHAPTER 4

Molecular Biology of the Cell

Chapter Contents

Molecular biology is the study of living things at the level of the molecules which control them and make them up. While traditional biology concentrated on studying whole living organisms and how they interact within populations (a "top down" approach), molecular biology strives to understand living things by examining the components that make them up (a "bottom up" approach). Both approaches to biology are equally valid, although improvements to technology have permitted scientists to concentrate more on the molecules of life in recent years.

INTRODUCTION

In cell biology, mitosis is a part of the cell cycle when replicated chromosomes are separated into two new nuclei. In general, mitosis (division of the nucleus) is preceded by the S stage of interphase (during which the DNA is replicated) and is often accompanied or followed by cytokinesis, which divides the cytoplasm, organelles and cell membrane into two new cells containing roughly equal shares of these cellular components.Mitosis and cytokinesis together define the mitotic (M) phase of an animal cell cycle—the division of the mother cell into two daughter cells genetically identical to each other.

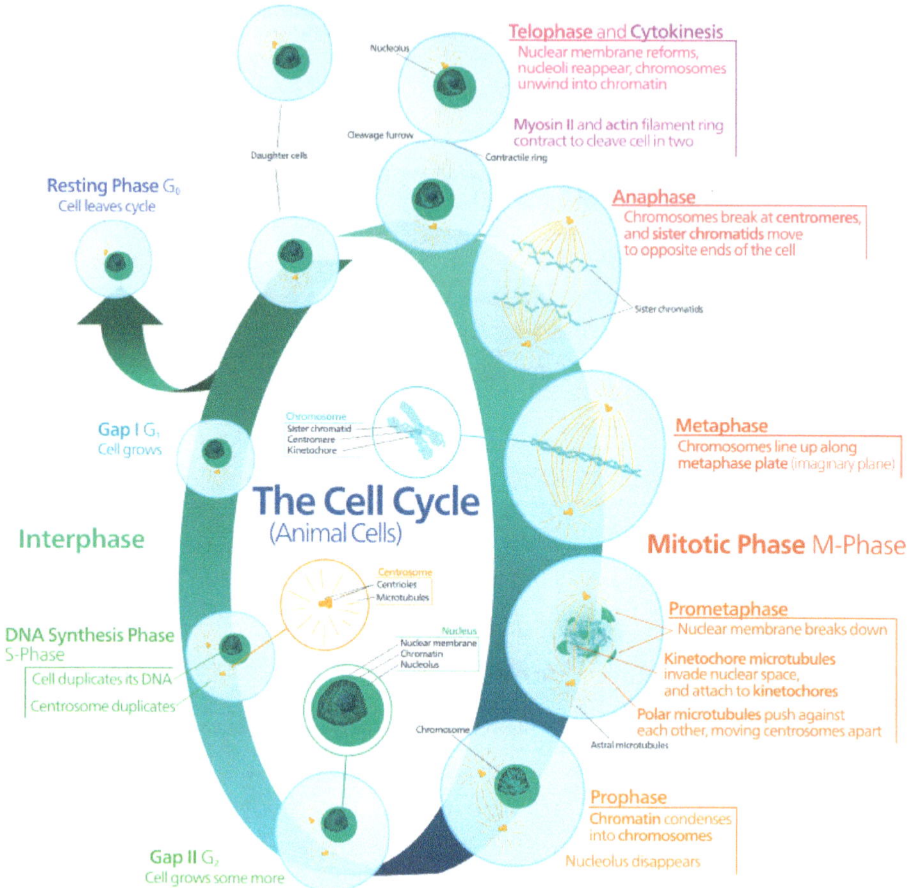

Mitosis in an animal cell (phases ordered counter-clockwise).

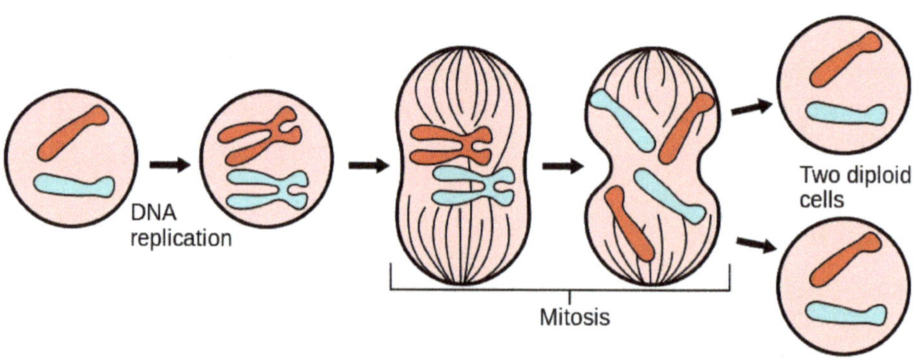

Mitosis divides the chromosomes in a cell nucleus.

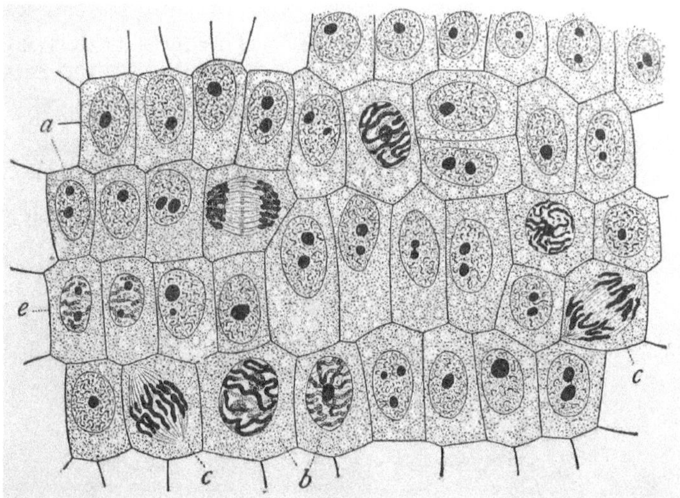

Onion (Allium) cells in different phases of the cell cycle enlarged 800 diameters.

a. non-dividing cells

b. nuclei preparing for division (spireme-stage)

c. dividing cells showing mitotic figures

d. pair of daughter-cells shortly after division

The process of mitosis is divided into stages corresponding to the completion of one set of activities and the start of the next. These stages are prophase, prometaphase, metaphase, anaphase, and telophase. During mitosis, the chromosomes, which have already duplicated, condense and attach to spindle fibers that pull one copy of each chromosome to opposite sides of the cell. The result is two genetically identical daughter nuclei. The rest of the cell may then continue to divide by cytokinesis to produce two daughter cells. Producing three or more daughter cells instead of normal two is a mitotic error called tripolar mitosis or multipolar mitosis (direct cell triplication / multiplication). Other errors during mitosis can induce apoptosis (programmed cell death) or cause mutations. Certain types of cancer can arise from such mutations.

Mitosis occurs only in eukaryotic cells and the process varies in different organisms. For example, animals undergo an "open" mitosis, where the nuclear envelope breaks down before the chromosomes separate, while fungi undergo a "closed" mitosis, where chromosomes divide within an intact cell nucleus.Furthermore, most animal cells undergo a shape change, known as mitotic cell rounding, to adopt a near spherical morphology at the start of mitosis. Prokaryotic cells, which lack a nucleus, divide by a different process called binary fission.

HISTORY

German zoologist Otto Bütschli might have claimed the discovery of the process presently known as "mitosis", a term coined by Walther Flemming in 1882.

Mitosis was discovered in frog, rabbit, and cat cornea cells in 1873 and described for the first time by the Polish histologistWacław Mayzel in 1875. The term is derived from the Greek word μίτος *mitos* "warp thread".

OVERVIEW

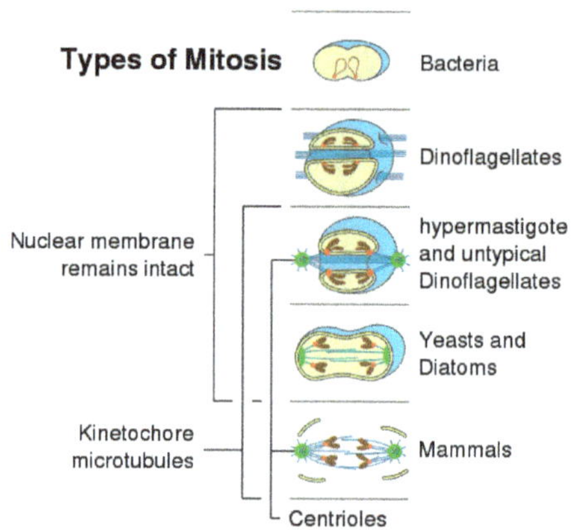

Types of mitosis

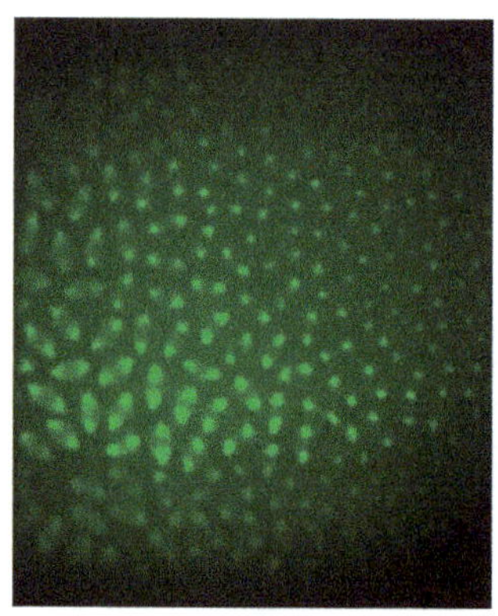

Time-lapse video of mitosis in a *Drosophila melanogaster* embryo

The primary result of mitosis and cytokinesis is the transfer of a parent cell's genome into two daughter cells. The genome is composed of a number of chromosomes—complexes of tightly coiled DNA that contain genetic information vital for proper cell function. Because each resultant daughter cell should be genetically identical to the parent cell, the parent cell must make a copy of each chromosome before mitosis. This occurs during the S phase of interphase.Chromosome duplication results in two identical *sister chromatids* bound together by cohesin proteins at the *centromere*.

When mitosis begins, the chromosomes condense and become visible. In some eukaryotes, for example animals, the nuclear envelope, which segregates the DNA from the cytoplasm, disintegrates into small vesicles. The nucleolus, which makes ribosomes in the cell, also disappears. Microtubules project from opposite ends of the cell, attach to the centromeres, and align the chromosomes centrally within the cell. The microtubules then contract to pull the sister chromatids of each chromosome apart. Sister chromatids at this point are called *daughter chromosomes*. As the cell elongates, corresponding daughter chromosomes are pulled toward opposite ends of the cell and condense maximally in late anaphase. A new nuclear envelope forms around the separated daughter chromosomes, which decondense to form interphase nuclei.

During mitotic progression, typically after anaphase onset, the cell may undergo cytokinesis. In animal cells, a cell membrane pinches inward between the two developing nuclei to produce two new cells. In plant cells, a cell plate forms between the two nuclei. Cytokinesis does not always occur; coenocytic (a type of multinucleate condition) cells undergo mitosis without cytokinesis.

PHASES OF CELL CYCLE AND MITOSIS

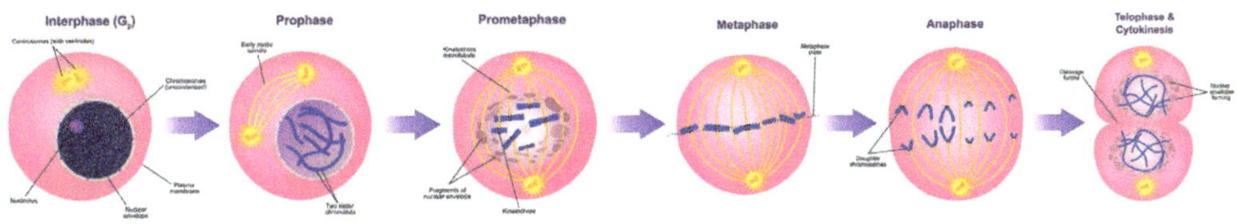

Interphase

The mitotic phase is a relatively short period of the cell cycle. It alternates with the much longer *interphase*, where the cell prepares itself for the process of cell division. Interphase is divided into three phases: G_1 (first gap), S (synthesis), and G_2(second gap). During all three phases, the cell grows by producing proteins and cytoplasmic organelles. However, chromosomes are replicated only during the S phase. Thus, a cell grows (G_1), continues to grow as it duplicates its chromosomes (S), grows more and prepares for mitosis (G_2), and finally divides (M) before restarting the cycle. All these phases in the cell cycle are highly regulated by cyclins, cyclin-dependent kinases, and other cell cycle proteins. The phases follow one another in strict order and there are "checkpoints" that give the cell cues to proceed from one phase to another. Cells may also temporarily or permanently leave the cell cycle and enter G_0 phase to stop dividing. This can occur when cells become overcrowded (density-dependent inhibition) or when they differentiate to carry out specific functions for the organism, as is the case for human heart muscle cells and neurons. Some G_0 cells have the ability to re-enter the cell cycle.

Preprophase (Plant Cells)

In plant cells only, prophase is preceded by a pre-prophase stage. In highly vacuolated plant cells, the nucleus has to migrate into the center of the cell before mitosis can begin. This is achieved through the formation of a phragmosome, a transverse sheet of cytoplasm that bisects the cell along the future plane of cell division. In addition to phragmosome formation, preprophase is characterized by the formation of a ring of microtubules and actin filaments (called preprophase band) underneath the plasma membrane around the equatorial plane of the future mitotic spindle. This band marks the position where the cell will eventually divide. The cells of higher plants (such as the flowering plants) lack centrioles; instead, microtubules form a spindle on the surface of the nucleus and are then organized into a spindle by the chromosomes themselves, after the nuclear envelope breaks down. The preprophase band disappears during nuclear envelope breakdown and spindle formation in prometaphase.

Prophase

During prophase, which occurs after G_2 interphase, the cell prepares to divide by tightly condensing its chromosomes and initiating mitotic spindle formation, this process is called chromosome condensation. During interphase, the genetic material in the nucleus consists of loosely packed chromatin. At the onset of prophase, chromatin fibers condense into discrete chromosomes that are typically visible at high magnification through a light microscope. In this stage, chromosomes are long, thin and thread-like. Each chromosome has two chromatids. The two chromatids are joined at a place called centromere.

Gene transcription ceases during prophase and does not resume until late anaphase to early G1 phase. The nucleolus also disappears during early prophase.

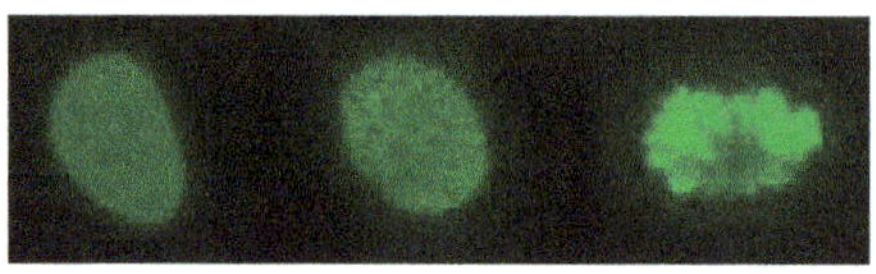

Condensing chromosomes. Interphase nucleus (left), condensing chromosomes (middle) and condensed chromosomes (right).

Close to the nucleus of animal cells are structures called centrosomes, consisting of a pair of centrioles surrounded by a loose collection of proteins. The centrosome is the coordinating center for the cell's microtubules. A cell inherits a single centrosome at cell division, which is duplicated by the cell before a new round of mitosis begins, giving a pair of centrosomes. The two centrosomes polymerize tubulin to help form a microtubule spindle apparatus. Motor proteins then push the centrosomes along these microtubules to opposite sides of the cell. Although centrosomes help organize microtubule assembly, they are not essential for the formation of the spindle apparatus, since they are absent from plants, and are not absolutely required for animal cell mitosis.

Prometaphase

At the beginning of prometaphase in animal cells, phosphorylation of nuclear lamins causes the nuclear envelope to disintegrate into small membrane vesicles. As this happens, microtubules invade the nuclear space. This is called *open mitosis*, and it occurs in some multicellular organisms. Fungi and some protists, such as algae or trichomonads, undergo a variation called *closed mitosis* where the spindle forms inside the nucleus, or the microtubules penetrate the intact nuclear envelope.

In late prometaphase, *kinetochore microtubules* begin to search for and attach to chromosomal kinetochores. A *kinetochore* is a proteinaceous microtubule-binding structure that forms on the chromosomal centromere during late prophase. A number of *polar microtubules* find and interact with corresponding polar microtubules from the opposite centrosome to form the mitotic spindle. Although the kinetochore structure and function are not fully understood, it is known that it contains some form of molecular motor. When a microtubule connects with the kinetochore, the motor activates, using energy from ATP to "crawl" up the tube toward the originating centrosome. This motor activity, coupled with polymerisation and depolymerisation of microtubules, provides the pulling force necessary to later separate the chromosome's two chromatids.

Metaphase

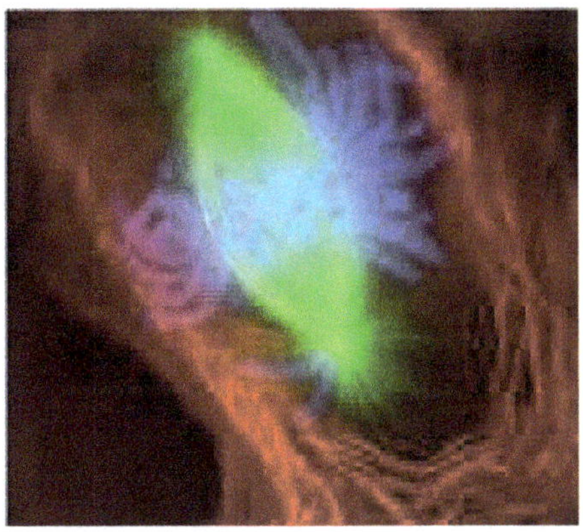

A cell in late metaphase. All chromosomes (blue) but one have arrived at the metaphase plate.

After the microtubules have located and attached to the kinetochores in prometaphase, the two centrosomes begin pulling the chromosomes towards opposite ends of the cell. The resulting tension causes the chromosomes to align along the *metaphase plate* or *equatorial plane*, an imaginary line that is centrally located between the two centrosomes (at approximately the midline of the cell). To ensure equitable distribution of chromosomes at the end of mitosis, the *metaphase checkpoint* guarantees that kinetochores are properly attached to the mitotic spindle and that the chromosomes are aligned along the metaphase plate. If the cell successfully passes through the metaphase checkpoint, it proceeds to anaphase.

Anaphase

During *anaphase A*, the cohesins that bind sister chromatids together are cleaved, forming two identical daughter chromosomes. Shortening of the kinetochore microtubules pulls the newly formed daughter chromosomes to opposite ends of the cell. During *anaphase B*, polar microtubules push against each other, causing the cell to elongate. In late anaphase, chromosomes also reach their overall maximal condensation level, to help chromosome segregation and the re-formation of the nucleus. In most animal cells, anaphase A precedes anaphase B, but some vertebrate egg cells demonstrate the opposite order of events.

Telophase

Telophase (from the Greek word τελος meaning «end») is a reversal of prophase and prometaphase events. At telophase, the polar microtubules continue to lengthen, elongating the cell even more. If the nuclear envelope has broken down, a new nuclear envelope forms using the membrane vesicles of the parent cell›s old nuclear envelope. The new envelope forms around each set of separated daughter chromosomes (though the membrane does not enclose the centrosomes) and the nucleolus reappears. Both sets of chromosomes, now surrounded by new nuclear membrane, begin to «relax» or decondense. Mitosis is complete. Each daughter nucleus has an identical set of chromosomes. Cell division may or may not occur at this time depending on the organism.

Cytokinesis

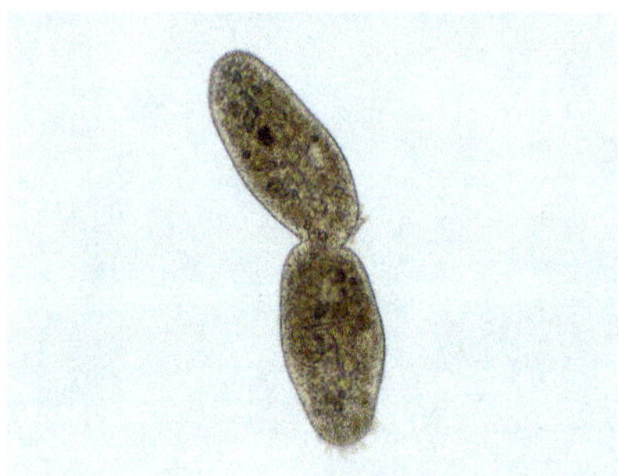

Cilliate undergoing cytokinesis, with the cleavage furrow being clearly visible

Cytokinesis is not a phase of mitosis but rather a separate process, necessary for completing cell division. In animal cells, a cleavage furrow (pinch) containing a contractile ring develops where the metaphase plate used to be, pinching off the separated nuclei. In both animal and plant cells, cell division is also driven by vesicles derived from the Golgi apparatus, which move along microtubules to the middle of the cell. In plants, this structure coalesces into a cell plate at the center of the phragmoplast and develops into a cell wall, separating the two nuclei. The phragmoplast is a microtubule structure typical for higher plants, whereas some green algae use a phycoplast microtubule array during cytokinesis. Each daughter cell has a complete copy of the genome of its parent cell. The end of cytokinesis marks the end of the M-phase.

There are many cells where mitosis and cytokinesis occur separately, forming single cells with multiple nuclei. The most notable occurrence of this is among the fungi, slime molds, and coenocytic algae, but the phenomenon is found in various other organisms. Even in animals, cytokinesis and mitosis may occur independently, for instance during certain stages of fruit fly embryonic development.

SIGNIFICANCE

Mitosis is important for the maintenance of the chromosomal set; each cell formed receives chromosomes that are alike in composition and equal in number to the chromosomes of the parent cell.

Mitosis occurs in the following circumstances:

Development and growth

The number of cells within an organism increases by mitosis. This is the basis of the development of a multicellular body from a single cell, i.e., zygote and also the basis of the growth of a multicellular body.

Cell Replacement

In some parts of body, e.g. skin and digestive tract, cells are constantly sloughed off and replaced by new ones. New cells are formed by mitosis and so are exact copies of the cells being replaced. In like manner, red blood cells have short lifespan (only about 4 months) and new RBCs are formed by mitosis.

Regeneration

Some organisms can regenerate body parts. The production of new cells in such instances is achieved by mitosis. For example, starfish regenerate lost arms through mitosis.

Asexual Reproduction

Some organisms produce genetically similar offspring through asexual reproduction. For example, the hydra reproduces asexually by budding. The cells at the surface of hydra undergo mitosis and form a mass called a bud. Mitosis continues in the cells of the bud and this grows into a new individual. The same division happens during asexual reproduction or vegetative propagation in plants.

CELL SHAPE CHANGES DURING MITOSIS

In animal tissue, most cells round up to a near-spherical shape during mitosis. In epithelia and epidermis, an efficient rounding process is correlated with proper mitotic spindle alignment and subsequent correct positioning of daughter cells. Moreover, researchers have found that if rounding is heavily suppressed it may result in spindle defects, primarily pole splitting and failure to efficiently capture chromosomes. Therefore, mitotic cell rounding is thought to play a protective role in ensuring accurate mitosis.

Rounding forces are driven by reorganization of F-actin and myosin (actomyosin) into a contractile homogeneous cell cortex that 1) rigidifies the cell peripheryand 2) facilitates generation of intracellular hydrostatic pressure (up to 10 fold higher than interphase). The generation of intracellular pressure is particularly critical under confinement, such as would be important in a tissue scenario, where outward forces must be produced to round up against surrounding cells and/or the extracellular matrix. Generation of pressure is dependent on formin-mediated F-actin nucleation and Rho kinase (ROCK)-mediated myosin II contraction, both of which are governed upstream by signaling pathways RhoA and ECT2 through the activity of Cdk1. Due to its importance in mitosis, the molecular components and dynamics of the mitotic actomyosin cortex is an area of active research.

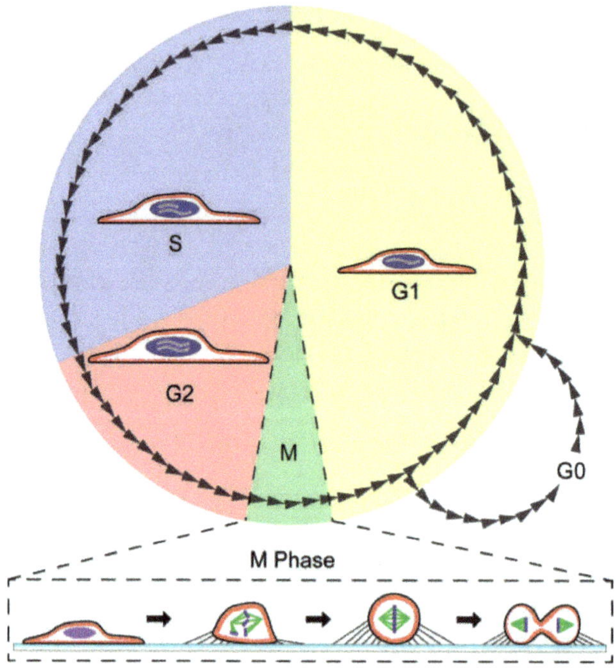

Cell shape changes through mitosis for a typical animal cell cultured on a flat surface. The cell undergoes mitotic cell rounding during spindle assembly and then divides via cytokinesis. The actomyosin cortex is depicted in red, DNA/chromosomes purple, microtubules green, and membrane and retraction fibers in black. Rounding also occurs in live tissue, as described in the text.

Errors and Variations of Mitosis

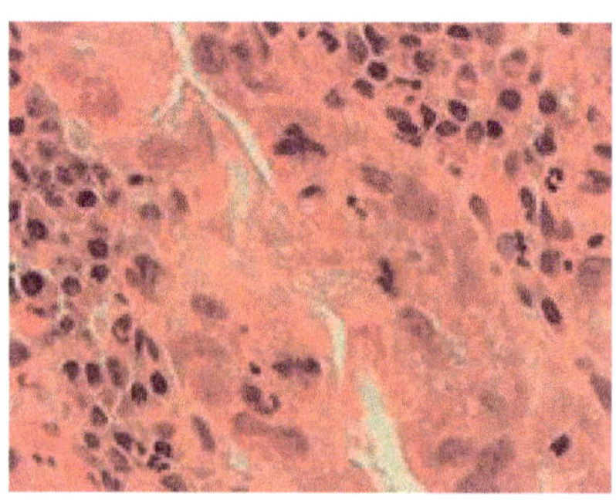

An abnormal (tripolar) mitosis (12 o'clock position) in a precancerous lesion of the stomach (H&E stain)

Errors can occur during mitosis, especially during early embryonic development in humans. Mitotic errors can create aneuploid cells that have too few or too many of one or more chromosomes, a condition associated with cancer. Early human embryos, cancer cells, infected or intoxicated cells can also suffer from pathological division into three or more daughter cells (tripolar or multipolar mitosis), resulting in severe errors in their chromosomal complements.

In *nondisjunction*, sister chromatids fail to separate during anaphase. One daughter cell receives both sister chromatids from the nondisjoining chromosome and the other cell receives none. As a result, the former cell gets three copies of the chromosome, a condition known as *trisomy*, and the latter will have only one copy, a condition known as *monosomy*. On occasion, when cells experience nondisjunction, they fail to complete cytokinesis and retain both nuclei in one cell, resulting in binucleated cells.

Anaphase lag occurs when the movement of one chromatid is impeded during anaphase. This may be caused by a failure of the mitotic spindle to properly attach to the chromosome. The lagging chromatid is excluded from both nuclei and is lost. Therefore, one of the daughter cells will be monosomic for that chromosome.

Endoreduplication (or endoreplication) occurs when chromosomes duplicate but the cell does not subsequently divide. This results in polyploid cells or, if the chromosomes duplicates repeatedly, polytene chromosomes. Endoreduplication is found in many species and appears to be a normal part of development. Endomitosis is a variant of endoreduplication in which cells replicate their chromosomes during S phase and enter, but prematurely terminate, mitosis. Instead of being divided into two new daughter nuclei, the replicated chromosomes are retained within the original nucleus. The cells then re-enter G_1 and S phase and replicate their chromosomes again. This may occur multiple times, increasing the chromosome number with each round of replication and endomitosis. Platelet-producing megakaryocytes go through endomitosis during cell differentiation.

Timeline in Pictures

Mitotic cells can be visualized microscopically by staining them with fluorescent antibodies and dyes.

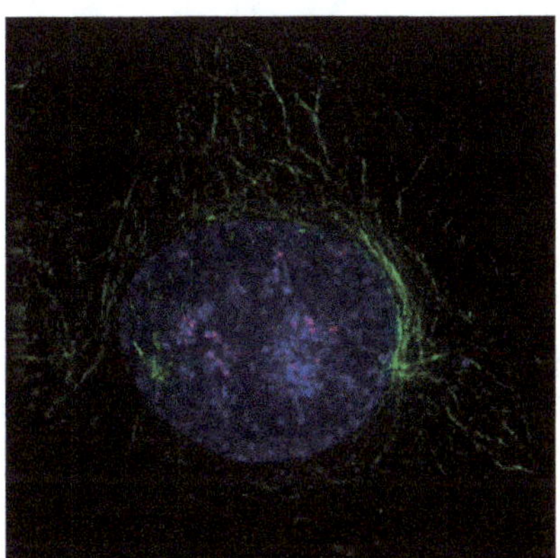

Early prophase: Nonkinetochore microtubules, shown as green strands, have established a matrix around the disassembling nucleus, with the condensing chromosomes in blue. The red nodules are the centromeres.

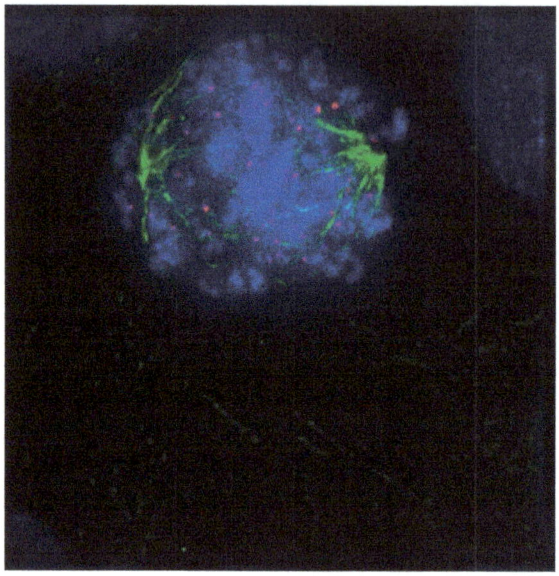

Early prometaphase: The nuclear envelope has just disassembled, allowing the microtubules to quickly interact with the kinetochores, which assemble on the centromeres of the condensing chromosomes.

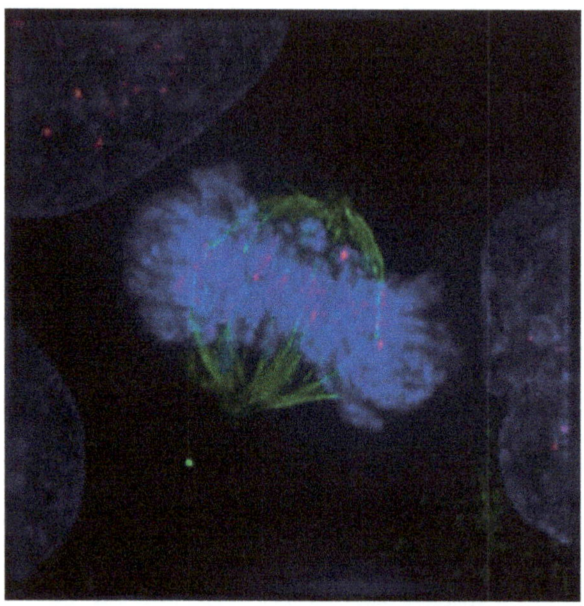

Metaphase: The centrosomes have moved to the poles of the cell and have established the mitotic spindle. The chromosomes have congressed at the metaphase plate.

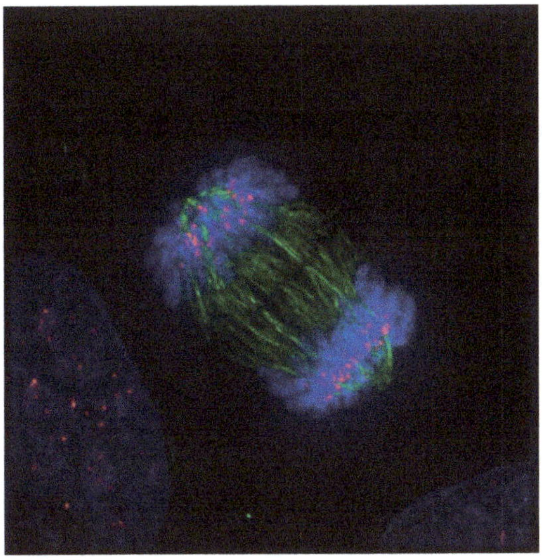

Anaphase: Kinetochore microtubules pull the two sets of chromosomes apart, and lengthening nonkinetochore microtubules push the halves of the dividing cell further apart, while chromosomes are condensed maximally.

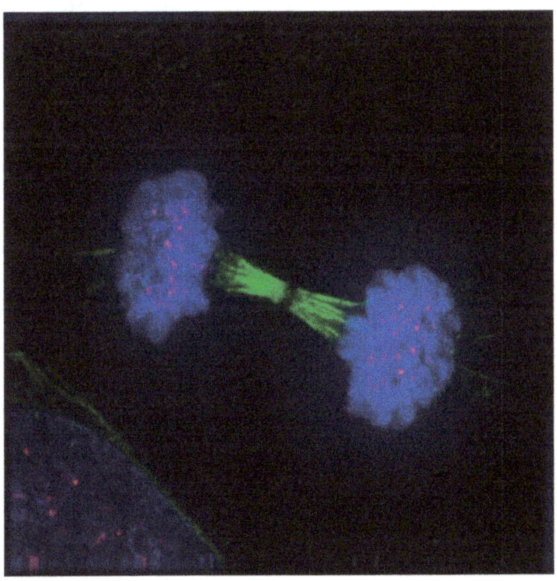

Telophase: Reversal of prophase and prometaphase events and thus completing the cell cycle.

CELL SIGNALING

Cell signaling (cell signalling in British English) is part of a complex system of communication that governs basic activities of cells and coordinates cell actions. The ability of cells to perceive and correctly respond to their microenvironment is the basis of development, tissue repair, and immunity as well as normal tissue homeostasis. Errors in cellular information processing are responsible for diseases such as cancer, autoimmunity, and diabetes. By understanding cell signaling, diseases may be treated more effectively and, theoretically, artificial tissues may be created.

Traditional work in biology has focused on studying individual parts of cell signaling pathways. Systems biology research helps us to understand the underlying structure of cell signaling networks and how changes in these networks may affect the transmission and flow of information (signal transduction). Such networks are complex systems in their organization and may exhibit a number of emergent properties including bistability and ultrasensitivity. Analysis of cell signaling networks requires a combination of experimental and theoretical approaches including the development and analysis of simulations and modeling. Long-range allostery is often a significant component of cell signaling events.

Signaling between cells of one organism and multiple organisms

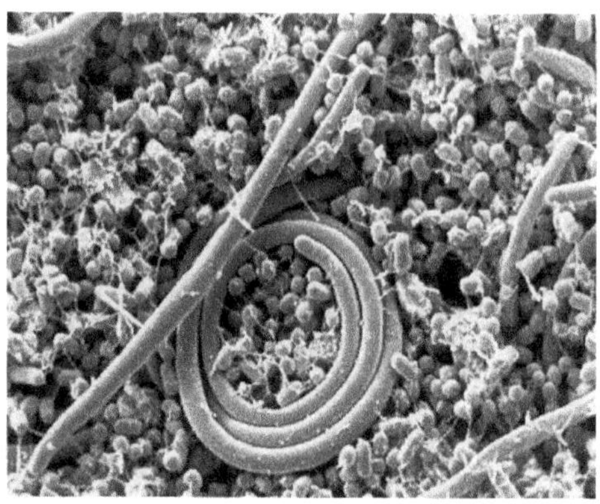

Figure 1. Example of signaling between bacteria. *Salmonella enteritidis* uses N-Acyl homoserine lactone for Quorum sensing (see: Inter-Bacterial Communication).

Cell signaling has been most extensively studied in the context of human diseases and signaling between cells of a single organism. However, cell signaling may also occur between the cells of two different organisms. In many mammals, early embryo cells exchange signals with cells of the uterus. In the human gastrointestinal tract, bacteria exchange signals with each other and with human epithelial and immune system cells. For the yeast *Saccharomyces cerevisiae* during mating, some cells send a peptide signal (mating factor pheromones) into their environment. The mating factor peptide may bind to a cell surface receptor on other yeast cells and induce them to prepare for mating.

Classification

Cell signaling can be classified to be mechanical and biochemical based on the type of the signal. Mechanical signals are the forces exerted on the cell and the forces produced by the cell. These forces can both be sensed and responded by the cells. Biochemical signals are the biochemical molecules such as proteins, lipids, ions and gases. These signals can be categorized based on the distance between signaling and responder cells. Signaling within, between, and among cells is subdivided into the following classifications:

- *Intracrine* signals are produced by the target cell that stay within the target cell.
- *Autocrine* signals are produced by the target cell, are secreted, and affect the target cell itself via receptors. Sometimes autocrine cells can target cells close by if they are the same type of cell as the emitting cell. An example of this are immune cells.
- *Juxtacrine* signals target adjacent (touching) cells. These signals are transmitted along cell membranes via protein or lipid components integral to the membrane and are capable of affecting either the emitting cell or cells immediately adjacent.
- *Paracrine* signals target cells in the vicinity of the emitting cell. Neurotransmitters represent an example.
- *Endocrine* signals target distant cells. Endocrine cells produce hormones that travel through the blood to reach all parts of the body.

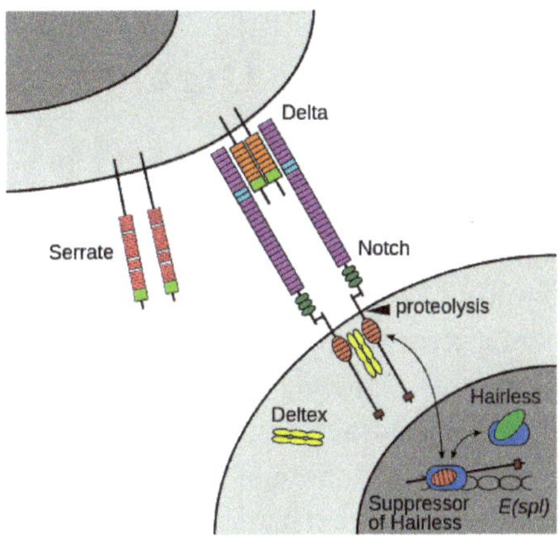

Figure 2: Notch-mediated juxtacrine signal between adjacent cells.

Cells communicate with each other via direct contact (juxtacrine signaling), over short distances (paracrine signaling), or over large distances and/or scales (endocrine signaling).

Some cell–cell communication requires direct cell–cell contact. Some cells can form gap junctions that connect their cytoplasm to the cytoplasm of adjacent cells. In cardiac muscle, gap junctions between adjacent cells allows for action potential propagation from the cardiac pacemaker region of the heart to spread and coordinately cause contraction of the heart.

The notch signaling mechanism is an example of juxtacrine signaling (also known as contact-dependent signaling) in which two adjacent cells must make physical contact in order to communicate. This requirement for direct contact allows for very precise control of cell differentiation during embryonic development. In the worm *Caenorhabditis elegans*, two cells of the developing gonad each have an equal chance of terminally differentiating or becoming a uterine precursor cell that continues to divide. The choice of which cell continues to divide is controlled by competition of cell surface signals. One cell will happen to produce more of a cell surface protein that activates the Notch receptor on the adjacent cell. This activates a feedback loop or system that reduces Notch expression in the cell that will differentiate and that increases Notch on the surface of the cell that continues as a stem cell.

Many cell signals are carried by molecules that are released by one cell and move to make contact with another cell. *Endocrine* signals are called hormones. Hormones are produced by endocrine cells and they travel through the blood to reach all parts of the body. Specificity of signaling can be controlled if only some cells can respond to a particular hormone. *Paracrine* signals such as retinoic acid target only cells in the vicinity of the emitting cell. Neurotransmitters represent another example of a paracrine signal. Some signaling molecules can function as both a hormone and a neurotransmitter. For example, epinephrine and norepinephrine can function as hormones when released from the adrenal gland and are transported to the heart by way of the blood stream. Norepinephrine can also be produced by neurons to function as a neurotransmitter within the brain. Estrogen can be released by the ovary and function as a hormone or act locally via paracrine or autocrine signaling. Active species of oxygen and nitric oxide can also act as cellular messengers. This process is dubbed redox signaling.

Cell Signaling in Multicellular Organisms

In a multicellular organism, signaling between cells occurs either through release into the extracellular space, divided in paracrine signaling (over short distances) and endocrine signaling (over long distances), or by direct contact, known as juxtacrine signaling. Autocrine signaling is a special case of paracrine signaling where the secreting cell has the ability to respond to the secreted signaling molecule. Synaptic signaling is a special case of paracrine signaling (for chemical synapses) or juxtacrine signaling (for electrical synapses) between neurons and target cells. Signaling molecules interact with a target cell as a ligand to cell surface receptors, and/or by entering into the cell through its membrane or endocytosis for intracrine signaling. This generally results in the activation of second messengers, leading to various physiological effects.

A particular molecule is generally used in diverse modes of signaling, and therefore a classification by mode of signaling is not possible. At least three important classes of signaling molecules are widely recognized, although non-exhaustive and with imprecise boundaries, as such membership is non-exclusive and depends on the context:

- Hormones are the major signaling molecules of the endocrine system, though they often regulate each other's secretion via local signaling (e.g. islet of Langerhans cells), and most are also expressed in tissues for local purposes (e.g. angiotensin) or failing that, structurally related molecules are (e.g. PTHrP).
- Neurotransmitters are signaling molecules of the nervous system, also including neuropeptides and neuromodulators. Neurotransmitters like the catecholamines are also secreted by the endocrine system into the systemic circulation.

- Cytokines are signaling molecules of the immune system, with a primary paracrine or juxtacrine role, though they can during significant immune responses have a strong presence in the circulation, with systemic effect (altering iron metabolism or body temperature). Growth factors can be considered as cytokines or a different class.

Signaling molecules can belong to several chemical classes: lipids, phospholipids, amino acids, monoamines, proteins, glycoproteins, or gases. Signaling molecules binding surface receptors are generally large and hydrophilic (e.g. TRH, Vasopressin, Acetylcholine), while those entering the cell are generally small and hydrophobic (e.g. glucocorticoids, thyroid hormones, cholecalciferol, retinoic acid), but important exceptions to both are numerous, and a same molecule can act both via surface receptor or in an intracrine manner to different effects. In intracrine signaling, once inside the cell, a signaling molecule can bind to intracellular receptors, other elements, or stimulate enzyme activity (e.g. gasses). The intracrine action of peptide hormones remains a subject of debate.

Hydrogen sulfide is produced in small amounts by some cells of the human body and has a number of biological signaling functions. Only two other such gases are currently known to act as signaling molecules in the human body: nitric oxide and carbon monoxide.

Receptors for Cell Motility and Differentiation

Cells receive information from their neighbors through a class of proteins known as receptors. Notch is a cell surface protein that functions as a receptor. Animals have a small set of genes that code for signaling proteins that interact specifically with Notch receptors and stimulate a response in cells that express Notch on their surface. Molecules that activate (or, in some cases, inhibit) receptors can be classified as hormones, neurotransmitters, cytokines, and growth factors, in general called receptor ligands. Ligand receptor interactions such as that of the Notch receptor interaction, are known to be main interactions responsible for cell signaling mechanisms and communication.

As shown in Figure 2 (above; left), notch acts as a receptor for ligands that are expressed on adjacent cells. While some receptors are cell surface proteins, others are found inside cells. For example, estrogen is a hydrophobic molecule that can pass through the lipid bilayer of the membranes. As part of the endocrine system, intracellular estrogen receptors from a variety of cell types can be activated by estrogen produced in the ovaries.

A number of transmembrane receptors for small molecules and peptide hormones as well as intracellular receptors for steroid hormones exist, giving cells the ability to respond to a great number of hormonal and pharmacological stimuli. In diseases, often, proteins that interact with receptors are aberrantly activated, resulting in constitutively activated downstream signals.

For several types of intercellular signaling molecules that are unable to permeate the hydrophobic cell membrane due to their hydrophilic nature, the target receptor is expressed on the membrane. When such signaling molecule activates its receptor, the signal is carried into the cell usually by means of a second messenger such as cAMP.\

Signaling Pathways

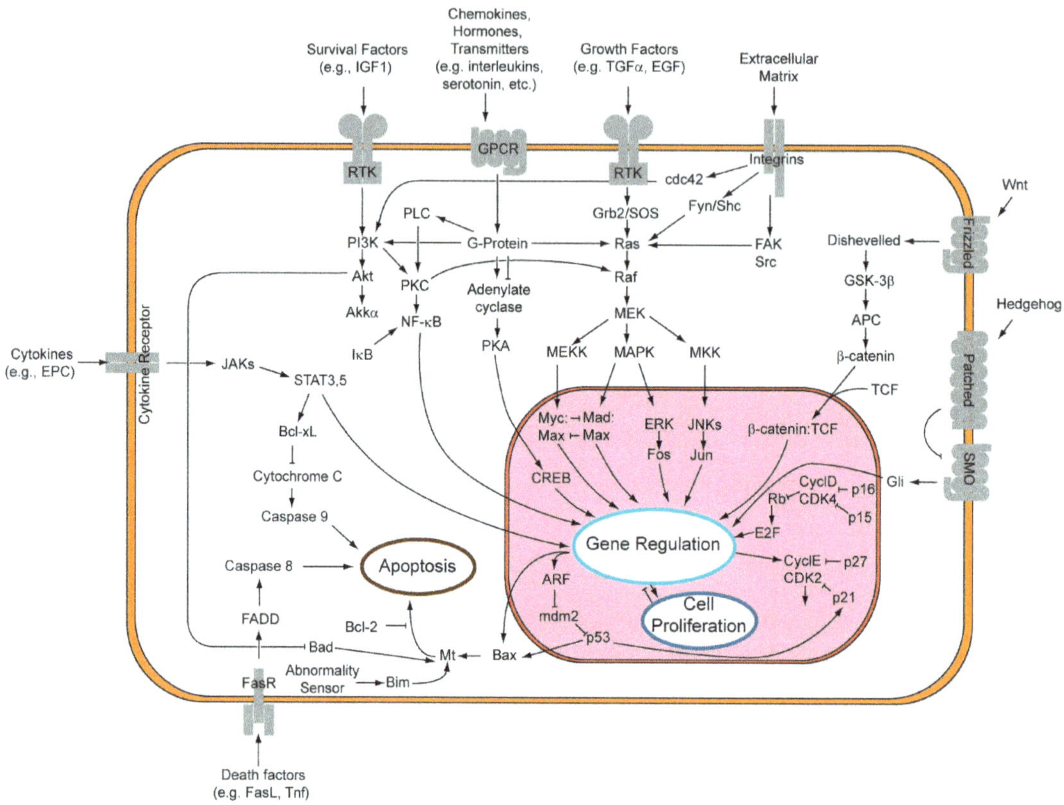

Overview of signal transduction pathways

In some cases, receptor activation caused by ligand binding to a receptor is directly coupled to the cell's response to the ligand. For example, the neurotransmitter GABA can activate a cell surface receptor that is part of an ion channel. GABA binding to a $GABA_A$ receptor on a neuron opens a chloride-selective ion channel that is part of the receptor. $GABA_A$ receptor activation allows negatively charged chloride ions to move into the neuron, which inhibits the ability of the neuron to produce action potentials. However, for many cell surface receptors, ligand-receptor interactions are not directly linked to the cell's response. The activated receptor must first interact with other proteins inside the cell before the ultimate physiological effect of the ligand on the cell's behavior is produced. Often, the behavior of a chain of several interacting cell proteins is altered following receptor activation. The entire set of cell changes induced by receptor activation is called a signal transduction mechanism or pathway.

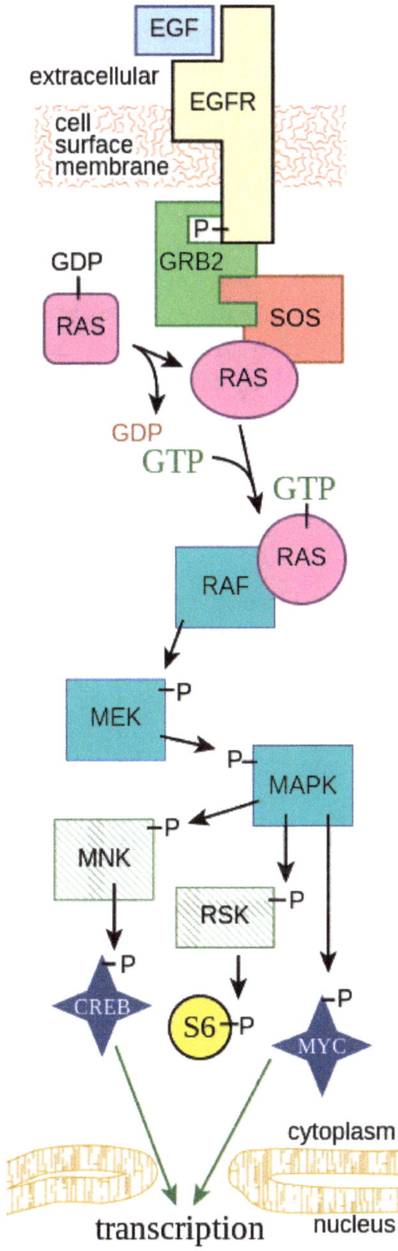

Figure 3. Key components of a signal transduction pathway (MAPK/ERK pathway shown).

In the case of Notch-mediated signaling, the signal transduction mechanism can be relatively simple. As shown in Figure 2, activation of Notch can cause the Notch protein to be altered by a protease. Part of the Notch protein is released from the cell surface membrane and takes part in gene regulation. Cell signaling research involves studying the spatial and temporal dynamics of both receptors and the components of signaling pathways that are activated by receptors in various cell types.

A more complex signal transduction pathway is shown in Figure 3. This pathway involves changes of protein–protein interactions inside the cell, induced by an external signal. Many growth factors bind to receptors at the cell surface and stimulate cells to progress through the cell cycle and divide. Several of these receptors are kinases that start to phosphorylate themselves and other proteins when binding to a ligand. This phosphorylation can generate a binding site for a different protein and thus induce protein–protein interaction. In Figure 3, the ligand (called epidermal growth factor (EGF)) binds to the receptor (called EGFR). This activates the receptor to phosphorylate itself. The phosphorylated receptor binds to an adaptor protein (GRB2), which couples the signal to further downstream signaling processes. For example, one of the signal transduction pathways that are activated is called the mitogen-activated protein kinase (MAPK) pathway. The signal transduction component labeled as "MAPK" in the pathway was originally called "ERK," so the pathway is called the MAPK/ERK pathway. The MAPK protein is an enzyme, a protein kinase that can attach phosphate to target proteins such as the transcription factor MYC and, thus, alter gene transcription and, ultimately, cell cycle progression. Many cellular proteins are activated downstream of the growth factor receptors (such as EGFR) that initiate this signal transduction pathway.

Some signaling transduction pathways respond differently, depending on the amount of signaling received by the cell. For instance, the hedgehog protein activates different genes, depending on the amount of hedgehog protein present.

Complex multi-component signal transduction pathways provide opportunities for feedback, signal amplification, and interactions inside one cell between multiple signals and signaling pathways.

Intraspecies and Interspecies Signaling

Molecular signaling can occur between different organisms, whether unicellular or multicellular, the emitting organism produces the signaling molecule, secrete it into the environment, where it diffuses, and it is sensed or internalized by the receiving organism. In some cases of interspecies signaling, the emitting organism can actually be a host of the receiving organism, or vice versa.

Intraspecies signaling occurs especially in bacteria, yeast, social insects, but also many vertebrates. The signaling molecules used by multicellular organisms are often called pheromones, they can have such purposes as alerting against danger, indicating food supply, or assisting in reproduction. In unicellular organisms such as bacteria, signaling can be used to ‹activate' peers from a dormant state, enhance virulence, defend against bacteriophages, etc. In quorum sensing, which is also found in social insects, the multiplicity of individual signals has the potentiality to create a positive feedback loop, generating coordinated response, in this context the signaling molecules are called autoinducers. This signaling mechanism may have been involved in evolution from unicellular to multicellular organisms. Bacteria also use contact-dependent signaling, notably to limit their growth.

Molecular signaling can also occur between individuals of different species, this has been particularly studied in bacteria. Different bacterial species can coordinate to colonize a host and participate in common quorum sensing.Therapeutic strategies to disrupt this phenomenon are being investigated. Interactions mediated through signaling molecules are also thought to occur between the gut flora and their host, as part of their commensal or symbiotic relationship.

Gram negative microbes deploy bacterial outer membrane vesicles for intra- and inter-species signaling in natural environments and at the host-pathogen interface.

Additionally, interspecies signaling occurs between multicellular organisms. In *Vespa mandarinia*, individuals release a scent that directs the colony to a food source.

REFERENCES

1. "Dynamic aberrant NF-κB spurs tumorigenesis: A new model encompassing the microenvironment". Cytokine Growth Factor Rev. 26: 389–403. Aug 2015. doi:10.1016/j.cytogfr.2015.06.001. PMID 26119834.
2. Wang, K; Grivennikov, SI; Karin, M (2013). "Implications of anti-cytokine therapy in colorectal cancer and autoimmune diseases". Ann. Rheum. Dis. 72 Suppl 2: ii100–3. doi:10.1136/annrheumdis-2012-202201. PMID 23253923.
3. Solinas G, Vilcu C, Neels JG, et al. (2007). "JNK1 in hematopoietically derived cells contributes to diet-induced inflammation and insulin resistance without affecting obesity". Cell Metab. 6 *(5): 386–97. doi:10.1016/j.cmet.2007.09.011. PMID 17983584.*
4. Smith, RJ; Koobatian, MT; Shahini, A; Swartz, DD; Andreadis, ST (2015). "Capture of endothelial cells under flow using immobilized vascular endothelial growth factor". Biomaterials. 51: 303–12. doi:10.1016/j.biomaterials.2015.02.025. PMC 4361797. PMID 25771020.
5. Eungdamrong, Narat J.; Iyengar, Ravi (2004-06-01). "Modeling Cell Signaling Networks". Biology of the Cell. 96 *(5): 355–362. doi:10.1016/j.biolcel.2004.03.004. ISSN 0248-4900. PMC 3620715. PMID 15207904.*
6. Bu Z, Callaway DJ (2011). "Proteins MOVE! Protein dynamics and long-range allostery in cell signaling". Advances in Protein Chemistry and Structural Biology. Advances in Protein Chemistry and Structural Biology. 83: 163–221. doi:10.1016/B978-0-12-381262-9.00005-7. ISBN 978-0-12-381262-9. PMID 21570668.
7. Mohamed OA, Jonnaert M, Labelle-Dumais C, Kuroda K, Clarke HJ, Dufort D (June 2005). "Uterine Wnt/β-catenin signaling is required for implantation". Proc Natl Acad Sci USA. 102 *(24): 8579–84. doi:10.1073/pnas.0500612102. PMC 1150820. PMID 15930138.*
8. Clarke MB, Sperandio V (June 2005). "Events at the host-microbial interface of the gastrointestinal tract III. Cell-to-cell signaling among microbial flora, host, and pathogens: there is a whole lot of talking going on". Am. J. Physiol. Gastrointest. Liver Physiol. 288 *(6): G1105–9. doi:10.1152/ajpgi.00572.2004. PMID 15890712.*
9. Lin JC, Duell K, Konopka JB (March 2004). "A Microdomain Formed by the Extracellular Ends of the Transmembrane Domains Promotes Activation of the G Protein-Coupled α-Factor Receptor". Mol Cell Biol. 24 *(5): 2041–51. doi:10.1128/MCB.24.5.2041-2051.2004. PMC 350546. PMID 14966283.*
10. Miller, Callie Johnson; Davidson, Lance A. (2013). "The interplay between cell signalling and mechanics in developmental processes". Nature Reviews Genetics. 14 *(10): 733–744.* doi:10.1038/nrg3513. ISSN 1471-0056. PMC 4056017. PMID 24045690.
11. Greenwald I (June 1998). "LIN-12/Notch signaling: lessons from worms and flies". Genes Dev. 12 *(12): 1751–62. doi:10.1101/gad.12.12.1751. PMID 9637676.*

12. Duester G (September 2008). "Retinoic Acid Synthesis and Signaling during Early Organogenesis". Cell. 134 *(6): 921–31. doi:10.1016/j.cell.2008.09.002. PMC 2632951. PMID 18805086.*

13. Cartford MC, Samec A, Fister M, Bickford PC (2004). "Cerebellar norepinephrine modulates learning of delay classical eyeblink conditioning: Evidence for post-synaptic signaling via PKA". Learn Mem. 11 *(6): 732–7. doi:10.1101/lm.83104. PMC 534701. PMID 15537737.*

14. Jesmin S, Mowa CN, Sakuma I, et al. (October 2004). "Aromatase is abundantly expressed by neonatal rat penis but downregulated in adulthood". J Mol Endocrinol. 33 *(2): 343–59. doi:10.1677/jme.1.01548. PMID 15525594.*

15. Gilbert, Scott F. (2000). "Juxtacrine Signaling". In NCBI bookshelf. Developmental biology (6. ed.). Sunderland, Mass.: Sinauer Assoc. ISBN 0878932437.

16. to:[a] [b] Bruce Alberts; et al. (2002). "General Principles of Cell Communication". In NCBI bookshelf. Molecular biology of the cell (4th ed.). New York: Garland Science. ISBN 0815332181.

17. Re, R (October 1999). "The nature of intracrine peptide hormone action". Hypertension. 34 *(4 Pt 1): 534–8. doi:10.1161/01.HYP.34.4.534. PMID 10523322.*

18. Hausman, Geoffrey M. Cooper, Robert E. (2000). "Signaling Molecules and Their Receptors". In NCBI bookshelf. The cell : a molecular approach (2nd ed.). Washington, D.C.: ASM Press. ISBN 087893300X.

19. Domazet I, Holleran BJ, Martin SS, Lavigne P, Leduc R, Escher E, Guillemette G (2009-05-01). "The Second Transmembrane Domain of the Human Type 1 Angiotensin II Receptor Participates in the Formation of the Ligand Binding Pocket and Undergoes Integral Pivoting Movement during the Process of Receptor Activation". J Biol Chem. 284 *(18): 11922–9. doi:10.1074/jbc.M808113200. PMC 2673261. PMID 19276075.*

20. Hislop JN, Henry AG, Marchese A, von Zastrow M (2009-07-17). "Ubiquitination Regulates Proteolytic Processing of G Protein-coupled Receptors after Their Sorting to Lysosomes". J Biol Chem. 284 *(29): 19361–70. doi:10.1074/jbc.M109.001644. PMC 2740561. PMID 19433584.*

21. Meng H, Zhang X, Hankenson KD, Wang MM (2009-03-20). "Thrombospondin 2 Potentiates Notch3/Jagged1 Signaling". J Biol Chem. 284 *(12): 7866–74. doi:10.1074/jbc.M803650200. PMC 2658079.* PMID 19147503.

22. Copland JA, Sheffield-Moore M, Koldzic-Zivanovic N, Gentry S, Lamprou G, Tzortzatou-Stathopoulou F, Zoumpourlis V, Urban RJ, Vlahopoulos SA (June 2009). "Sex steroid receptors in skeletal differentiation and epithelial neoplasia: is tissue-specific intervention possible?". BioEssays. 31 *(6): 629–41. doi:10.1002/bies.200800138. PMID 19382224.*

23. Goh SL, Looi Y, Shen H, Fang J, Bodner C, Houle M, Ng AC, Screaton RA, Featherstone M (2009-07-10). "Transcriptional Activation by MEIS1A in Response to Protein Kinase A Signaling Requires the Transducers of Regulated CREB Family of CREB Co-activators". J Biol Chem. 284 *(28): 18904–12. doi:10.1074/jbc.M109.005090. PMC 2707216. PMID 19473990.*

24. Wojtal KA, Hoekstra D, Van IJzendoorn SC (February 2008). "cAMP-dependent protein kinase A and the dynamics of epithelial cell surface domains: moving membranes to keep

in shape". BioEssays. 30 *(2): 146–55. doi:10.1002/bies.20705. PMID 18200529.*

25. Dinasarapu AR; et al. (2011). "Signaling gateway molecule pages—a data model perspective". Bioinformatics. 27 *(12): 1736–1738. doi:10.1093/bioinformatics/btr190. PMC 3106186. PMID 21505029.*

26. Tirindelli, R; Dibattista, M; Pifferi, S; Menini, A (July 2009). "From pheromones to behavior". Physiological reviews. 89 *(3): 921–56. doi:10.1152/physrev.00037.2008. PMID 19584317.*

27. Mukamolova, GV; Kaprelyants, AS; Young, DI; Young, M; Kell, DB (Jul 21, 1998). "A bacterial cytokine". Proceedings of the National Academy of Sciences of the United States of America. 95 *(15): 8916–21. doi:10.1073/pnas.95.15.8916. PMC 21177. PMID 9671779.*

28. Miller, Melissa B.; Bassler, Bonnie L. (1 October 2001). "Quorum sensing in bacteria". Annual Review of Microbiology. 55 *(1): 165–199. doi:10.1146/annurev.micro.55.1.165. PMID 11544353.*

29. Kaper, JB; Sperandio, V (June 2005). "Bacterial cell-to-cell signaling in the gastrointestinal tract". Infection and immunity. 73 *(6): 3197–209. doi:10.1128/IAI.73.6.3197-3209.2005. PMC 1111840. PMID 15908344.*

30. Camilli, A.; Bonnie L. Bassler (24 February 2006). "Bacterial Small-Molecule Signaling Pathways". Science. 311 *(5764): 1113–1116. doi:10.1126/science.1121357. PMC 2776824. PMID 16497924.*

31. Stoka, AM (June 1999). "Phylogeny and evolution of chemical communication: an endocrine approach" (PDF). Journal of molecular endocrinology. 22 *(3): 207–25. doi:10.1677/jme.0.0220207. PMID 10343281.*

32. Blango, Matthew G; Mulvey, Matthew A (31 March 2009). "Bacterial landlines: contact-dependent signaling in bacterial populations". Current Opinion in Microbiology. 12 *(2): 177–181. doi:10.1016/j.mib.2009.01.011. PMC 2668724. PMID 19246237.*

33. Shank, Elizabeth Anne; Kolter, Roberto (31 March 2009). "New developments in microbial interspecies signaling". Current Opinion in Microbiology. 12 *(2): 205–214. doi:10.1016/j.mib.2009.01.003. PMC 2709175. PMID 19251475.*

34. Ryan, R. P.; Dow, J. M. (1 July 2008). "Diffusible signals and interspecies communication in bacteria" (PDF). Microbiology. 154 *(7): 1845–1858. doi:10.1099/mic.0.2008/017871-0.*

35. Ryan, Robert P.; Dow, J. Maxwell (28 February 2011). "Communication with a growing family: diffusible signal factor (DSF) signaling in bacteria". Trends in Microbiology. 19 *(3): 145–152. doi:10.1016/j.tim.2010.12.003. PMID 21227698.*

36. Déziel, E; Lépine, F; Milot, S; He, J; Mindrinos, MN; Tompkins, RG; Rahme, LG (Feb 3, 2004). "Analysis of Pseudomonas aeruginosa 4-hydroxy-2-alkylquinolines (HAQs) reveals a role for 4-hydroxy-2-heptylquinoline in cell-to-cell communication". Proceedings of the National Academy of Sciences of the United States of America. 101 *(5): 1339–44. doi:10.1073/pnas.0307694100. PMC 337054. PMID 14739337.*

37. Federle, Michael J.; Bassler, Bonnie L. (31 October 2003). "Interspecies communication in bacteria". Journal of Clinical Investigation. 112 *(9): 1291–1299.* doi:10.1172/JCI20195. PMC 228483. PMID 14597753.

38. to:[a] [b] Sperandio, V.; et al. (14 July 2003). "Bacteria-host communication: The language of hormones". Proceedings of the National Academy of Sciences. 100 *(15): 8951–8956. doi:10.1073/pnas.1537100100. PMC 166419. PMID 12847292.*

39. Hooper, LV; Gordon, JI (May 11, 2001). "Commensal host-bacterial relationships in the gut". Science. 292 *(5519): 1115–1118. doi:10.1126/science.1058709. PMID 11352068.*

40. Schueller, Teresa (10 June 2010). "The cues have it; nest-based, cue-mediated recruitment to carbohydrate resources in a swarm-founding social wasp". Naturwissenschaften (Springer-Verlag).

CHAPTER 5

Techniques of Molecular Biology

Chapter Contents

Molecular biology is a specialised branch of biochemistry, the study of the chemistry of molecules which are specifically connected to living processes. Of particular importance to molecular biology are the nucleic acids (DNA and RNA) and the proteins which are constructed using the genetic instructions encoded in those molecules. Other biomolecules, such as carbohydrates and lipids may also be studied for the interactions they have with nucleic acids and proteins. Molecular biology is often separated from the field of cell biology, which concentrates on cellular structures (organelles and the like), molecular pathways within cells and cell life cycles.

INTRODUCTION

Since the late 1950s and early 1960s, molecular biologists have learned to characterize, isolate, and manipulate the molecular components of cells and organisms.

These components include DNA, the repository of genetic information; RNA, a close relative of DNA whose functions range from serving as a temporary working copy of DNA to actual structural and enzymatic functions as well as a functional and structural part of the translational apparatus, the ribosome; and proteins, the major structural and enzymatic type of molecule in cells.

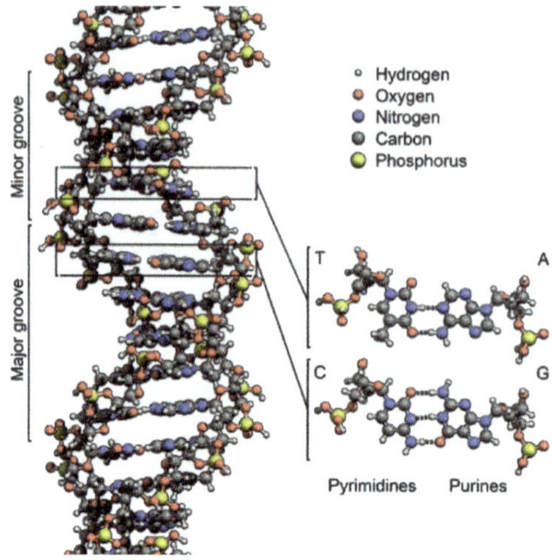

DNA Structure

DRY LAB

A **dry lab** is a laboratory where computational or applied mathematical analyses are done on a computer-generated model to simulate a phenomenon in the physical realm. Examples of such phenomena include a molecule changing quantum states, the event horizon of a black hole or anything that otherwise might be impossible or too dangerous to observe under normal laboratory conditions. This term may also refer to a lab that uses primarily electronic equipment, for example, a robotics lab. A dry lab can also refer to a laboratory space for the storage of dry materials. The term dry lab is also used in the photo printing industry to refer to photo printing systems that do not employ the use of "wet" photographic chemicals.

Dry labbing can also refer to supplying fictional (yet plausible) results in lieu of performing an assigned experiment.

In Silico Chemistry

As computing power has grown exponentially this approach to research, often referred to as *in silico* (as opposed to *in vitro*), has amassed more attention especially in the area of bioinformatics. More specifically, within bioinformatics, is the study of proteins or proteomics, which is the elucidation of their unknown structures and folding patterns. The general approach in the elucidation of protein structure has been to first purify a protein, crystallize it and then send X-rays through such a purified protein crystal to observe how these x-rays diffract into specific pattern—a process referred to as X-ray crystallography. However, many proteins, especially those embedded in cellular membranes, are nearly impossible to crystallize due to their hydrophobic nature. Although other techniques exists, such as ramachandran plotting and mass spectrometry, these alone generally do not lead to the full elucidation of protein structure or folding mechanisms.

Distributed Computing

As a means of surpassing the limitations of these techniques, projects such as Folding@home and Rosetta@home are aimed at resolving this problem using computational analysis, this means of resolving protein structure is referred to as protein structure prediction. Although many labs have a slightly different approach, the main concept is to find, from a myriad of protein conformations, which conformation has the lowest energy or, in the case of Folding@Home, to find relatively low energies of proteins that could cause the protein to misfold and aggregate other proteins to itself—like in the case of sickle cell anemia. The general scheme in these projects is that a small number of computations are parsed to, or sent to be calculated on, a computer, generally a home computer, and then that computer analyzes the likelihood that a specific protein will take a certain shape or conformation based on the amount of energy required for that protein to stay in that shape, this way of processing data is what is generally referred to as distributed computing. This analysis is done on an extraordinarily large number of different conformations, owing to the support of hundreds of thousands of home-based computers, in hopes to find the conformation of lowest possible energy or set of conformations of lowest possible energy relative to any conformations that are just slightly different. Although doing so is quite difficult, one can, by observing the energy distribution of a large number of conformations, despite the almost infinite number of different protein conformations possible for any given protein (see Levinthal Paradox), with a reasonably large number of protein energy samplings, predict relatively closely what conformation, within a range of conformations, has the expected lowest energy using methods in statistical inference. There are other factors such as salt concentration, pH, ambient temperature or chaperonins, which are proteins that assist in the folding process of other proteins, that can greatly affect how a protein folds. However, if the given protein is shown to fold on its own, especially in vitro, these findings can be further supported. Once we can see how a protein folds then we can see how it works as a catalyst, or in intracellular communication, e.g. neuroreceptor-neurotransmitter interaction. How certain compounds may be used to enhance or prevent the function of these proteins and how an elucidated protein overall plays a role in diseases such as Alzheimer's Disease or Huntington's Disease can also be much better understood.

Of course, there are many other avenues of research in which the dry lab approach has been implemented. Other physical phenomena, such as sound, properties of newly discovered or hypothetical compounds and quantum mechanics models have recently received more attention in this area of approach.

Wet Laboratory

Wet laboratories are laboratories where chemicals, drugs, or other material or biological matter are handled in liquid solutions or volatile phases, requiring direct ventilation, and specialized piped utilities (typically water and various gases). Wet laboratory space types do not include biohazards in Levels BL-2, BL-3, and BL-4 as defined by the 1999 NIH/CDC guideline "Biosafety in Microbiological and Biomedical Laboratories." The wet laboratory space types are typically located within a building specifically designed to house them.

In biology, chemistry, genetics or biochemistry, the terms wet lab or wet laboratory distinguish classical benchtop experiments handling biological material from in silico (computer analysis) or other theoretical work.

MOLECULAR CLONING

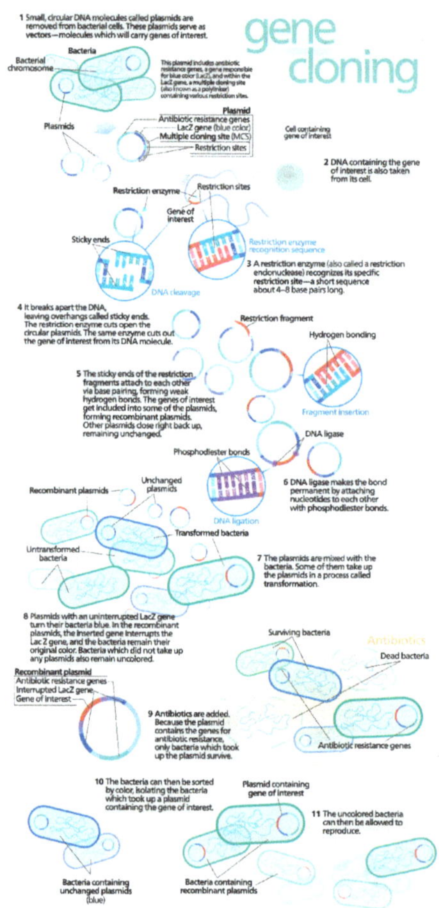

Diagram of molecular cloning using bacteria and plasmids.

Molecular cloning is a set of experimental methods in molecular biology that are used to assemble recombinant DNA molecules and to direct their replication within host organisms. The use of the word *cloning* refers to the fact that the method involves the replication of one molecule to produce a population of cells with identical DNA molecules. Molecular cloning generally uses DNA sequences from two different organisms: the species that is the source of the DNA to be cloned, and the species that will serve as the living host for replication of the recombinant DNA. Molecular cloning methods are central to many contemporary areas of modern biology and medicine.

In a conventional molecular cloning experiment, the DNA to be cloned is obtained from an organism of interest, then treated with enzymes in the test tube to generate smaller DNA fragments. Subsequently, these fragments are then combined with vector DNA to generate recombinant DNA molecules. The recombinant DNA is then introduced into a host organism (typically an easy-to-grow, benign, laboratory strain of *E. coli* bacteria). This will generate a population of organisms in which recombinant DNA molecules are replicated along with the host DNA. Because they contain foreign DNA fragments, these are transgenic or genetically modified microorganisms (GMO). This process takes advantage of the fact that a single bacterial cell can be induced to take up and replicate a single recombinant DNA molecule. This single cell can then be expanded exponentially to generate a large amount of bacteria, each of which contain copies of the original recombinant molecule. Thus, both the resulting bacterial population, and the recombinant DNA molecule, are commonly referred to as "clones". Strictly speaking, *recombinant DNA* refers to DNA molecules, while *molecular cloning* refers to the experimental methods used to assemble them.

History of Molecular Cloning

Prior to the 1970s, our understanding of genetics and molecular biology was severely hampered by an inability to isolate and study individual genes from complex organisms. This changed dramatically with the advent of molecular cloning methods. Microbiologists, seeking to understand the molecular mechanisms through which bacteria restricted the growth of bacteriophage, isolated restriction endonucleases, enzymes that could cleave DNA molecules only when specific DNA sequences were encountered. They showed that restriction enzymes cleaved chromosome-length DNA molecules at specific locations, and that specific sections of the larger molecule could be purified by size fractionation. Using a second enzyme, DNA ligase, fragments generated by restriction enzymes could be joined in new combinations, termed recombinant DNA. By recombining DNA segments of interest with vector DNA, such as bacteriophage or plasmids, which naturally replicate inside bacteria, large quantities of purified recombinant DNA molecules could be produced in bacterial cultures. The first recombinant DNA molecules were generated and studied in 1972.

Molecular cloning takes advantage of the fact that the chemical structure of DNA is fundamentally the same in all living organisms. Therefore, if any segment of DNA from any organism is inserted into a DNA segment containing the molecular sequences required for DNA replication, and the resulting recombinant DNA is introduced into the organism from which the replication sequences were obtained, then the foreign DNA will be replicated along with the host cell's DNA in the transgenic organism.

Molecular cloning is similar to polymerase chain reaction (PCR) in that it permits the replication of DNA sequence. The fundamental difference between the two methods is that molecular cloning involves replication of the DNA in a living microorganism, while PCR replicates DNA in an *in vitro* solution, free of living cells.

Steps in Molecular Cloning

In standard molecular cloning experiments, the cloning of any DNA fragment essentially involves seven steps: (1) Choice of host organism and cloning vector, (2) Preparation of vector DNA, (3) Preparation of DNA to be cloned, (4) Creation of recombinant DNA, (5) Introduction of recombinant DNA into host organism, (6) Selection of organisms containing recombinant DNA, (7) Screening for clones with desired DNA inserts and biological properties.

Although the detailed planning of the cloning can be done in any text editor, together with online utilities for e.g. PCR primer design, dedicated software exist for the purpose. Software for the purpose include for example ApE (open source), DNAStrider (open source), Serial Cloner (gratis) and Collagene (open source).

Choice of Host Organism and Cloning Vector

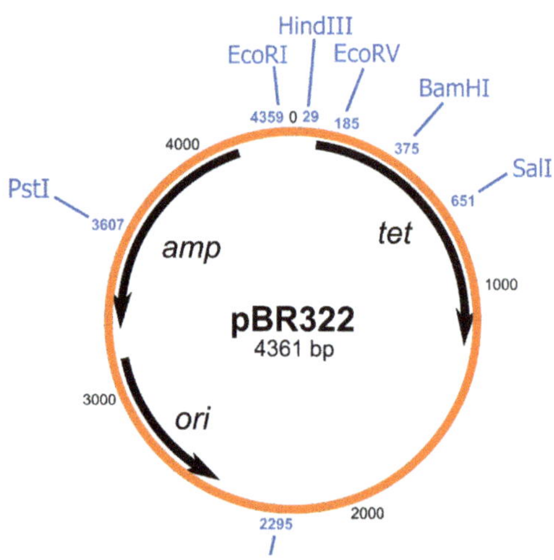

Diagram of a commonly used cloning plasmid; pBR322. It's a circular piece of DNA 4361 bases long. Two antibiotic resistancegenes are present, conferring resistance to ampicillin and tetracycline, and an origin of replication that the host uses to replicate the DNA.

Although a very large number of host organisms and molecular cloning vectors are in use, the great majority of molecular cloning experiments begin with a laboratory strain of the bacterium *E. coli* (*Escherichia coli*) and a plasmid cloning vector. *E. coli* and plasmid vectors are in common use because they are technically sophisticated, versatile, widely available, and offer rapid growth of recombinant organisms with minimal equipment. If the DNA to be cloned is exceptionally large (hundreds of thousands to millions of base pairs), then a bacterial artificial chromosome or yeast artificial chromosome vector is often chosen.

Specialized applications may call for specialized host-vector systems. For example, if the experimentalists wish to harvest a particular protein from the recombinant organism, then an expression vector is chosen that contains appropriate signals for transcription and translation in the desired host organism. Alternatively, if replication of the DNA in different species is desired (for example, transfer of DNA from bacteria to plants), then a multiple host range vector (also termed shuttle vector) may be selected. In practice, however, specialized molecular cloning experiments usually begin with cloning into a bacterial plasmid, followed by subcloning into a specialized vector.

Whatever combination of host and vector are used, the vector almost always contains four DNA segments that are critically important to its function and experimental utility:

- DNA *replication origin* is necessary for the vector (and its linked recombinant sequences) to replicate inside the host organism
- one or more unique *restriction endonuclease recognition sites* to serves as sites where foreign DNA may be introduced
- a *selectable genetic marker* gene that can be used to enable the survival of cells that have taken up vector sequences
- a *tag* gene that can be used to screen for cells containing the foreign DNA

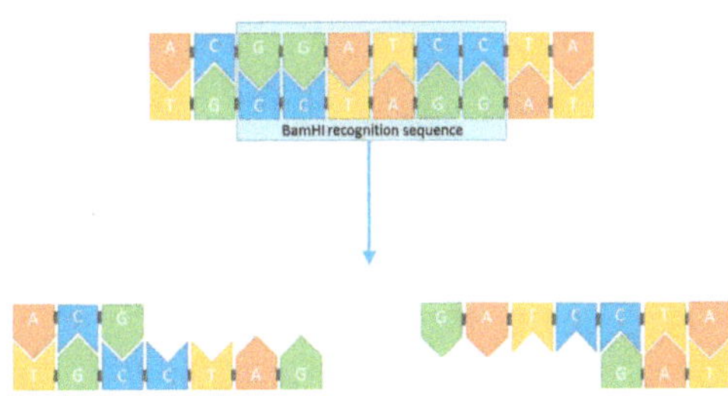

Cleavage of a DNA sequence containing the BamHI restriction site. The DNA is cleaved at the palindromic sequence to produce ‹sticky ends'.

Preparation of Vector DNA

The cloning vector is treated with a restriction endonuclease to cleave the DNA at the site where foreign DNA will be inserted. The restriction enzyme is chosen to generate a configuration at the cleavage site that is compatible with the ends of the foreign DNA. Typically, this is done by cleaving the vector DNA and foreign DNA with the same restriction enzyme, for example EcoRI. Most modern vectors contain a variety of convenient cleavage sites that are unique within the vector molecule (so that the vector can only be cleaved at a single site) and are located within a gene (frequently beta-galactosidase) whose inactivation can be used to distinguish recombinant from non-recombinant organisms at a later step in the process. To improve the ratio of recombinant to non-recombinant organisms, the cleaved vector may be treated with an enzyme (alkaline phosphatase) that dephosphorylates the vector ends. Vector molecules with dephosphorylated ends are unable to replicate, and replication can only be restored if foreign DNA is integrated into the cleavage site.

Preparation of DNA to be Cloned

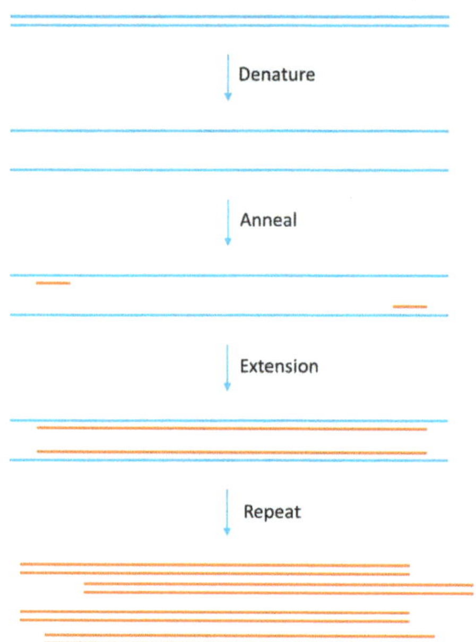

DNA for cloning is most commonly produced using PCR. Template DNA is mixed with bases (the building blocks of DNA), primers (short pieces of complementary single stranded DNA) and a DNA polymerase enzyme that builds the DNA chain. The mix goes through cycles of heating and cooling to produce large quantities of copied DNA.

For cloning of genomic DNA, the DNA to be cloned is extracted from the organism of interest. Virtually any tissue source can be used (even tissues from extinct animals), as long as the DNA is not extensively degraded. The DNA is then purified using simple methods to remove contaminating proteins (extraction with phenol), RNA (ribonuclease) and smaller molecules (precipitation and/or chromatography). Polymerase chain reaction (PCR) methods are often used for amplification of specific DNA or RNA (RT-PCR) sequences prior to molecular cloning.

DNA for cloning experiments may also be obtained from RNA using reverse transcriptase (complementary DNA or cDNA cloning), or in the form of synthetic DNA (artificial gene synthesis). cDNA cloning is usually used to obtain clones representative of the mRNA population of the cells of interest, while synthetic DNA is used to obtain any precise sequence defined by the designer.

The purified DNA is then treated with a restriction enzyme to generate fragments with ends capable of being linked to those of the vector. If necessary, short double-stranded segments of DNA (*linkers*) containing desired restriction sites may be added to create end structures that are compatible with the vector.

Creation of Recombinant DNA with DNA ligase

The creation of recombinant DNA is in many ways the simplest step of the molecular cloning process. DNA prepared from the vector and foreign source are simply mixed together at appropriate concentrations and exposed to an enzyme (DNA ligase) that covalently links the ends together. This joining reaction is often termed ligation. The resulting DNA mixture containing randomly joined ends is then ready for introduction into the host organism.

DNA ligase only recognizes and acts on the ends of linear DNA molecules, usually resulting in a complex mixture of DNA molecules with randomly joined ends. The desired products (vector DNA covalently linked to foreign DNA) will be present, but other sequences (e.g. foreign DNA linked to itself, vector DNA linked to itself and higher-order combinations of vector and foreign DNA) are also usually present. This complex mixture is sorted out in subsequent steps of the cloning process, after the DNA mixture is introduced into cells.

Introduction of Recombinant DNA into host Organism

The DNA mixture, previously manipulated in vitro, is moved back into a living cell, referred to as the host organism. The methods used to get DNA into cells are varied, and the name applied to this step in the molecular cloning process will often depend upon the experimental method that is chosen (e.g. transformation, transduction, transfection, electroporation).

When microorganisms are able to take up and replicate DNA from their local environment, the process is termed transformation, and cells that are in a physiological state such that they can take up DNA are said to be competent. In mammalian cell culture, the analogous process of introducing DNA into cells is commonly termed transfection. Both transformation and transfection usually require preparation of the cells through a special growth regime and chemical treatment process that will vary with the specific species and cell types that are used.

Electroporation uses high voltage electrical pulses to translocate DNA across the cell membrane (and cell wall, if present). In contrast, transduction involves the packaging of DNA into virus-derived particles, and using these virus-like particles to introduce the encapsulated DNA into the cell through a process resembling viral infection. Although electroporation and transduction are highly specialized methods, they may be the most efficient methods to move DNA into cells.

Selection of Organisms Containing Vector Sequences

Whichever method is used, the introduction of recombinant DNA into the chosen host organism is usually a low efficiency process; that is, only a small fraction of the cells will actually take up DNA. Experimental scientists deal with this issue through a step of artificial genetic selection, in which cells that have not taken up DNA are selectively killed, and only those cells that can actively replicate DNA containing the selectable marker gene encoded by the vector are able to survive.

When bacterial cells are used as host organisms, the selectable marker is usually a gene that confers resistance to an antibiotic that would otherwise kill the cells, typically ampicillin. Cells harboring the plasmid will survive when exposed to the antibiotic, while those that have failed to take up plasmid sequences will die. When mammalian cells (e.g. human or mouse cells) are used, a similar strategy is used, except that the marker gene (in this case typically encoded as part of the kanMX cassette) confers resistance to the antibiotic Geneticin.

Screening for clones with Desired DNA Inserts and Biological Properties

Modern bacterial cloning vectors (e.g. pUC19 and later derivatives including the pGEM vectors) use the blue-white screening system to distinguish colonies (clones) of transgenic cells from those that contain the parental vector (i.e. vector DNA with no recombinant sequence inserted). In these vectors, foreign DNA is inserted into a sequence that encodes an essential part of beta-galactosidase, an enzyme whose activity results in formation of a blue-colored colony on the culture medium that is used for this work. Insertion of the foreign DNA into the beta-galactosidase coding sequence disables the function of the enzyme, so that colonies containing transformed DNA remain colorless (white). Therefore, experimentalists are easily able to identify and conduct further studies on transgenic bacterial clones, while ignoring those that do not contain recombinant DNA.

The total population of individual clones obtained in a molecular cloning experiment is often termed a DNA library. Libraries may be highly complex (as when cloning complete genomic DNA from an organism) or relatively simple (as when moving a previously cloned DNA fragment into a different plasmid), but it is almost always necessary to examine a number of different clones to be sure that the desired DNA construct is obtained. This may be accomplished through a very wide range of experimental methods, including the use of nucleic acid hybridizations, antibody probes, polymerase chain reaction, restriction fragment analysis and/or DNA sequencing.

Applications of Molecular Cloning

Molecular cloning provides scientists with an essentially unlimited quantity of any individual DNA segments derived from any genome. This material can be used for a wide range of purposes, including those in both basic and applied biological science. A few of the more important applications are summarized here.

Genome Organization and Gene Expression

Molecular cloning has led directly to the elucidation of the complete DNA sequence of the genomes of a very large number of species and to an exploration of genetic diversity within individual species, work that has been done mostly by determining the DNA sequence of large numbers of randomly cloned fragments of the genome, and assembling the overlapping sequences.

At the level of individual genes, molecular clones are used to generate probes that are used for examining how genes are expressed, and how that expression is related to other processes in biology, including the metabolic environment, extracellular signals, development, learning, senescence and cell death. Cloned genes can also provide tools to examine the biological function and importance of individual genes, by allowing investigators to inactivate the genes, or make more subtle mutations using regional mutagenesis or site-directed mutagenesis.

Production of Recombinant Proteins

Obtaining the molecular clone of a gene can lead to the development of organisms that produce the protein product of the cloned genes, termed a recombinant protein. In practice, it is frequently more difficult to develop an organism that produces an active form of the recombinant protein in desirable quantities than it is to clone the gene. This is because the molecular signals for gene expression are complex and variable, and because protein folding, stability and transport can be very challenging.

Many useful proteins are currently available as recombinant products. These include--(1) medically useful proteins whose administration can correct a defective or poorly expressed gene (e.g. recombinant factor VIII, a blood-clotting factor deficient in some forms of hemophilia, and recombinant insulin, used to treat some forms of diabetes), (2) proteins that can be administered to assist in a life-threatening emergency (e.g. tissue plasminogen activator, used to treat strokes), (3) recombinant subunit vaccines, in which a purified protein can be used to immunize patients against infectious diseases, without exposing them to the infectious agent itself (e.g. hepatitis B vaccine), and (4) recombinant proteins as standard material for diagnostic laboratory tests.

Transgenic Organisms

Once characterized and manipulated to provide signals for appropriate expression, cloned genes may be inserted into organisms, generating transgenic organisms, also termed genetically modified organisms (GMOs). Although most GMOs are generated for purposes of basic biological research (see for example, transgenic mouse), a number of GMOs have been developed for commercial use, ranging from animals and plants that produce pharmaceuticals or other compounds (pharming), herbicide-resistant crop plants, and fluorescent tropical fish (GloFish) for home entertainment.

Gene Therapy

Gene therapy involves supplying a functional gene to cells lacking that function, with the aim of correcting a genetic disorder or acquired disease. Gene therapy can be broadly divided into two categories. The first is alteration of germ cells, that is, sperm or eggs, which results in a permanent genetic change for the whole organism and subsequent generations. This "germ line gene therapy" is considered by many to be unethical in human beings. The second type of gene therapy, "somatic cell gene therapy", is analogous to an organ transplant. In this case, one or more specific tissues are targeted by direct treatment or by removal of the tissue, addition of the therapeutic gene or genes in the laboratory, and return of the treated cells to the patient. Clinical trials of somatic cell gene therapy began in the late 1990s, mostly for the treatment of cancers and blood, liver, and lung disorders.

Despite a great deal of publicity and promises, the history of human gene therapy has been characterized by relatively limited success. The effect of introducing a gene into cells often promotes only partial and/or transient relief from the symptoms of the disease being treated. Some gene therapy trial patients have suffered adverse consequences of the treatment itself, including deaths. In some cases, the adverse effects result from disruption of essential genes within the patient's genome by insertional inactivation. In others, viral vectors used for gene therapy have been contaminated with infectious virus. Nevertheless, gene therapy is still held to be a promising future area of medicine, and is an area where there is a significant level of research and development activity.

POLYMERASE CHAIN REACTION

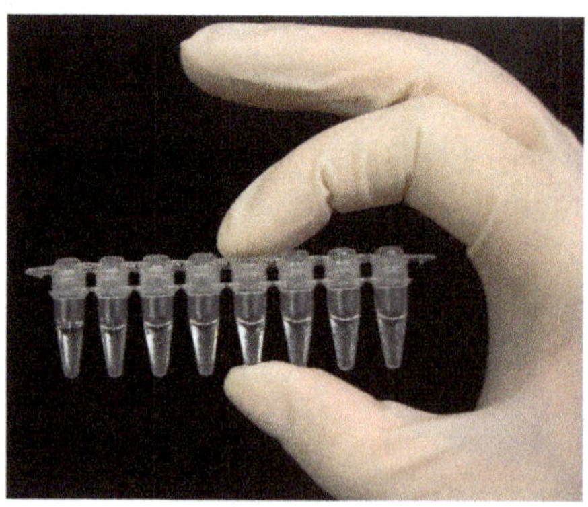

A strip of eight PCR tubes, each containing a 100 μl reaction mixture

Polymerase chain reaction (PCR) is a technique used in molecular biology to amplify a single copy or a few copies of a piece of DNA across several orders of magnitude, generating thousands to millions of copies of a particular DNA sequence. It is an easy and cheap tool to amplify a focused segment of DNA, useful for such purposes as the diagnosis and monitoring of genetic diseases, identification of criminals (in the field of forensics), and studying the function of a targeted segment of DNA.

Developed in 1983 by Kary Mullis, PCR is now a common and often indispensable technique used in clinical laboratories and research laboratories for a variety of applications. These include DNA cloning for sequencing, DNA-based phylogeny, or functional analysis of genes; the diagnosis of hereditary diseases; the identification of genetic fingerprints (used in forensic sciences and DNA paternity testing); and the detection of pathogens in nucleic acid tests for the diagnosis of infectious diseases. In 1993, Mullis was awarded the Nobel Prize in Chemistry along with Michael Smith for his work on PCR.

The method relies on thermal cycling, consisting of cycles of repeated heating and cooling of the reaction for DNA melting and enzymatic replication of the DNA. Primers (short DNA fragments) containing sequences complementary to the target region along with a DNA polymerase, which the method is named after, are key components to enable selective and repeated amplification. As PCR progresses, the DNA generated is itself used as a template for replication, setting in motion a chain reaction in which the DNA template is exponentially amplified. PCR can be extensively modified to perform a wide array of genetic manipulations. PCR is not generally considered to be a recombinant DNA method, as it does not involve cutting and pasting DNA, only amplification of existing sequences.

Almost all PCR applications employ a heat-stable DNA polymerase, such as Taq polymerase (an enzyme originally isolated from the bacterium *Thermus aquaticus*). This DNA polymerase enzymatically assembles a new DNA strand from DNA building-blocks, the nucleotides, by using single-stranded DNA as a template and DNA oligonucleotides (also called DNA primers), which are required for initiation of DNA synthesis. The vast majority of PCR methods use thermal cycling, i.e., alternately heating and cooling the PCR sample through a defined series of temperature steps.

In the first step, the two strands of the DNA double helix are physically separated at a high temperature in a process called DNA melting. In the second step, the temperature is lowered and the two DNA strands become templates for DNA polymerase to selectively amplify the target DNA. The selectivity of PCR results from the use of primers that are complementary to the DNA region targeted for amplification under specific thermal cycling conditions.

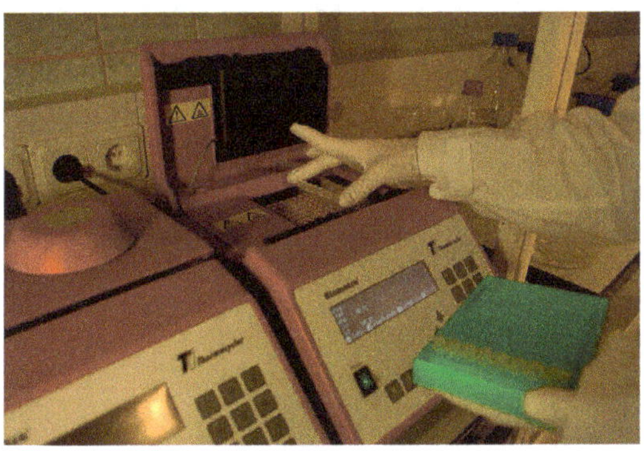

Placing a strip of eight PCR tubes into a thermal cycler

Principles

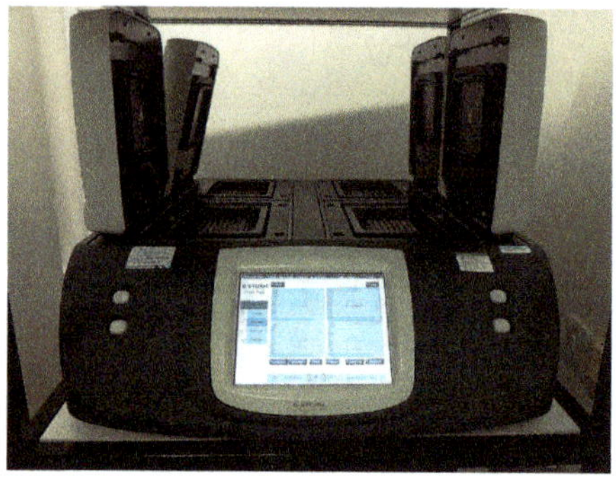

A thermal cycler for PCR

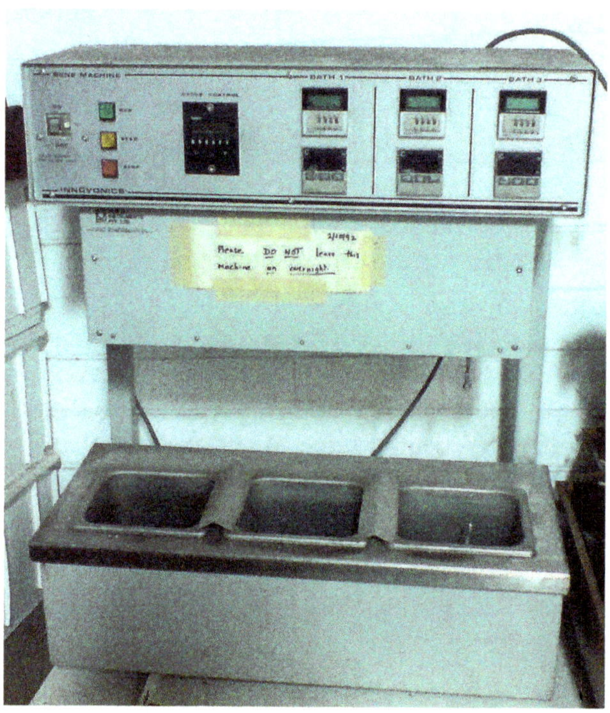

An older model three-temperature thermal cycler for PCR

PCR amplifies a specific region of a DNA strand (the DNA target). Most PCR methods typically amplify DNA fragments of between 0.1 and 10 kilo base pairs (kbp), although some techniques allow for amplification of fragments up to 40 kbp in size. The amount of amplified product is determined by the available substrates in the reaction, which become limiting as the reaction progresses.

A basic PCR set up requires several components and reagents. These components include:

- *DNA template* that contains the DNA region (target) to amplify
- Two *primers* that are complementary to the 3' (three prime) ends of each of the sense and anti-sense strand of the DNA target. Primers can be custom made in a laboratory that are complementary to the DNA segment to be amplified.
- *Taq polymerase* (as the DNA Polymerase can not attach to a DNA strand and elongate on its own. It should also be heat resistant, so that it can withstand the denaturation process.
- *Deoxynucleoside triphosphates* (dNTPs, sometimes called "deoxynucleotide triphosphates"; nucleotides containing triphosphate groups), the building-blocks from which the DNA polymerase synthesizes a new DNA strand.
- *Buffer solution*, providing a suitable chemical environment for optimum activity and stability of the DNA polymerase
- *Bivalent cations*, magnesium or manganese ions; generally Mg^{2+} is used, but Mn^{2+} can be used for PCR-mediated DNA mutagenesis, as higher Mn^{2+} concentration increases the error rate during DNA synthesis
- *Monovalent cation* potassium ions

The PCR is commonly carried out in a reaction volume of 10–200 μl in small reaction tubes (0.2–0.5 ml volumes) in a thermal cycler. The thermal cycler heats and cools the reaction tubes to achieve the temperatures required at each step of the reaction (see below). Many modern thermal cyclers make use of the Peltier effect, which permits both heating and cooling of the block holding the PCR tubes simply by reversing the electric current. Thin-walled reaction tubes permit favorable thermal conductivity to allow for rapid thermal equilibration. Most thermal cyclers have heated lids to prevent condensation at the top of the reaction tube. Older thermocyclers lacking a heated lid require a layer of oil on top of the reaction mixture or a ball of wax inside the tube.

Procedure

Typically, PCR consists of a series of 20–40 repeated temperature changes, called cycles, with each cycle commonly consisting of 2–3 discrete temperature steps, usually three (Figure below). The cycling is often preceded by a single temperature step at a high temperature (>90 °C), and followed by one hold at the end for final product extension or brief storage. The temperatures used and the length of time they are applied in each cycle depend on a variety of parameters. These include the enzyme used for DNA synthesis, the concentration of divalent ions and dNTPs in the reaction, and the melting temperature (Tm) of the primers.

- *Initialization step* (Only required for DNA polymerases that require heat activation by hot-start PCR.): This step consists of heating the reaction to a temperature of 94–96 °C (or 98 °C if extremely thermostable polymerases are used), which is held for 1–9 minutes.

- *Denaturation step*: This step is the first regular cycling event and consists of heating the reaction to 94–98 °C for 20–30 seconds. It causes DNA melting of the DNA template by disrupting the hydrogen bonds between complementary bases, yielding single-stranded DNA molecules.
- *Annealing step*: The reaction temperature is lowered to 50–65 °C for 20–40 seconds allowing annealing of the primers to the single-stranded DNA template. This temperature must be low enough to allow for hybridization of the primer to the strand, but high enough for the hybridization to be specific, i.e., the primer should only bind to a perfectly complementary part of the template. If the temperature is too low, the primer could bind imperfectly. If it is too high, the primer might not bind. Typically the annealing temperature is about 3–5 °C below the Tm of the primers used. Stable DNA–DNA hydrogen bonds are only formed when the primer sequence very closely matches the template sequence. The polymerase binds to the primer-template hybrid and begins DNA formation. It is very vital to determine the annealing temperature in PCR. This is because in PCR, efficiency and specificity are affected by the annealing temperature. An incorrect annealing temperature will cause an error in the test.
- *Extension/elongation step*: The temperature at this step depends on the DNA polymerase used; Taq polymerase has its optimum activity temperature at 75–80 °C, and commonly a temperature of 72 °C is used with this enzyme. At this step the DNA polymerase synthesizes a new DNA strand complementary to the DNA template strand by adding dNTPs that are complementary to the template in 5' to 3' direction, condensing the 5'-phosphate group of the dNTPs with the 3'-hydroxyl group at the end of the nascent (extending) DNA strand. The extension time depends both on the DNA polymerase used and on the length of the DNA fragment to amplify. As a rule-of-thumb, at its optimum temperature, the DNA polymerase polymerizes a thousand bases per minute. Under optimum conditions, i.e., if there are no limitations due to limiting substrates or reagents, at each extension step, the amount of DNA target is doubled, leading to exponential (geometric) amplification of the specific DNA fragment.

The processes of denaturation, annealing and elongation constitute of one cycle. Multiple cycles are required to amplify DNA. The formula used to calculate the number of DNA copies formed after a given # of cycles repeated is, 2# of cycles.

- *Final elongation*: This single step is occasionally performed at a temperature of 70–74 °C (this is the temperature needed for optimal activity for most polymerases used in PCR) for 5–15 minutes after the last PCR cycle to ensure that any remaining single-stranded DNA is fully extended.
- *Final hold*: This step at 4–15 °C for an indefinite time may be employed for short-term storage of the reaction.

Stages

The PCR process can be divided into three stages:

Exponential amplification: At every cycle, the amount of product is doubled (assuming

100% reaction efficiency). The reaction is very sensitive: only minute quantities of DNA must be present.

Leveling off stage: The reaction slows as the DNA polymerase loses activity and as consumption of reagents such as dNTPs and primers causes them to become limiting.

Plateau: No more product accumulates due to exhaustion of reagents and enzyme.

Optimization

In practice, PCR can fail for various reasons, in part due to its sensitivity to contamination causing amplification of spurious DNA products. Because of this, a number of techniques and procedures have been developed for optimizing PCR conditions. Contamination with extraneous DNA is addressed with lab protocols and procedures that separate pre-PCR mixtures from potential DNA contaminants. This usually involves spatial separation of PCR-setup areas from areas for analysis or purification of PCR products, use of disposable plasticware, and thoroughly cleaning the work surface between reaction setups. Primer-design techniques are important in improving PCR product yield and in avoiding the formation of spurious products, and the usage of alternate buffer components or polymerase enzymes can help with amplification of long or otherwise problematic regions of DNA. Addition of reagents, such as formamide, in buffer systems may increase the specificity and yield of PCR. Computer simulations of theoretical PCR results (Electronic PCR) may be performed to assist in primer design.

Applications

Selective DNA Isolation

PCR allows isolation of DNA fragments from genomic DNA by selective amplification of a specific region of DNA. This use of PCR augments many methods, such as generating hybridization probes for Southern or northern hybridization and DNA cloning, which require larger amounts of DNA, representing a specific DNA region. PCR supplies these techniques with high amounts of pure DNA, enabling analysis of DNA samples even from very small amounts of starting material.

Other applications of PCR include DNA sequencing to determine unknown PCR-amplified sequences in which one of the amplification primers may be used in Sanger sequencing, isolation of a DNA sequence to expedite recombinant DNA technologies involving the insertion of a DNA sequence into a plasmid, phage, or cosmid (depending on size) or the genetic material of another organism. Bacterial colonies *(such as E. coli)* can be rapidly screened by PCR for correct DNA vector constructs. PCR may also be used for genetic fingerprinting; a forensic technique used to identify a person or organism by comparing experimental DNAs through different PCR-based methods.

Some PCR ‹fingerprints' methods have high discriminative power and can be used to identify genetic relationships between individuals, such as parent-child or between siblings, and are used in paternity testing (Fig. 4). This technique may also be used to determine evolutionary relationships among organisms when certain molecular clocks are used (i.e., the 16S rRNA and recA genes of microorganisms).

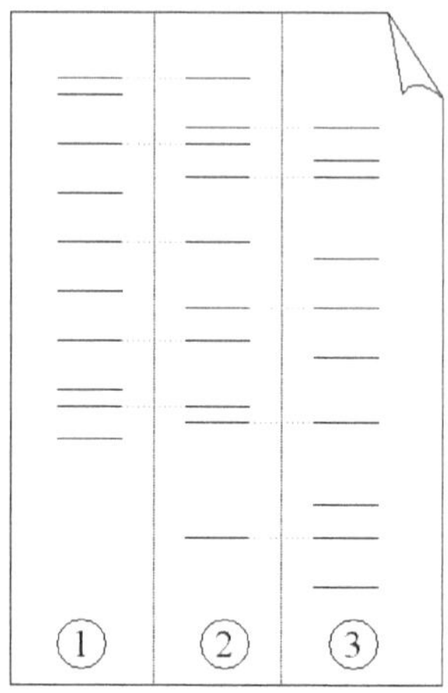

Electrophoresis of PCR-amplified DNA fragments. (1) Father. (2) Child. (3) Mother. The child has inherited some, but not all of the fingerprint of each of its parents, giving it a new, unique fingerprint.

Amplification and Quantification of DNA

Because PCR amplifies the regions of DNA that it targets, PCR can be used to analyze extremely small amounts of sample. This is often critical for forensic analysis, when only a trace amount of DNA is available as evidence. PCR may also be used in the analysis of ancient DNA that is tens of thousands of years old. These PCR-based techniques have been successfully used on animals, such as a forty-thousand-year-old mammoth, and also on human DNA, in applications ranging from the analysis of Egyptian mummies to the identification of a Russian tsar and the body of English king Richard III.

Quantitative PCR methods allow the estimation of the amount of a given sequence present in a sample—a technique often applied to quantitatively determine levels of gene expression. Quantitative PCR is an established tool for DNA quantification that measures the accumulation of DNA product after each round of PCR amplification.

Medical Applications

PCR has been applied to a large number of medical procedures:

- The first application of PCR was for genetic testing, where a sample of DNA is analyzed for the presence of genetic disease mutations. Prospective parents can be tested for being genetic carriers, or their children might be tested for actually being affected by a disease. DNA samples for prenatal testing can be obtained by amniocentesis, chorionic villus sampling, or even by the analysis of rare fetal cells circulating in the mother's bloodstream. PCR analysis is also essential to preimplantation genetic diagnosis, where individual cells of a developing embryo are tested for mutations.

- PCR can also be used as part of a sensitive test for tissue typing, vital to organ transplantation. As of 2008, there is even a proposal to replace the traditional antibody-based tests for blood type with PCR-based tests.
- Many forms of cancer involve alterations to oncogenes. By using PCR-based tests to study these mutations, therapy regimens can sometimes be individually customized to a patient. PCR permits early diagnosis of malignant diseases such as leukemia and lymphomas, which is currently the highest-developed in cancer research and is already being used routinely. PCR assays can be performed directly on genomic DNA samples to detect translocation-specific malignant cells at a sensitivity that is at least 10,000 fold higher than that of other methods.

Infectious Disease Applications

PCR allows for rapid and highly specific diagnosis of infectious diseases, including those caused by bacteria or viruses. PCR also permits identification of non-cultivatable or slow-growing microorganisms such as mycobacteria, anaerobic bacteria, or viruses from tissue culture assays and animal models. The basis for PCR diagnostic applications in microbiology is the detection of infectious agents and the discrimination of non-pathogenic from pathogenic strains by virtue of specific genes.

Characterization and detection of infectious disease organisms have been revolutionized by PCR in the follow ways:

- The human immunodeficiency virus (or HIV), is a difficult target to find and eradicate. The earliest tests for infection relied on the presence of antibodies to the virus circulating in the bloodstream. However, antibodies don't appear until many weeks after infection, maternal antibodies mask the infection of a newborn, and therapeutic agents to fight the infection don't affect the antibodies. PCR tests have been developed that can detect as little as one viral genome among the DNA of over 50,000 host cells. Infections can be detected earlier, donated blood can be screened directly for the virus, newborns can be immediately tested for infection, and the effects of antiviral treatments can be quantified.
- Some disease organisms, such as that for tuberculosis, are difficult to sample from patients and slow to be grown in the laboratory. PCR-based tests have allowed detection of small numbers of disease organisms (both live or dead), in convenient samples. Detailed genetic analysis can also be used to detect antibiotic resistance, allowing immediate and effective therapy. The effects of therapy can also be immediately evaluated.
- The spread of a disease organism through populations of domestic or wild animals can be monitored by PCR testing. In many cases, the appearance of new virulent sub-types can be detected and monitored. The sub-types of an organism that were responsible for earlier epidemics can also be determined by PCR analysis.

- Viral DNA can be detected by PCR. The primers used must be specific to the targeted sequences in the DNA of a virus, and PCR can be used for diagnostic analyses or DNA sequencing of the viral genome. The high sensitivity of PCR permits virus detection soon after infection and even before the onset of disease. Such early detection may give physicians a significant lead time in treatment. The amount of virus ("viral load") in a patient can also be quantified by PCR-based DNA quantitation techniques (see below).

Forensic Applications

The development of PCR-based genetic (or DNA) fingerprinting protocols has seen widespread application in forensics:

- In its most discriminating form, **genetic fingerprinting** can uniquely discriminate any one person from the entire population of the world. Minute samples of DNA can be isolated from a crime scene, and compared to that from suspects, or from a DNA database of earlier evidence or convicts. Simpler versions of these tests are often used to rapidly rule out suspects during a criminal investigation. Evidence from decades-old crimes can be tested, confirming or exonerating the people originally convicted.
- Less discriminating forms of DNA fingerprinting can help in **DNA paternity testing**, where an individual is matched with their close relatives. DNA from unidentified human remains can be tested, and compared with that from possible parents, siblings, or children. Similar testing can be used to confirm the biological parents of an adopted (or kidnapped) child. The actual biological father of a newborn can also be confirmed (or ruled out).

Research Applications

PCR has been applied to many areas of research in molecular genetics:

- PCR allows rapid production of short pieces of DNA, even when not more than the sequence of the two primers is known. This ability of PCR augments many methods, such as generating hybridization probes for Southern or northern blot hybridization. PCR supplies these techniques with large amounts of pure DNA, sometimes as a single strand, enabling analysis even from very small amounts of starting material.
- The task of DNA sequencing can also be assisted by PCR. Known segments of DNA can easily be produced from a patient with a genetic disease mutation. Modifications to the amplification technique can extract segments from a completely unknown genome, or can generate just a single strand of an area of interest.
- PCR has numerous applications to the more traditional process of DNA cloning. It can extract segments for insertion into a vector from a larger genome, which may be only available in small quantities. Using a single set of ‹vector primers', it can also analyze or extract fragments that have already been inserted into vectors. Some alterations to the PCR protocol can generate mutations (general

or site-directed) of an inserted fragment.

- Sequence-tagged sites is a process where PCR is used as an indicator that a particular segment of a genome is present in a particular clone. The Human Genome Project found this application vital to mapping the cosmid clones they were sequencing, and to coordinating the results from different laboratories.
- An exciting application of PCR is the phylogenic analysis of DNA from ancient sources, such as that found in the recovered bones of Neanderthals, or from frozen tissues of mammoths. In some cases the highly degraded DNA from these sources might be reassembled during the early stages of amplification.
- A common application of PCR is the study of patterns of gene expression. Tissues (or even individual cells) can be analyzed at different stages to see which genes have become active, or which have been switched off. This application can also use quantitative PCR to quantitate the actual levels of expression
- The ability of PCR to simultaneously amplify several loci from individual sperm has greatly enhanced the more traditional task of genetic mapping by studying chromosomal crossovers after meiosis. Rare crossover events between very close loci have been directly observed by analyzing thousands of individual sperms. Similarly, unusual deletions, insertions, translocations, or inversions can be analyzed, all without having to wait (or pay for) the long and laborious processes of fertilization, embryogenesis, etc.

Limitations

DNA polymerase is prone to error, which in turn causes mutations in the PCR fragments that are made. Additionally, the specificity of the PCR fragments can mutate to the template DNA, due to nonspecific binding of primers. Furthermore, prior information on the sequence is necessary in order to generate the primers.

Variations

- *Allele-specific PCR*: a diagnostic or cloning technique based on single-nucleotide variations (SNVs not to be confused with SNPs) (single-base differences in a patient). It requires prior knowledge of a DNA sequence, including differences between alleles, and uses primers whose 3' ends encompass the SNV (base pair buffer around SNV usually incorporated). PCR amplification under stringent conditions is much less efficient in the presence of a mismatch between template and primer, so successful amplification with an SNP-specific primer signals presence of the specific SNP in a sequence. See SNP genotyping for more information.
- *Assembly PCR* or *Polymerase Cycling Assembly (PCA)*: artificial synthesis of long DNA sequences by performing PCR on a pool of long oligonucleotides with short overlapping segments. The oligonucleotides alternate between sense and antisense directions, and the overlapping segments determine the order of the PCR fragments, thereby selectively producing the final long DNA product.

- *Asymmetric PCR*: preferentially amplifies one DNA strand in a double-stranded DNA template. It is used in sequencing and hybridization probing where amplification of only one of the two complementary strands is required. PCR is carried out as usual, but with a great excess of the primer for the strand targeted for amplification. Because of the slow (arithmetic) amplification later in the reaction after the limiting primer has been used up, extra cycles of PCR are required. A recent modification on this process, known as *L*inear-*A*fter-*T*he-*E*xponential-PCR (LATE-PCR), uses a limiting primer with a higher melting temperature (Tm) than the excess primer to maintain reaction efficiency as the limiting primer concentration decreases mid-reaction.
- *Dial-out PCR*: a highly parallel method for retrieving accurate DNA molecules for gene synthesis. A complex library of DNA molecules is modified with unique flanking tags before massively parallel sequencing. Tag-directed primers then enable the retrieval of molecules with desired sequences by PCR.
- *Digital PCR (dPCR)*: used to measure the quantity of a target DNA sequence in a DNA sample. The DNA sample is highly diluted so that after running many PCRs in parallel, some of them do not receive a single molecule of the target DNA. The target DNA concentration is calculated using the proportion of negative outcomes. Hence the name 'digital PCR'.
- *Helicase-dependent amplification*: similar to traditional PCR, but uses a constant temperature rather than cycling through denaturation and annealing/extension cycles. DNA helicase, an enzyme that unwinds DNA, is used in place of thermal denaturation.
- *Hot start PCR*: a technique that reduces non-specific amplification during the initial set up stages of the PCR. It may be performed manually by heating the reaction components to the denaturation temperature (e.g., 95 °C) before adding the polymerase. Specialized enzyme systems have been developed that inhibit the polymerase's activity at ambient temperature, either by the binding of an antibody or by the presence of covalently bound inhibitors that dissociate only after a high-temperature activation step. Hot-start/cold-finish PCR is achieved with new hybrid polymerases that are inactive at ambient temperature and are instantly activated at elongation temperature.
- *In silico PCR* (digital PCR, virtual PCR, electronic PCR, e-PCR) refers to computational tools used to calculate theoretical polymerase chain reaction results using a given set of primers (probes) to amplify DNA sequences from a sequenced genome or transcriptome. In silico PCR was proposed as an educational tool for molecular biology.
- *Intersequence-specific PCR* (ISSR): a PCR method for DNA fingerprinting that amplifies regions between simple sequence repeats to produce a unique fingerprint of amplified fragment lengths.
- *Inverse PCR*: is commonly used to identify the flanking sequences around genomic inserts. It involves a series of DNA digestions and self ligation, resulting in known sequences at either end of the unknown sequence.
- *Ligation-mediated PCR*: uses small DNA linkers ligated to the DNA of interest

and multiple primers annealing to the DNA linkers; it has been used for DNA sequencing, genome walking, and DNA footprinting.

- *Methylation-specific PCR* (MSP): developed by Stephen Baylin and Jim Herman at the Johns Hopkins School of Medicine, and is used to detect methylation of CpG islands in genomic DNA. DNA is first treated with sodium bisulfite, which converts unmethylated cytosine bases to uracil, which is recognized by PCR primers as thymine. Two PCRs are then carried out on the modified DNA, using primer sets identical except at any CpG islands within the primer sequences. At these points, one primer set recognizes DNA with cytosines to amplify methylated DNA, and one set recognizes DNA with uracil or thymine to amplify unmethylated DNA. MSP using qPCR can also be performed to obtain quantitative rather than qualitative information about methylation.
- *Miniprimer PCR*: uses a thermostable polymerase (S-Tbr) that can extend from short primers ("smalligos") as short as 9 or 10 nucleotides. This method permits PCR targeting to smaller primer binding regions, and is used to amplify conserved DNA sequences, such as the 16S (or eukaryotic 18S) rRNA gene.
- *Multiplex ligation-dependent probe amplification* (*MLPA*): permits amplifying multiple targets with a single primer pair, thus avoiding the resolution limitations of multiplex PCR (see below).
- *Multiplex-PCR*: consists of multiple primer sets within a single PCR mixture to produce amplicons of varying sizes that are specific to different DNA sequences. By targeting multiple genes at once, additional information may be gained from a single test-run that otherwise would require several times the reagents and more time to perform. Annealing temperatures for each of the primer sets must be optimized to work correctly within a single reaction, and amplicon sizes. That is, their base pair length should be different enough to form distinct bands when visualized by gel electrophoresis.
- *Nanoparticle-Assisted PCR (nanoPCR)*: In recent years, it has been reported that some nanoparticles (NPs) can enhance the efficiency of PCR (thus being called nanoPCR), and some even perform better than the original PCR enhancers. It was also found that quantum dots (QDs) can improve PCR specificity and efficiency. Single-walled carbon nanotubes (SWCNTs) and multi-walled carbon nanotubes (MWCNTs) are efficient in enhancing the amplification of long PCR. Carbon nanopowder (CNP) was reported be able to improve the efficiency of repeated PCR and long PCR. ZnO, TiO2, and Ag NPs were also found to increase PCR yield. Importantly, already known data has indicated that non-metallic NPs retained acceptable amplification fidelity. Given that many NPs are capable of enhancing PCR efficiency, it is clear that there is likely to be great potential for nanoPCR technology improvements and product development.
- *Nested PCR*: increases the specificity of DNA amplification, by reducing background due to non-specific amplification of DNA. Two sets of primers are used in two successive PCRs. In the first reaction, one pair of primers is used to generate DNA products, which besides the intended target, may still

consist of non-specifically amplified DNA fragments. The product(s) are then used in a second PCR with a set of primers whose binding sites are completely or partially different from and located 3' of each of the primers used in the first reaction. Nested PCR is often more successful in specifically amplifying long DNA fragments than conventional PCR, but it requires more detailed knowledge of the target sequences.

- *Overlap-extension PCR* or *Splicing by overlap extension (SOEing)* : a genetic engineering technique that is used to splice together two or more DNA fragments that contain complementary sequences. It is used to join DNA pieces containing genes, regulatory sequences, or mutations; the technique enables creation of specific and long DNA constructs. It can also introduce deletions, insertions or point mutations into a DNA sequence.
- *PAN-AC*: uses isothermal conditions for amplification, and may be used in living cells.
- *quantitative PCR* (qPCR): used to measure the quantity of a target sequence (commonly in real-time). It quantitatively measures starting amounts of DNA, cDNA, or RNA. quantitative PCR is commonly used to determine whether a DNA sequence is present in a sample and the number of its copies in the sample. *Quantitative PCR* has a very high degree of precision. Quantitative PCR methods use fluorescent dyes, such as Sybr Green, EvaGreen or fluorophore-containing DNA probes, such as TaqMan, to measure the amount of amplified product in real time. It is also sometimes abbreviated to RT-PCR (*real-time* PCR) but this abbreviation should be used only for reverse transcription PCR. qPCR is the appropriate contractions for quantitative PCR (real-time PCR).
- *Reverse Transcription PCR (RT-PCR)*: for amplifying DNA from RNA. Reverse transcriptase reverse transcribes RNA into cDNA, which is then amplified by PCR. RT-PCR is widely used in expression profiling, to determine the expression of a gene or to identify the sequence of an RNA transcript, including transcription start and termination sites. If the genomic DNA sequence of a gene is known, RT-PCR can be used to map the location of exons and introns in the gene. The 5' end of a gene (corresponding to the transcription start site) is typically identified by RACE-PCR (*Rapid Amplification of cDNA Ends*).
- *Solid Phase PCR*: encompasses multiple meanings, including Polony Amplification (where PCR colonies are derived in a gel matrix, for example), Bridge PCR(primers are covalently linked to a solid-support surface), conventional Solid Phase PCR (where Asymmetric PCR is applied in the presence of solid support bearing primer with sequence matching one of the aqueous primers) and Enhanced Solid Phase PCR (where conventional Solid Phase PCR can be improved by employing high Tm and nested solid support primer with optional application of a thermal 'step' to favour solid support priming).
- *Suicide PCR*: typically used in paleogenetics or other studies where avoiding false positives and ensuring the specificity of the amplified fragment is the highest priority. It was originally described in a study to verify the presence

of the microbe Yersinia pestis in dental samples obtained from 14th Century graves of people supposedly killed by plague during the medieval Black Death epidemic. The method prescribes the use of any primer combination only once in a PCR (hence the term "suicide"), which should never have been used in any positive control PCR reaction, and the primers should always target a genomic region never amplified before in the lab using this or any other set of primers. This ensures that no contaminating DNA from previous PCR reactions is present in the lab, which could otherwise generate false positives.

- *Thermal asymmetric interlaced PCR (TAIL-PCR)*: for isolation of an unknown sequence flanking a known sequence. Within the known sequence, TAIL-PCR uses a nested pair of primers with differing annealing temperatures; a degenerate primer is used to amplify in the other direction from the unknown sequence.
- *Touchdown PCR* (*Step-down PCR*): a variant of PCR that aims to reduce nonspecific background by gradually lowering the annealing temperature as PCR cycling progresses. The annealing temperature at the initial cycles is usually a few degrees (3–5 °C) above the T_m of the primers used, while at the later cycles, it is a few degrees (3–5 °C) below the primer T_m. The higher temperatures give greater specificity for primer binding, and the lower temperatures permit more efficient amplification from the specific products formed during the initial cycles.
- *Universal Fast Walking*: for genome walking and genetic fingerprinting using a more specific 'two-sided' PCR than conventional 'one-sided' approaches (using only one gene-specific primer and one general primer—which can lead to artefactual 'noise') by virtue of a mechanism involving lariat structure formation. Streamlined derivatives of UFW are LaNe RAGE (lariat-dependent nested PCR for rapid amplification of genomic DNA ends), 5'RACE LaNe and 3'RACE LaNe.

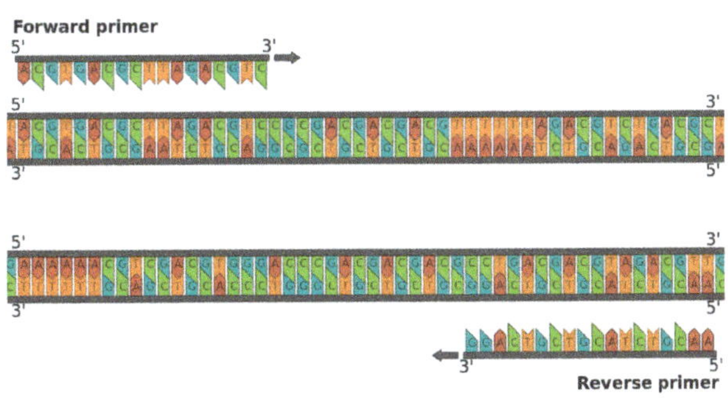

Diagrammatic representation of an example primer pair. The use of primers in an in vitro assay to allow DNA synthesis was a major innovation that allowed the development of PCR.

A 1971 paper in the Journal of Molecular Biology by Kjell Kleppe (no) and co-workers in the laboratory of H. Gobind Khorana first described a method using an enzymatic assay to replicate a short DNA template with primers *in vitro*. However, this early manifestation of the basic PCR principle did not receive much attention at the time, and the invention of the polymerase chain reaction in 1983 is generally credited to Kary Mullis.

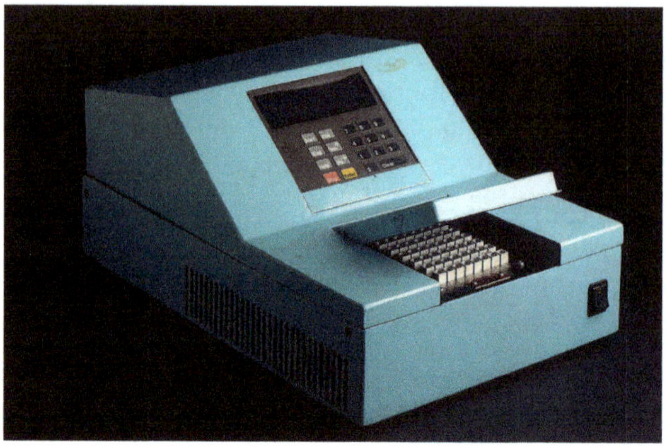

"Baby Blue", a 1986 prototype machine for doing PCR

When Mullis developed the PCR in 1983, he was working in Emeryville, California for Cetus Corporation, one of the first biotechnology companies. There, he was responsible for synthesizing short chains of DNA. Mullis has written that he conceived of PCR while cruising along the Pacific Coast Highway one night in his car. He was playing in his mind with a new way of analyzing changes (mutations) in DNA when he realized that he had instead invented a method of amplifying any DNA region through repeated cycles of duplication driven by DNA polymerase. In *Scientific American*, Mullis summarized the procedure: "Beginning with a single molecule of the genetic material DNA, the PCR can generate 100 billion similar molecules in an afternoon. The reaction is easy to execute. It requires no more than a test tube, a few simple reagents, and a source of heat." He was awarded the Nobel Prize in Chemistry in 1993 for his invention, seven years after he and his colleagues at Cetus first put his proposal to practice. However, some controversies have remained about the intellectual and practical contributions of other scientists to Mullis' work, and whether he had been the sole inventor of the PCR principle (see below).

At the core of the PCR method is the use of a suitable DNA polymerase able to withstand the high temperatures of >90 °C (194 °F) required for separation of the two DNA strands in the DNA double helix after each replication cycle. The DNA polymerases initially employed for in vitro experiments presaging PCR were unable to withstand these high temperatures.So the early procedures for DNA replication were very inefficient and time-consuming, and required large amounts of DNA polymerase and continuous handling throughout the process.

The discovery in 1976 of Taq polymerase — a DNA polymerase purified from the thermophilic bacterium, *Thermus aquaticus*, which naturally lives in hot (50 to 80 °C (122 to 176 °F)) environments such as hot springs — paved the way for dramatic improvements of the PCR method. The DNA polymerase isolated from *T. aquaticus* is stable at high temperatures remaining active even after DNA denaturation, thus obviating the need to add new DNA polymerase after each cycle.This allowed an automated thermocycler-based process for DNA amplification.

Patent Disputes

The PCR technique was patented by Kary Mullis and assigned to Cetus Corporation, where Mullis worked when he invented the technique in 1983. The *Taq* polymerase enzyme was also covered by patents. There have been several high-profile lawsuits related to the technique, including an unsuccessful lawsuit brought by DuPont. The pharmaceutical company Hoffmann-La Roche purchased the rights to the patents in 1992 and currently holds those that are still protected.

A related patent battle over the Taq polymerase enzyme is still ongoing in several jurisdictions around the world between Roche and Promega. The legal arguments have extended beyond the lives of the original PCR and Taq polymerase patents, which expired on March 28, 2005.

GEL ELECTROPHORESIS

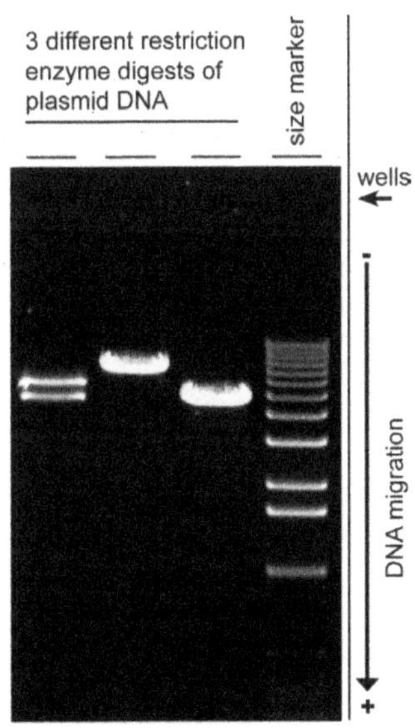

Digital image of 3 plasmid restriction digests run on a 1% w/v agarose gel, 3 volt/cm, stained with ethidium bromide. The DNA size marker is a commercial 1 kbp ladder. The position of the wells and direction of DNA migration is noted.

Gel electrophoresis is a method for separation and analysis of macromolecules (DNA, RNA and proteins) and their fragments, based on their size and charge. It is used in clinical chemistry to separate proteins by charge and/or size (IEF agarose, essentially size independent) and in biochemistry and molecular biology to separate a mixed population of DNA and RNA fragments by length, to estimate the size of DNA and RNA fragments or to separate proteins by charge.

Nucleic acid molecules are separated by applying an electric field to move the negatively charged molecules through a matrix of agarose or other substances. Shorter molecules move faster and migrate farther than longer ones because shorter molecules migrate more easily through the pores of the gel. This phenomenon is called sieving. Proteins are separated by charge in agarose because the pores of the gel are too large to sieve proteins. Gel electrophoresis can also be used for separation of nanoparticles.

Gel electrophoresis uses a gel as an anticonvective medium and/or sieving medium during electrophoresis, the movement of a charged particle in an electrical field. Gels suppress the thermal convection caused by application of the electric field, and can also act as a sieving medium, retarding the passage of molecules; gels can also simply serve to maintain the finished separation, so that a post electrophoresis stain can be applied. DNA Gel electrophoresis is usually performed for analytical purposes, often after amplification of DNA via PCR, but may be used as a preparative technique prior to use of other methods such as mass spectrometry, RFLP, PCR, cloning, DNA sequencing, or Southern blotting for further characterization.

Physical Basis

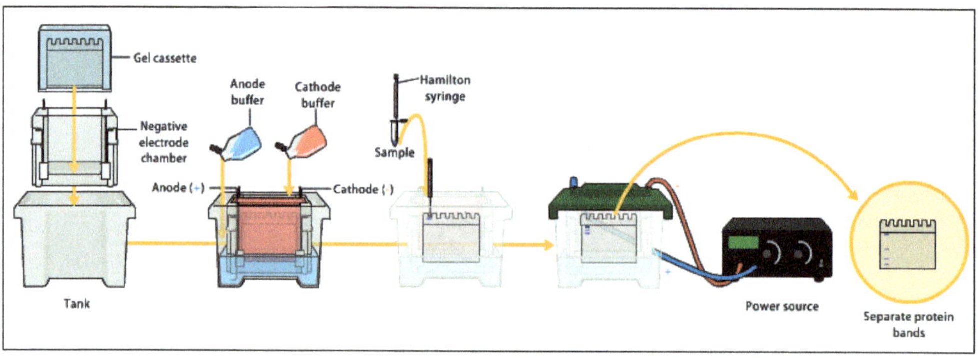

In simple terms, electrophoresis is a process which enables the sorting of molecules based on size. Using an electric field, molecules (such as DNA) can be made to move through a gel made of agar or polyacrylamide. The electric field consists of a negative charge at one end which pushes the molecules through the gel, and a positive charge at the other end that pulls the molecules through the gel. The molecules being sorted are dispensed into a well in the gel material. The gel is placed in an electrophoresis chamber, which is then connected to a power source. When the electric current is applied, the larger molecules move more slowly through the gel while the smaller molecules move faster. The different sized molecules form distinct bands on the gel.

The term "gel" in this instance refers to the matrix used to contain, then separate the target molecules. In most cases, the gel is a crosslinked polymer whose composition and porosity is chosen based on the specific weight and composition of the target to be analyzed.

When separating proteins or small nucleic acids (DNA, RNA, or oligonucleotides) the gel is usually composed of different concentrations of acrylamide and a cross-linker, producing different sized mesh networks of polyacrylamide. When separating larger nucleic acids (greater than a few hundred bases), the preferred matrix is purified agarose. In both cases, the gel forms a solid, yet porous matrix. Acrylamide, in contrast to polyacrylamide, is a neurotoxin and must be handled using appropriate safety precautions to avoid poisoning. Agarose is composed of long unbranched chains of uncharged carbohydrate without cross links resulting in a gel with large pores allowing for the separation of macromolecules and macromolecular complexes.

"Electrophoresis" refers to the electromotive force (EMF) that is used to move the molecules through the gel matrix. By placing the molecules in wells in the gel and applying an electric field, the molecules will move through the matrix at different rates, determined largely by their mass when the charge to mass ratio (Z) of all species is uniform. However, when charges are not all uniform then, the electrical field generated by the electrophoresis procedure will affect the species that have different charges and therefore will attract the species according to their charges being the opposite. Species that are positively charged will migrate towards the cathode which is negatively charged (because this is an electrolytic rather than galvanic cell). If the species are negatively charged they will migrate towards the positively charged anode.

If several samples have been loaded into adjacent wells in the gel, they will run parallel in individual lanes. Depending on the number of different molecules, each lane shows separation of the components from the original mixture as one or more distinct bands, one band per component. Incomplete separation of the components can lead to overlapping bands, or to indistinguishable smears representing multiple unresolved components. Bands in different lanes that end up at the same distance from the top contain molecules that passed through the gel with the same speed, which usually means they are approximately the same size. There are molecular weight size markers available that contain a mixture of molecules of known sizes. If such a marker was run on one lane in the gel parallel to the unknown samples, the bands observed can be compared to those of the unknown in order to determine their size. The distance a band travels is approximately inversely proportional to the logarithm of the size of the molecule.

There are limits to electrophoretic techniques. Since passing current through a gel causes heating, gels may melt during electrophoresis. Electrophoresis is performed in buffer solutions to reduce pH changes due to the electric field, which is important because the charge of DNA and RNA depends on pH, but running for too long can exhaust the buffering capacity of the solution. Further, different preparations of genetic material may not migrate consistently with each other, for morphological or other reasons.

Types of Gel

The types of gel most typically used are agarose and polyacrylamide gels. Each type of gel is well-suited to different types and sizes of analyte. Polyacrylamide gels are usually used for proteins, and have very high resolving power for small fragments of DNA (5-500 bp). Agarose gels on the other hand have lower resolving power for DNA but have greater range of separation, and are therefore used for DNA fragments of usually 50-20,000 bp in size, but resolution of over 6 Mb is possible with pulsed field gel electrophoresis (PFGE). Polyacrylamide gels are run in a vertical configuration while agarose gels are typically run horizontally in a submarine mode. They also differ in their casting methodology, as agarose sets thermally, while polyacrylamide forms in a chemical polymerization reaction.

Agarose

Agarose gels are made from the natural polysaccharide polymers extracted from seaweed. Agarose gels are easily cast and handled compared to other matrices, because the gel setting is a physical rather than chemical change. Samples are also easily recovered. After the experiment is finished, the resulting gel can be stored in a plastic bag in a refrigerator.

Agarose gels do not have a uniform pore size, but are optimal for electrophoresis of proteins that are larger than 200 kDa. Agarose gel electrophoresis can also be used for the separation of DNA fragments ranging from 50 base pair to several megabases (millions of bases), the largest of which require specialized apparatus. The distance between DNA bands of different lengths is influenced by the percent agarose in the gel, with higher percentages requiring longer run times, sometimes days. Instead high percentage agarose gels should be run with a pulsed field electrophoresis (PFE), or field inversion electrophoresis.

"Most agarose gels are made with between 0.7% (good separation or resolution of large 5–10kb DNA fragments) and 2% (good resolution for small 0.2–1kb fragments) agarose dissolved in electrophoresis buffer. Up to 3% can be used for separating very tiny fragments but a vertical polyacrylamide gel is more appropriate in this case. Low percentage gels are very weak and may break when you try to lift them. High percentage gels are often brittle and do not set evenly. 1% gels are common for many applications."

Polyacrylamide

Polyacrylamide gel electrophoresis (PAGE) is used for separating proteins ranging in size from 5 to 2,000 kDa due to the uniform pore size provided by the polyacrylamide gel. Pore size is controlled by modulating the concentrations of acrylamide and bis-acrylamide powder used in creating a gel. Care must be used when creating this type of gel, as acrylamide is a potent neurotoxin in its liquid and powdered forms.

Traditional DNA sequencing techniques such as Maxam-Gilbert or Sanger methods used polyacrylamide gels to separate DNA fragments differing by a single base-pair in length so the sequence could be read. Most modern DNA separation methods now use agarose gels, except for particularly small DNA fragments. It is currently most often used in the field of immunology and protein analysis, often used to separate different proteins or isoforms of the same protein into separate bands. These can be transferred onto a nitrocellulose or PVDF membrane to be probed with antibodies and corresponding markers, such as in a western blot.

Typically resolving gels are made in 6%, 8%, 10%, 12% or 15%. Stacking gel (5%) is poured on top of the resolving gel and a gel comb (which forms the wells and defines the lanes where proteins, sample buffer and ladders will be placed) is inserted. The percentage chosen depends on the size of the protein that one wishes to identify or probe in the sample. The smaller the known weight, the higher the percentage that should be used. Changes on the buffer system of the gel can help to further resolve proteins of very small sizes.

Starch

Partially hydrolysed potato starch makes for another non-toxic medium for protein electrophoresis. The gels are slightly more opaque than acrylamide or agarose. Non-denatured proteins can be separated according to charge and size. They are visualised using Napthal Black or Amido Black staining. Typical starch gel concentrations are 5% to 10%.

Gel Conditions

Denaturing

TTGE profiles representing the bifidobacterial diversity of fecal samples from two healthy volunteers (A and B) before and after AMC (Oral Amoxicillin-Clavulanic Acid) treatment

Denaturing gels are run under conditions that disrupt the natural structure of the analyte, causing it to unfold into a linear chain. Thus, the mobility of each macromolecule depends only on its linear length and its mass-to-charge ratio. Thus, the secondary, tertiary, and quaternary levels of biomolecular structure are disrupted, leaving only the primary structure to be analyzed.

Nucleic acids are often denatured by including urea in the buffer, while proteins are denatured using sodium dodecyl sulfate, usually as part of the SDS-PAGE process. For full denaturation of proteins, it is also necessary to reduce the covalent disulfide bonds that stabilize their tertiary and quaternary structure, a method called reducing PAGE. Reducing conditions are usually maintained by the addition of beta-mercaptoethanol or dithiothreitol. For general analysis of protein samples, reducing PAGE is the most common form of protein electrophoresis.

Denaturing conditions are necessary for proper estimation of molecular weight of RNA. RNA is able to form more intramolecular interactions than DNA which may result in change of its electrophoretic mobility. Urea, DMSO and glyoxal are the most often used denaturing agents to disrupt RNA structure. Originally, highly toxic methylmercury hydroxide was often used in denaturing RNA electrophoresis, but it may be method of choice for some samples.

Denaturing gel electrophoresis is used in the DNA and RNA banding pattern-based methods DGGE (denaturing gradient gel electrophoresis), TGGE (temperature gradient gel electrophoresis), and TTGE (temporal temperature gradient electrophoresis).

Native

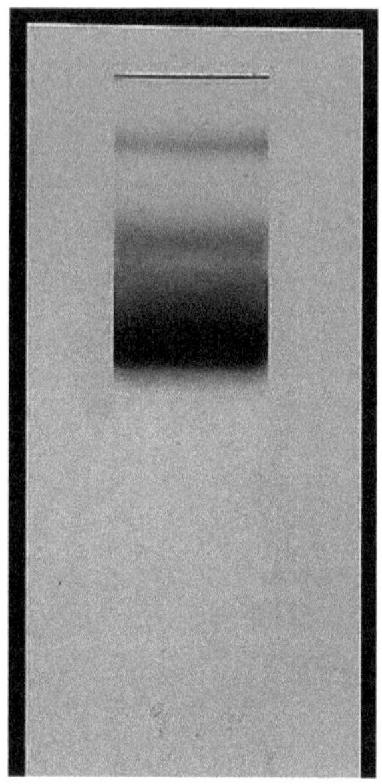

Specific enzyme-linked staining: Glucose-6-Phosphate Dehydrogenaseisoenzymes in *Plasmodium falciparum* infected Red blood cells

Native gels are run in non-denaturing conditions, so that the analyte's natural structure is maintained. This allows the physical size of the folded or assembled complex to affect the mobility, allowing for analysis of all four levels of the biomolecular structure. For biological samples, detergents are used only to the extent that they are necessary to lyse lipid membranes in the cell. Complexes remain—for the most part—associated and folded as they would be in the cell. One downside, however, is that complexes may not separate cleanly or predictably, as it is difficult to predict how the molecule's shape and size will affect its mobility. Addressing and solving this problem is a major aim of quantitative native PAGE.

Unlike denaturing methods, native gel electrophoresis does not use a charged denaturing agent. The molecules being separated (usually proteins or nucleic acids) therefore differ not only in molecular mass and intrinsic charge, but also the cross-sectional area, and thus experience different electrophoretic forces dependent on the shape of the overall structure. For proteins, since they remain in the native state they may be visualised not only by general protein staining reagents but also by specific enzyme-linked staining.

Native gel electrophoresis is typically used in proteomics and metallomics. However, native PAGE is also used to scan genes (DNA) for unknown mutations as in Single-strand conformation polymorphism.

Buffers

Buffers in gel electrophoresis are used to provide ions that carry a current and to maintain the pH at a relatively constant value. There are a number of buffers used for electrophoresis. The most common being, for nucleic acids Tris/Acetate/EDTA (TAE), Tris/Borate/EDTA (TBE). Many other buffers have been proposed, e.g. lithium borate, which is almost never used, based on Pubmed citations (LB), iso electric histidine, pK matched goods buffers, etc.; in most cases the purported rationale is lower current (less heat) and or matched ion mobilities, which leads to longer buffer life. Borate is problematic; Borate can polymerize, and/or interact with cis diols such as those found in RNA. TAE has the lowest buffering capacity but provides the best resolution for larger DNA. This means a lower voltage and more time, but a better product. LB is relatively new and is ineffective in resolving fragments larger than 5 kbp; However, with its low conductivity, a much higher voltage could be used (up to 35 V/cm), which means a shorter analysis time for routine electrophoresis. As low as one base pair size difference could be resolved in 3% agarose gel with an extremely low conductivity medium (1 mM Lithium borate).

Most SDS-PAGE protein separations are performed using a "discontinuous" (or DISC) buffer system that significantly enhances the sharpness of the bands within the gel. During electrophoresis in a discontinuous gel system, an ion gradient is formed in the early stage of electrophoresis that causes all of the proteins to focus into a single sharp band in a process called isotachophoresis. Separation of the proteins by size is achieved in the lower, "resolving" region of the gel. The resolving gel typically has a much smaller pore size, which leads to a sieving effect that now determines the electrophoretic mobility of the proteins.

Visualization

After the electrophoresis is complete, the molecules in the gel can be stained to make them visible. DNA may be visualized using ethidium bromide which, when intercalated into DNA, fluoresce under ultraviolet light, while protein may be visualised using silver stain or Coomassie Brilliant Blue dye. Other methods may also be used to visualize the separation of the mixture's components on the gel. If the molecules to be separated contain radioactivity, for example in a DNA sequencing gel, an autoradiogram can be recorded of the gel. Photographs can be taken of gels, often using a Gel Doc system.

Downstream Processing

After separation, an additional separation method may then be used, such as isoelectric focusing or SDS-PAGE. The gel will then be physically cut, and the protein complexes extracted from each portion separately. Each extract may then be analysed, such as by peptide mass fingerprinting or de novo peptide sequencing after in-gel digestion. This can provide a great deal of information about the identities of the proteins in a complex.

Applications

- Estimation of the size of DNA molecules following restriction enzyme digestion, e.g. in restriction mapping of cloned DNA.
- Analysis of PCR products, e.g. in molecular genetic diagnosis or genetic fingerprinting
- Separation of restricted genomic DNA prior to Southern transfer, or of RNA prior to Northern transfer.

Gel electrophoresis is used in forensics, molecular biology, genetics, microbiology and biochemistry. The results can be analyzed quantitatively by visualizing the gel with UV light and a gel imaging device. The image is recorded with a computer operated camera, and the intensity of the band or spot of interest is measured and compared against standard or markers loaded on the same gel. The measurement and analysis are mostly done with specialized software.

Depending on the type of analysis being performed, other techniques are often implemented in conjunction with the results of gel electrophoresis, providing a wide range of field-specific applications.

Nucleic Acids

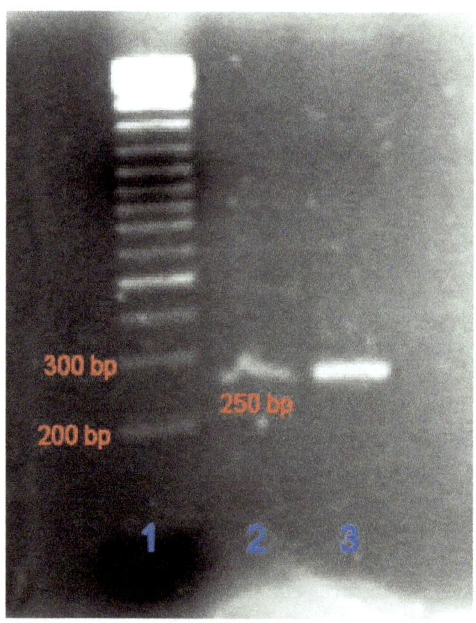

An agarose gel of a PCR product compared to a DNA ladder.

In the case of nucleic acids, the direction of migration, from negative to positive electrodes, is due to the naturally occurring negative charge carried by their sugar-phosphate backbone.

Double-stranded DNA fragments naturally behave as long rods, so their migration through the gel is relative to their size or, for cyclic fragments, their radius of gyration. Circular DNA such as plasmids, however, may show multiple bands, the speed of migration may depend on whether it is relaxed or supercoiled. Single-stranded DNA or RNA tend to fold up into molecules with complex shapes and migrate through the gel in a complicated manner based on their tertiary structure. Therefore, agents that disrupt the hydrogen bonds, such as sodium hydroxide or formamide, are used to denature the nucleic acids and cause them to behave as long rods again.

Gel electrophoresis of large DNA or RNA is usually done by agarose gel electrophoresis. See the "Chain termination method" page for an example of a polyacrylamide DNA sequencing gel. Characterization through ligand interaction of nucleic acids or fragments may be performed by mobility shift affinity electrophoresis.

Electrophoresis of RNA samples can be used to check for genomic DNA contamination and also for RNA degradation. RNA from eukaryotic organisms shows distinct bands of 28s and 18s rRNA, the 28s band being approximately twice as intense as the 18s band. Degraded RNA has less sharply defined bands, has a smeared appearance, and intensity ratio is less than 2:1.

Proteins

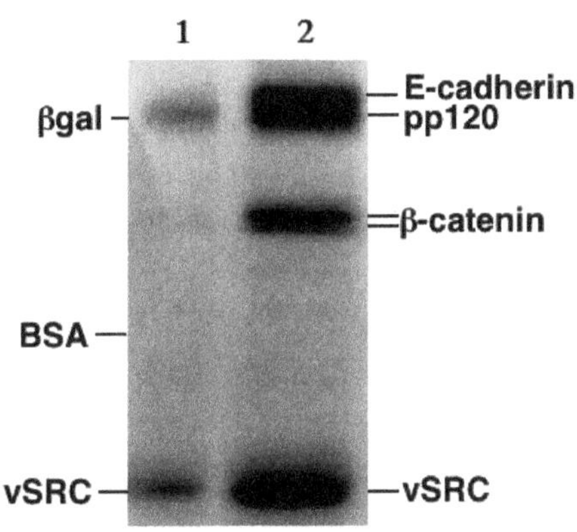

SDS-PAGE autoradiography – The indicated proteins are present in different concentrations in the two samples.

Proteins, unlike nucleic acids, can have varying charges and complex shapes, therefore they may not migrate into the polyacrylamide gel at similar rates, or at all, when placing a negative to positive EMF on the sample. Proteins therefore, are usually denatured in the presence of a detergent such as sodium dodecyl sulfate (SDS) that coats the proteins with a negative charge. Generally, the amount of SDS bound is relative to the size of the protein (usually 1.4g SDS per gram of protein), so that the resulting denatured proteins have an overall negative charge, and all the proteins have a similar charge to mass ratio. Since denatured proteins act like long rods instead of having a complex tertiary shape, the rate at which the resulting SDS coated proteins migrate in the gel is relative only to its size and not its charge or shape.

Proteins are usually analyzed by sodium dodecyl sulfate polyacrylamide gel electrophoresis (SDS-PAGE), by native gel electrophoresis, by preparative gel electrophoresis (QPNC-PAGE), or by 2-D electrophoresis.

Characterization through ligand interaction may be performed by electroblotting or by affinity electrophoresis in agarose or by capillary electrophoresis as for estimation of binding constants and determination of structural features like glycan content through lectin binding.

Macromolecule Blotting and Probing

The terms *northern*, *western* and *eastern* blotting are derived from what initially was a molecular biology joke that played on the term *Southern blotting*, after the technique described by Edwin Southern for the hybridisation of blotted DNA. Patricia Thomas, developer of the RNA blot which then became known as the *northern blot*, actually didn't use the term. Further combinations of these techniques produced such terms as *southwesterns* (protein-DNA hybridizations), *northwesterns* (to detect protein-RNA interactions) and *farwesterns*(protein-protein interactions), all of which are presently found in the literature.

Southern Blotting

Named after its inventor, biologist Edwin Southern, the Southern blot is a method for probing for the presence of a specific DNA sequence within a DNA sample. DNA samples before or after restriction enzyme (restriction endonuclease) digestion are separated by gel electrophoresis and then transferred to a membrane by blotting via capillary action. The membrane is then exposed to a labeled DNA probe that has a complement base sequence to the sequence on the DNA of interest. Most original protocols used radioactive labels; however, non-radioactive alternatives are now available. Southern blotting is less commonly used in laboratory science due to the capacity of other techniques, such as PCR, to detect specific DNA sequences from DNA samples. These blots are still used for some applications, however, such as measuring transgene copy number in transgenic mice or in the engineering of gene knockout embryonic stem cell lines.

Northern Blotting

The northern blot is used to study the expression patterns of a specific type of RNA molecule as relative comparison among a set of different samples of RNA. It is essentially a combination of denaturing RNA gel electrophoresis, and a blot. In this process RNA is separated based on size and is then transferred to a membrane that is then probed with a labeled complement of a sequence of interest. The results may be visualized through a variety of ways depending on the label used; however, most result in the revelation of bands representing the sizes of the RNA detected in sample. The intensity of these bands is related to the amount of the target RNA in the samples analyzed. The procedure is commonly used to study when and how much gene expression is occurring by measuring how much of that RNA is present in different samples. It is one of the most basic tools for determining at what time, and under what conditions, certain genes are expressed in living tissues.

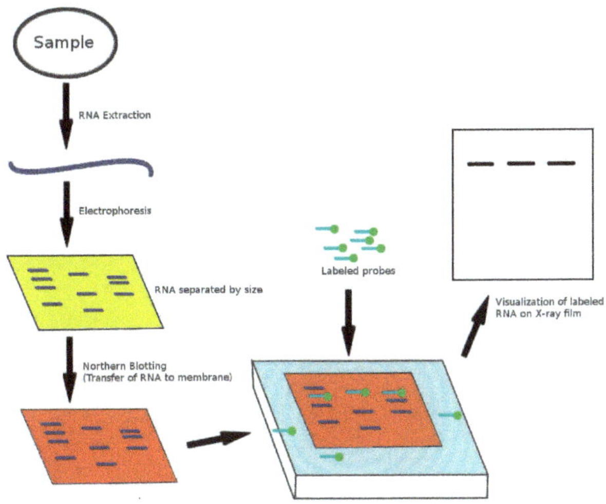

Western Blotting

Antibodies to most proteins can be created by injecting small amounts of the protein into an animal such as a mouse, rabbit, sheep, or donkey (polyclonal antibodies) or produced in cell culture (monoclonal antibodies). These antibodies can be used for a variety of analytical and preparative techniques.

In western blotting, proteins are first separated by size, in a thin gel sandwiched between two glass plates in a technique known as SDS-PAGE (sodium dodecyl sulfatepolyacrylamide gel electrophoresis). The proteins in the gel are then transferred to a polyvinylidene fluoride (PVDF), nitrocellulose, nylon, or other support membrane. This membrane can then be probed with solutions of antibodies. Antibodies that specifically bind to the protein of interest can then be visualized by a variety of techniques, including colored products, chemiluminescence, or autoradiography. Often, the antibodies are labeled with enzymes. When a chemiluminescent substrate is exposed to the enzyme it allows detection. Using western blotting techniques allows not only detection but also quantitative analysis. Analogous methods to western blotting can be used to directly stain specific proteins in live cells or tissue sections. However, these *immunostaining* methods, such as FISH, are used more often in cell biology research.

Eastern Blotting

The Eastern blotting technique is used to detect post-translational modification of proteins. Proteins blotted on to the PVDF or nitrocellulose membrane are probed for modifications using specific substrates.

DNA MICROARRAY

A **DNA microarray** (also commonly known as DNA chip or biochip) is a collection of microscopic DNA spots attached to a solid surface. Scientists use DNA microarrays to measure the expression levels of large numbers of genes simultaneously or to genotype multiple regions of a genome. Each DNA spot contains picomoles (10^{-12} moles) of a specific DNA sequence, known as *probes* (or *reporters* or *oligos*). These can be a short section of a gene or other DNA element that are used to hybridize a cDNA or cRNA (also called anti-sense RNA) sample (called *target*) under high-stringency conditions. Probe-target hybridization is usually detected and quantified by detection of fluorophore-, silver-, or chemiluminescence-labeled targets to determine relative abundance of nucleic acid sequences in the target. The original nucleic acid arrays were macro arrays approximately 9 x 12 cm and the first computerized image based analysis was published in 1981

Principle

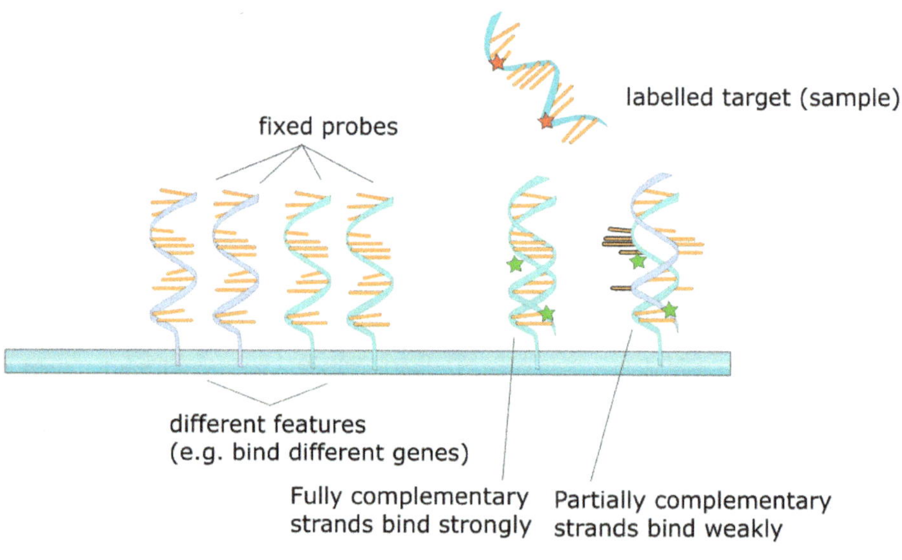

Hybridization of the target to the probe.

The core principle behind microarrays is hybridization between two DNA strands, the property of complementary nucleic acid sequences to specifically pair with each other by forming hydrogen bonds between complementary nucleotide base pairs. A high number of complementary base pairs in a nucleotide sequence means tighter non-covalent bonding between the two strands. After washing off non-specific bonding sequences, only strongly paired strands will remain hybridized. Fluorescently labeled target sequences that bind to a probe sequence generate a signal that depends on the hybridization conditions (such as temperature), and washing after hybridization. Total strength of the signal, from a spot (feature), depends upon the amount of target sample binding to the probes present on that spot. Microarrays use relative quantitation in which the intensity of a feature is compared to the intensity of the same feature under a different condition, and the identity of the feature is known by its position.

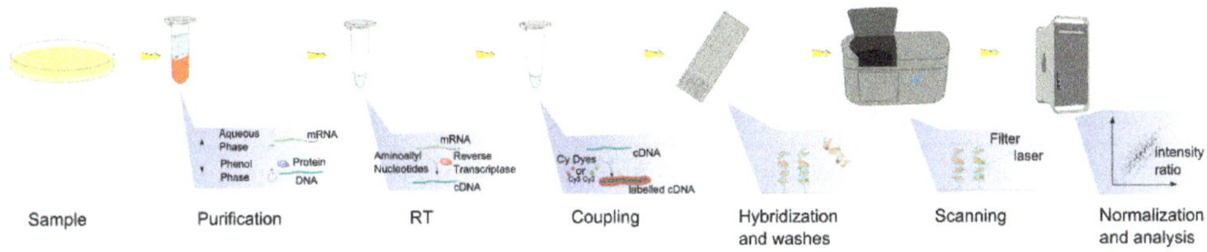

The steps required in a microarray experiment.

Uses and Types

Two Affymetrix chips. A match is shown at bottom left for size comparison.

Many types of arrays exist and the broadest distinction is whether they are spatially arranged on a surface or on coded beads:

- The traditional solid-phase array is a collection of orderly microscopic "spots", called features, each with thousands of identical and specific probes attached to a solid surface, such as glass, plastic or silicon biochip (commonly known as a *genome chip*, *DNA chip* or *gene array*). Thousands of these features can be placed in known locations on a single DNA microarray.
- The alternative bead array is a collection of microscopic polystyrene beads, each with a specific probe and a ratio of two or more dyes, which do not interfere with the fluorescent dyes used on the target sequence.

DNA microarrays can be used to detect DNA (as in comparative genomic hybridization), or detect RNA (most commonly as cDNA after reverse transcription) that may or may not be translated into proteins. The process of measuring gene expression via cDNA is called expression analysis or expression profiling.

Applications Include:

Application or technology	Synopsis
Gene expression profiling	In an mRNA or gene expression profiling experiment the expression levels of thousands of genes are simultaneously monitored to study the effects of certain treatments, diseases, and developmental stages on gene expression. For example, microarray-based gene expression profiling can be used to identify genes whose expression is changed in response to pathogens or other organisms by comparing gene expression in infected to that in uninfected cells or tissues.
Comparative genomic hybridization	Assessing genome content in different cells or closely related organisms.
GeneID	Small microarrays to check IDs of organisms in food and feed (like GMO), mycoplasms in cell culture, or pathogens for disease detection, mostly combining PCR and microarray technology.
Chromatin immunoprecipitation on Chip	DNA sequences bound to a particular protein can be isolated by immunoprecipitating that protein (ChIP), these fragments can be then hybridized to a microarray (such as a tiling array) allowing the determination of protein binding site occupancy throughout the genome. Example protein to immunoprecipitate are histone modifications (H3K27me3, H3K4me2, H3K9me3, etc.), Polycomb-group protein (PRC2:Suz12, PRC1:YY1) and trithorax-group protein (Ash1) to study the epigenetic landscape or RNA Polymerase II to study the transcription landscape.
DamID	Analogously to ChIP, genomic regions bound by a protein of interest can be isolated and used to probe a microarray to determine binding site occupancy. Unlike ChIP, DamID does not require antibodies but makes use of adenine methylation near the protein's binding sites to selectively amplify those regions, introduced by expressing minute amounts of protein of interest fused to bacterial DNA adenine methyltransferase.
SNP detection	Identifying single nucleotide polymorphism among alleles within or between populations. Several applications of microarrays make use of SNP detection, including genotyping, forensic analysis, measuring predisposition to disease, identifying drug-candidates, evaluating germline mutations in individuals or somatic mutations in cancers, assessing loss of heterozygosity, or genetic linkage analysis.
Alternative splicing detection	An *exon junction array* design uses probes specific to the expected or potential splice sites of predicted exons for a gene. It is of intermediate density, or coverage, to a typical gene expression array (with 1-3 probes per gene) and a genomic tiling array (with hundreds or thousands of probes per gene). It is used to assay the expression of alternative splice forms of a gene. Exon arrays have a different design, employing probes designed to detect each individual exon for known or predicted genes, and can be used for detecting different splicing isoforms.
Fusion genes microarray	A Fusion gene microarray can detect fusion transcripts, *e.g.* from cancer specimens. The principle behind this is building on the alternative splicing microarrays. The oligo design strategy enables combined measurements of chimeric transcript junctions with exon-wise measurements of individual fusion partners.

Tiling array	Genome tiling arrays consist of overlapping probes designed to densely represent a genomic region of interest, sometimes as large as an entire human chromosome. The purpose is to empirically detect expression of transcripts or alternatively spliced forms which may not have been previously known or predicted.
Double-stranded B-DNA microarrays	Right-handed double-stranded B-DNA microarrays can be used to characterize novel drugs and biologicals that can be employed to bind specific regions of immobilized, intact, double-stranded DNA. This approach can be used to inhibit gene expression. They also allow for characterization of their structure under different environmental conditions.
Double-stranded Z-DNA microarrays	Left-handed double-stranded Z-DNA microarrays can be used to identify short sequences of the alternative Z-DNA structure located within longer stretches of right-handed B-DNA genes (e.g., transcriptional enhancement, recombination, RNA editing). The microarrays also allow for characterization of their structure under different environmental conditions.
Multi-stranded DNA microarrays (triplex-DNA microarrays and quadruplex-DNA microarrays)	Multi-stranded DNA and RNA microarrays can be used to identify novel drugs that bind to these multi-stranded nucleic acid sequences. This approach can be used to discover new drugs and biologicals that have the ability to inhibit gene expression. These microarrays also allow for characterization of their structure under different environmental conditions.

Fabrication

Microarrays can be manufactured in different ways, depending on the number of probes under examination, costs, customization requirements, and the type of scientific question being asked. Arrays may have as few as 10 probes or up to 2.1 million micrometre-scale probes from commercial vendors.

Spotted vs. *in situ* Synthesised Arrays

A DNA microarray being printed by a robot at the University of Delaware

Microarrays can be fabricated using a variety of technologies, including printing with fine-pointed pins onto glass slides, photolithography using pre-made masks, photolithography using dynamic micromirror devices, ink-jet printing, or electrochemistry on microelectrode arrays.

In *spotted microarrays*, the probes are oligonucleotides, cDNA or small fragments of PCR products that correspond to mRNAs. The probes are synthesized prior to deposition on the array surface and are then "spotted" onto glass. A common approach utilizes an array of fine pins or needles controlled by a robotic arm that is dipped into wells containing DNA probes and then depositing each probe at designated locations on the array surface. The resulting "grid" of probes represents the nucleic acid profiles of the prepared probes and is ready to receive complementary cDNA or cRNA "targets" derived from experimental or clinical samples. This technique is used by research scientists around the world to produce "in-house" printed microarrays from their own labs. These arrays may be easily customized for each experiment, because researchers can choose the probes and printing locations on the arrays, synthesize the probes in their own lab (or collaborating facility), and spot the arrays. They can then generate their own labeled samples for hybridization, hybridize the samples to the array, and finally scan the arrays with their own equipment. This provides a relatively low-cost microarray that may be customized for each study, and avoids the costs of purchasing often more expensive commercial arrays that may represent vast numbers of genes that are not of interest to the investigator. Publications exist which indicate in-house spotted microarrays may not provide the same level of sensitivity compared to commercial oligonucleotide arrays, possibly owing to the small batch sizes and reduced printing efficiencies when compared to industrial manufactures of oligo arrays.

In *oligonucleotide microarrays*, the probes are short sequences designed to match parts of the sequence of known or predicted open reading frames. Although oligonucleotide probes are often used in "spotted" microarrays, the term "oligonucleotide array" most often refers to a specific technique of manufacturing. Oligonucleotide arrays are produced by printing short oligonucleotide sequences designed to represent a single gene or family of gene splice-variants by synthesizing this sequence directly onto the array surface instead of depositing intact sequences. Sequences may be longer (60-mer probes such as the Agilent design) or shorter (25-mer probes produced by Affymetrix) depending on the desired purpose; longer probes are more specific to individual target genes, shorter probes may be spotted in higher density across the array and are cheaper to manufacture. One technique used to produce oligonucleotide arrays include photolithographic synthesis (Affymetrix) on a silica substrate where light and light-sensitive masking agents are used to "build" a sequence one nucleotide at a time across the entire array. Each applicable probe is selectively "unmasked" prior to bathing the array in a solution of a single nucleotide, then a masking reaction takes place and the next set of probes are unmasked in preparation for a different nucleotide exposure. After many repetitions, the sequences of every probe become fully constructed. More recently, Maskless Array Synthesis from NimbleGen Systems has combined flexibility with large numbers of probes.

Two-Channel vs. one-channel Detection

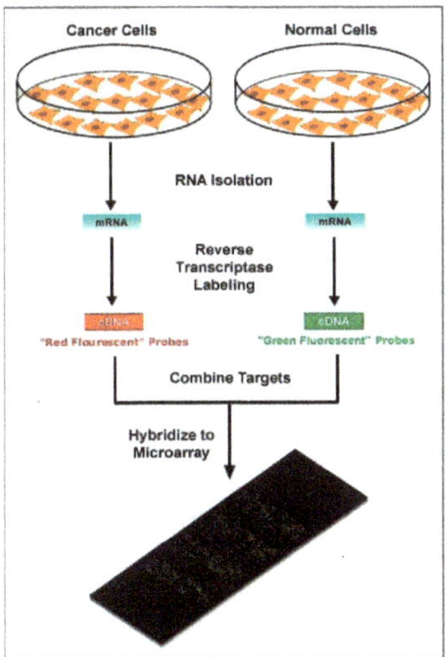

Diagram of typical dual-colour microarray experiment.

Two-color microarrays or *two-channel microarrays* are typically hybridized with cDNA prepared from two samples to be compared (e.g. diseased tissue versus healthy tissue) and that are labeled with two different fluorophores. Fluorescent dyes commonly used for cDNA labeling include Cy3, which has a fluorescence emission wavelength of 570 nm (corresponding to the green part of the light spectrum), and Cy5 with a fluorescence emission wavelength of 670 nm (corresponding to the red part of the light spectrum). The two Cy-labeled cDNA samples are mixed and hybridized to a single microarray that is then scanned in a microarray scanner to visualize fluorescence of the two fluorophores after excitation with a laser beam of a defined wavelength. Relative intensities of each fluorophore may then be used in ratio-based analysis to identify up-regulated and down-regulated genes.

Oligonucleotide microarrays often carry control probes designed to hybridize with RNA spike-ins. The degree of hybridization between the spike-ins and the control probes is used to normalize the hybridization measurements for the target probes. Although absolute levels of gene expression may be determined in the two-color array in rare instances, the relative differences in expression among different spots within a sample and between samples is the preferred method of data analysis for the two-color system. Examples of providers for such microarrays includes Agilent with their Dual-Mode platform, Eppendorf with their DualChip platform for colorimetric Silverquant labeling, and TeleChem International with Arrayit.

In *single-channel microarrays* or *one-color microarrays*, the arrays provide intensity data for each probe or probe set indicating a relative level of hybridization with the labeled target. However, they do not truly indicate abundance levels of a gene but rather relative abundance when compared to other samples or conditions when processed in the same experiment. Each RNA molecule encounters protocol and batch-specific bias during amplification, labeling, and hybridization phases of the experiment making comparisons between genes for the same microarray uninformative. The comparison of two conditions for the same gene requires two separate single-dye hybridizations. Several popular single-channel systems are the Affymetrix "Gene Chip", Illumina "Bead Chip", Agilent single-channel arrays, the Applied Microarrays "CodeLink" arrays, and the Eppendorf "DualChip & Silverquant". One strength of the single-dye system lies in the fact that an aberrant sample cannot affect the raw data derived from other samples, because each array chip is exposed to only one sample (as opposed to a two-color system in which a single low-quality sample may drastically impinge on overall data precision even if the other sample was of high quality). Another benefit is that data are more easily compared to arrays from different experiments as long as batch effects have been accounted for.

One channel microarray may be the only choice in some situations. Suppose i samples need to be compared: then the number of experiments required using the two channel arrays quickly becomes unfeasible, unless a sample is used as a reference.

number of samples	one-channel microarray	two channel microarray	two channel microarray (with reference)
1	1	1	1
2	2	1	1
3	3	3	2
4	4	6	3
i	i	i	i

A Typical Protocol

This is an example of a **DNA microarray experiment**, detailing a particular case to better explain DNA microarray experiments, while enumerating possible alternatives.

- The two samples to be compared (pairwise comparison) are grown/acquired. In this example treated sample (case) and untreated sample (control).
- The nucleic acid of interest is purified: this can be all RNA for expression profiling, DNA for comparative hybridization, or DNA/RNA bound to a particular protein which is immunoprecipitated (ChIP-on-chip) for epigenetic or regulation studies. In this example total RNA is isolated (total as it is nuclear and cytoplasmic) by Guanidinium thiocyanate-phenol-chloroform extraction (e.g. Trizol) which isolates most RNA (whereas column methods have a cut off of 200 nucleotides) and if done correctly has a better purity.
- The purified RNA is analysed for quality (by capillary electrophoresis) and quantity (for example, by using a NanoDrop or NanoPhotometer spectrometer). If the material is of acceptable quality and sufficient quantity is present (e.g., >1 μg, although the required amount varies by microarray platform), the experiment can proceed.

- The labelled product is generated via reverse transcription and sometimes with an optional PCR amplification. The RNA is reverse transcribed with either polyT primers (which amplify only mRNA) or random primers (which amplify all RNA, most of which is rRNA); miRNA microarrays ligate an oligonucleotide to the purified small RNA (isolated with a fractionator), which is then reverse transcribed and amplified. The label is added either during the reverse transcription step, or following amplification if it is performed. The sense labelling is dependent on the microarray; e.g. if the label is added with the RT mix, the cDNA is antisense and the microarray probe is sense, except in the case of negative controls. The label is typically fluorescent; only one machine uses radiolabels. The labelling can be direct (not used) or indirect (requires a coupling stage). For two-channel arrays, the coupling stage occurs before hybridization, using aminoallyl uridine triphosphate (aminoallyl-UTP, or aaUTP) and NHS amino-reactive dyes (such as cyanine dyes); for single-channel arrays, the coupling stage occurs after hybridization, using biotin and labelled streptavin. The modified nucleotides (usually in a ratio of 1 aaUTP: 4 TTP (thymidine triphosphate)) are added enzymatically in a low ratio to normal nucleotides, typically resulting in 1 every 60 bases. The aaDNA is then purified with a column (using a phosphate buffer solution, as Tris contains amine groups). The aminoallyl group is an amine group on a long linker attached to the nucleobase, which reacts with a reactive dye. A form of replicate known as a dye flipcan be performed to remove any dye effects in two-channel experiments; for a dye flip, a second slide is used, with the labels swapped (the sample that was labeled with Cy3 in the first slide is labeled with Cy5, and vice versa). In this example, aminoallyl-UTP is present in the reverse-transcribed mixture.
- The labeled samples are then mixed with a propriety hybridization solution which can consist of SDS, SSC, dextran sulfate, a blocking agent (such as COT1 DNA, salmon sperm DNA, calf thymus DNA, PolyA or PolyT), Denhardt's solution, or formamine.
- The mixture is denatured and added to the pinholes of the microarray. The holes are sealed and the microarray hybridized, either in a hyb oven, where the microarray is mixed by rotation, or in a mixer, where the microarray is mixed by alternating pressure at the pinholes.
- After an overnight hybridization, all nonspecific binding is washed off (SDS and SSC).
- The microarray is dried and scanned by a machine that uses a laser to excite the dye and measures the emission levels with a detector.
- The image is gridded with a template and the intensities of each feature (composed of several pixels) is quantified.
- The raw data is normalized; the simplest normalization method is to subtract background intensity and scale so that the total intensities of the features of the two channels are equal, or to use the intensity of a reference gene to calculate the t-value for all of the intensities. More sophisticated methods include z-ratio,

loess and lowess regression and RMA (robust multichip analysis) for Affymetrix chips (single-channel, silicon chip, in situ synthesised short oligonucleotides).

Microarrays and Bioinformatics

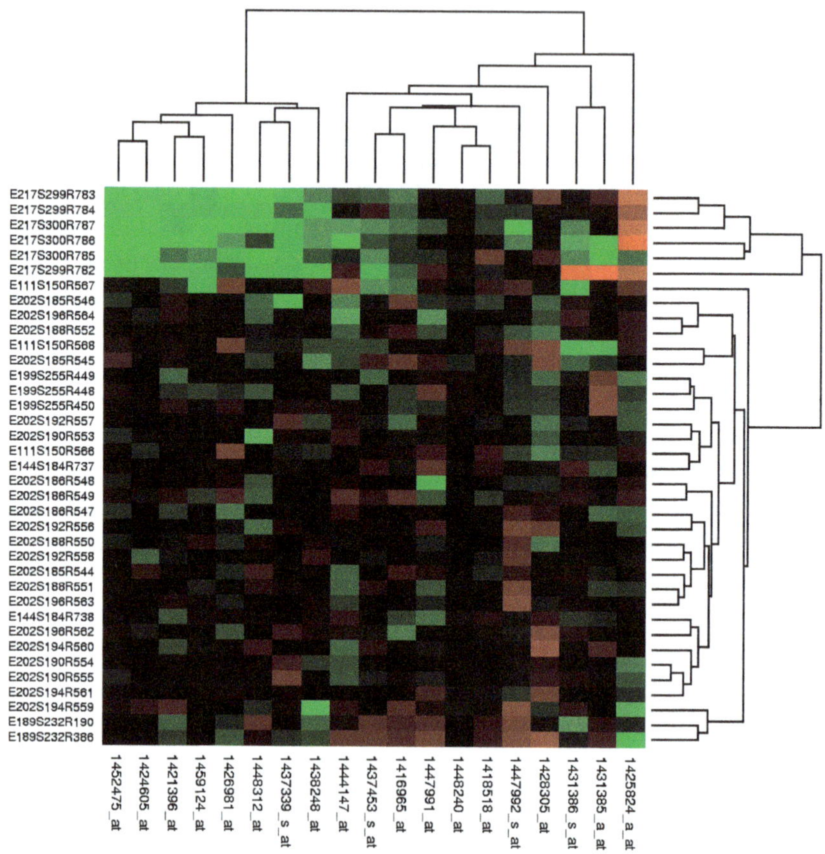

Gene expression values from microarray experiments can be represented as heat maps to visualize the result of data analysis.

The advent of inexpensive microarray experiments created several specific bioinformatics challenges:

- the multiple levels of replication in experimental design (Experimental design)
- the number of platforms and independent groups and data format (Standardization)
- the statistical treatment of the data (Statistical analysis)
- mapping each probe to the mRNA transcript that it measures (Annotation)
- the sheer volume of data and the ability to share it (Data warehousing)

Experimental Design

Due to the biological complexity of gene expression, the considerations of experimental design that are discussed in the expression profiling article are of critical importance if statistically and biologically valid conclusions are to be drawn from the data.

There are three main elements to consider when designing a microarray experiment. First, replication of the biological samples is essential for drawing conclusions from the experiment. Second, technical replicates (two RNA samples obtained from each experimental unit) help to ensure precision and allow for testing differences within treatment groups. The biological replicates include independent RNA extractions and technical replicates may be two aliquots of the same extraction. Third, spots of each cDNA clone or oligonucleotide are present as replicates (at least duplicates) on the microarray slide, to provide a measure of technical precision in each hybridization. It is critical that information about the sample preparation and handling is discussed, in order to help identify the independent units in the experiment and to avoid inflated estimates of statistical significance.

Standardization

Microarray data is difficult to exchange due to the lack of standardization in platform fabrication, assay protocols, and analysis methods. This presents an interoperability problem in bioinformatics. Various grass-roots open-source projects are trying to ease the exchange and analysis of data produced with non-proprietary chips:

- For example, the "Minimum Information About a Microarray Experiment" (MIAME) checklist helps define the level of detail that should exist and is being adopted by many journals as a requirement for the submission of papers incorporating microarray results. But MIAME does not describe the format for the information, so while many formats can support the MIAME requirements, as of 2007 no format permits verification of complete semantic compliance.
- The "MicroArray Quality Control (MAQC) Project" is being conducted by the US Food and Drug Administration (FDA) to develop standards and quality control metrics which will eventually allow the use of MicroArray data in drug discovery, clinical practice and regulatory decision-making.
- The MGED Society has developed standards for the representation of gene expression experiment results and relevant annotations.

Data analysis

Microarray data sets are commonly very large, and analytical precision is influenced by a number of variables. Statistical challenges include taking into account effects of background noise and appropriate normalization of the data. Normalization methods may be suited to specific platforms and, in the case of commercial platforms, the analysis may be proprietary. Algorithms that affect statistical analysis include:

- Image analysis: gridding, spot recognition of the scanned image (segmentation algorithm), removal or marking of poor-quality and low-intensity features (called *flagging*).

- Data processing: background subtraction (based on global or local background), determination of spot intensities and intensity ratios, visualisation of data (e.g. see MA plot), and log-transformation of ratios, global or local normalization of intensity ratios, and segmentation into different copy number regions using step detection algorithms.
- Class discovery analysis: This analytic approach, sometimes called unsupervised classification or knowledge discovery, tries to identify whether microarrays (objects, patients, mice, etc.) or genes cluster together in groups. Identifying naturally existing groups of objects (microarrays or genes) which cluster together can enable the discovery of new groups that otherwise were not previously known to exist. During knowledge discovery analysis, various unsupervised classification techniques can be employed with DNA microarray data to identify novel clusters (classes) of arrays. This type of approach is not hypothesis-driven, but rather is based on iterative pattern recognition or statistical learning methods to find an "optimal" number of clusters in the data. Examples of unsupervised analyses methods include self-organizing maps, neural gas, k-means cluster analyses, hierarchical cluster analysis, Genomic Signal Processing based clustering and model-based cluster analysis. For some of these methods the user also has to define a distance measure between pairs of objects. Although the Pearson correlation coefficient is usually employed, several other measures have been proposed and evaluated in the literature. The input data used in class discovery analyses are commonly based on lists of genes having high informativeness (low noise) based on low values of the coefficient of variation or high values of Shannon entropy, etc. The determination of the most likely or optimal number of clusters obtained from an unsupervised analysis is called cluster validity. Some commonly used metrics for cluster validity are the silhouette index, Davies-Bouldin index, Dunn's index, or Hubert's γ statistic.
- Class prediction analysis: This approach, called supervised classification, establishes the basis for developing a predictive model into which future unknown test objects can be input in order to predict the most likely class membership of the test objects. Supervised analysis for class prediction involves use of techniques such as linear regression, k-nearest neighbor, learning vector quantization, decision tree analysis, random forests, naive Bayes, logistic regression, kernel regression, artificial neural networks, support vector machines, mixture of experts, and supervised neural gas. In addition, various metaheuristic methods are employed, such as genetic algorithms, covariance matrix self-adaptation, particle swarm optimization, and ant colony optimization. Input data for class prediction are usually based on filtered lists of genes which are predictive of class, determined using classical hypothesis tests (next section), Gini diversity index, or information gain (entropy).
- Hypothesis-driven statistical analysis: Identification of statistically significant changes in gene expression are commonly identified using the t-test, ANOVA, Bayesian method Mann–Whitney test methods tailored to microarray data

sets, which take into account multiple comparisons or cluster analysis. These methods assess statistical power based on the variation present in the data and the number of experimental replicates, and can help minimize Type I and type II errors in the analyses.

- Dimensional reduction: Analysts often reduce the number of dimensions (genes) prior to data analysis. This may involve linear approaches such as principal components analysis (PCA), or non-linear manifold learning (distance metric learning) using kernel PCA, diffusion maps, Laplacian eigenmaps, local linear embedding, locally preserving projections, and Sammon's mapping.
- Network-based methods: Statistical methods that take the underlying structure of gene networks into account, representing either associative or causative interactions or dependencies among gene products. Weighted gene co-expression network analysis is widely used for identifying co-expression modules and intramodular hub genes. Modules may corresponds to cell types or pathways. Highly connected intramodular hubs best represent their respective modules.

Microarray data may require further processing aimed at reducing the dimensionality of the data to aid comprehension and more focused analysis. Other methods permit analysis of data consisting of a low number of biological or technical replicates; for example, the Local Pooled Error (LPE) test pools standard deviations of genes with similar expression levels in an effort to compensate for insufficient replication.

Annotation

The relation between a probe and the mRNA that it is expected to detect is not trivial. Some mRNAs may cross-hybridize probes in the array that are supposed to detect another mRNA. In addition, mRNAs may experience amplification bias that is sequence or molecule-specific. Thirdly, probes that are designed to detect the mRNA of a particular gene may be relying on genomic EST information that is incorrectly associated with that gene.

Data Warehousing

Microarray data was found to be more useful when compared to other similar datasets. The sheer volume of data, specialized formats (such as MIAME), and curation efforts associated with the datasets require specialized databases to store the data. A number of open-source data warehousing solutions, such as InterMine and BioMart, have been created for the specific purpose of integrating diverse biological datasets, and also support analysis.

Alternative Technologies

Advances in massively parallel sequencing has led to the development of RNA-Seq technology, that enables a whole transcriptome shotgun approach to characterize and quantify gene expression. Unlike microarrays, which need a reference genome and transcriptome to be available before the microarray itself can be designed, RNA-Seq can also be used for new model organisms whose genome has not been sequenced yet.

Multi-stranded DNA Microarray

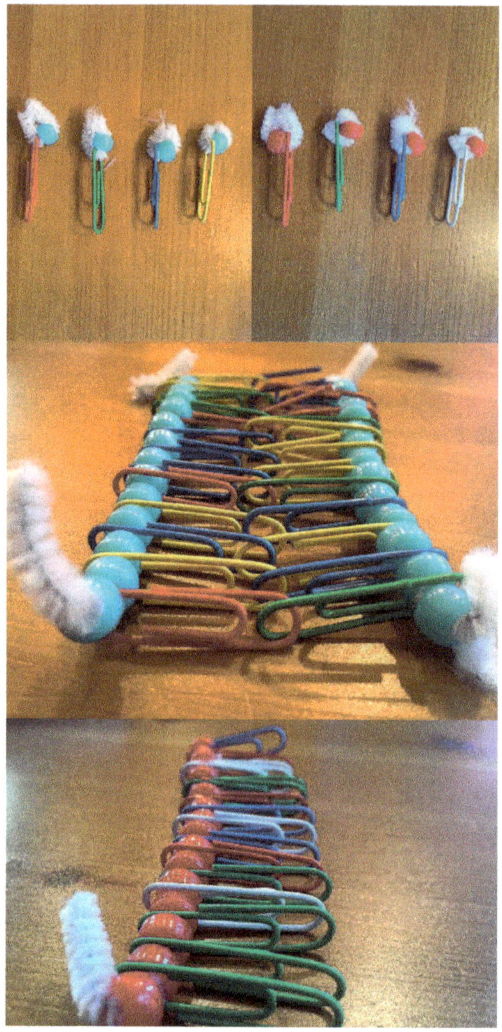

Allele-specific Oligonucleotide

Allele-specific oligonucleotide (ASO) is a technique that allows detection of single base mutations without the need for PCR or gel electrophoresis. Short (20-25 nucleotides in length), labeled probes are exposed to the non-fragmented target DNA. Hybridization occurs with high specificity due to the short length of the probes and even a single base change will hinder hybridization. The target DNA is then washed and the labeled probes that didn't hybridize are removed. The target DNA is then analyzed for the presence of the probe via radioactivity or fluorescence. In this experiment, as in most molecular biology techniques, a control must be used to ensure successful experimentation. The Illumina Methylation Assay is an example of a method that takes advantage of the ASO technique to measure one base pair differences in sequence.

Antiquated Technologies

In molecular biology, procedures and technologies are continually being developed and older technologies abandoned. For example, before the advent of DNA gel electrophoresis (agarose or polyacrylamide), the size of DNA molecules was typically determined by rate sedimentation in sucrose gradients, a slow and labor-intensive technique requiring expensive instrumentation; prior to sucrose gradients, viscometry was used. Aside from their historical interest, it is often worth knowing about older technology, as it is occasionally useful to solve another new problem for which the newer technique is inappropriate.

PROTEIN METHODS

Protein methods are the techniques used to study proteins. There are experimental methods for studying proteins (e.g., for detecting proteins, for isolating and purifying proteins, and for characterizing the structure and function of proteins, often requiring that the protein first be purified). Computational methods typically use computer programs to analyze proteins. However, many experimental methods (e.g., mass spectrometry) require computational analysis of the raw data.

REFERENCES

1. Bartlett, J. M. S.; Stirling, D. (2003). "A Short History of the Polymerase Chain Reaction". PCR Protocols. Methods in Molecular Biology. **226** *(2nd* ed.). pp. 3–6. doi:10.1385/1-59259-384-4:3. ISBN 1-59259-384-4.
2. Saiki, R.; Scharf, S.; Faloona, F.; Mullis, K.; Horn, G.; Erlich, H.; Arnheim, N. (1985). "Enzymatic amplification of beta-globin genomic sequences and restriction site analysis for diagnosis of sickle cell anemia". Science. **230** *(4732): 1350–1354. doi:10.1126/science.2999980. PMID 2999980.*
3. Saiki, R.; Gelfand, D.; Stoffel, S.; Scharf, S.; Higuchi, R.; Horn, G.; Mullis, K.; Erlich, H. (1988). "Primer-directed enzymatic amplification of DNA with a thermostable DNA polymerase". Science. **239** *(4839): 487–491. doi:10.1126/science.2448875. PMID 2448875.*
4. Cheng, S.; Fockler, C.; Barnes, W. M.; Higuchi, R. (1994). "Effective Amplification of Long Targets from Cloned Inserts and Human Genomic DNA". Proceedings of the National Academy of Sciences. **91** *(12): 5695–5699. doi:10.1073/pnas.91.12.5695. PMC 44063. PMID 8202550.*
5. Carr AC, Moore SD (2012). Lucia, Alejandro, ed. "Robust quantification of polymerase chain reactions using global fitting". PLOS ONE. **7** *(5): e37640. doi:10.1371/journal.pone.0037640. PMC 3365123. PMID 22701526.*
6. Pavlov, A. R.; Pavlova, N. V.; Kozyavkin, S. A.; Slesarev, A. I. (2004). "Recent developments in the optimization of thermostable DNA polymerases for efficient applications☆". Trends in Biotechnology. **22** *(5): 253–260. doi:10.1016/j.tibtech.2004.02.011. PMID 15109812.*

7. Rychlik W, Spencer WJ, Rhoads RE (1990). "Optimization of the annealing temperature for DNA amplification in vitro". Nucl Acids Res. **18** *(21): 6409–6412. doi:10.1093/*nar/18.21.6409. PMC 332522. PMID 2243783.

8. Sharkey, D. J.; Scalice, E. R.; Christy, K. G.; Atwood, S. M.; Daiss, J. L. (1994). "Antibodies as Thermolabile Switches: High Temperature Triggering for the Polymerase Chain Reaction". Bio/Technology. **12** *(5): 506–509.* doi:10.1038/nbt0594-506.

9. Chien A, Edgar DB, Trela JM (1976). "Deoxyribonucleic acid polymerase from the extreme thermophile Thermus aquaticus". J. Bacteriol. **127** *(3): 1550–1557. PMC 232952. PMID 8432.*

10. Lawyer, F.; Stoffel, S.; Saiki, R.; Chang, S.; Landre, P.; Abramson, R.; Gelfand, D. (1993). "High-level expression, purification, and enzymatic characterization of full-length Thermus aquaticus DNA polymerase and a truncated form deficient in 5' to 3' exonuclease activity". PCR methods and applications. **2** *(4): 275–287. doi:10.1101/gr.2.4.275. PMID 8324500.*

11. Sarkar, G.; Kapelner, S.; Sommer, S. (1990). "Formamide can dramatically improve the specificity of PCR". Nucleic Acids Research. **18** *(24): 7465. doi:10.1093/*nar/18.24.7465. PMC 332902. PMID 2259646.

12. Pavlov AR, Pavlova NV, Kozyavkin SA, Slesarev AI (2006). "Thermostable DNA Polymerases for a Wide Spectrum of Applications: Comparison of a Robust Hybrid TopoTaq to other enzymes". In Kieleczawa J. DNA Sequencing II: Optimizing Preparation and Cleanup. Jones and Bartlett. pp. 241–257. ISBN 0-7637-3383-0.

13. Pombert JF, Sistek V, Boissinot M, Frenette M (2009). "Evolutionary relationships among salivarius streptococci as inferred from multilocus phylogenies based on 16S rRNA-encoding, recA, secA, and secY gene sequences". BMC Microbiol. **9**: 232. doi:10.1186/1471-2180-9-232. PMC 2777182. PMID 19878555.

14. Chemical Synthesis, Sequencing, and Amplification of DNA (class notes on MBB/BIO 343)". Arizona State University. Retrieved 2007-10-29.

15. Quill E "Blood-Matching Goes Genetic" Science Magazine (14 March 2008) pp. 1478-1479.

16. Tomar, Rukam (2010). Molecular Markers and Plant Biotechnology. Pitman Pura, New Delhi: New India Publishing Agency. p. 188. ISBN 978-93-80235-25-7.

17. Cai, H; Caswell JL; Prescott JF (March 2014). "Nonculture Molecular Techniques for Diagnosis of Bacterial Disease in Animals: A Diagnostic Laboratory Perspective". Veterinary Pathology. **51** *(2): 341–350.* doi:10.1177/0300985813511132. PMID 24569613.

18. Salis AD (2009). "Applications in Clinical Microbiology". Real-Time PCR: Current Technology and Applications. Caister Academic Press. ISBN 978-1-904455-39-4.

19. Garibyan L, Avashia N (2013). "Polymerase Chain Reaction". Journal of Investigative Dermatology. **133**: 1–4. doi:10.1038/jid.2013.1.

20. Newton CR, Graham A, Heptinstall LE, Powell SJ, Summers C, Kalsheker N, Smith JC, Markham AF (1989). "Analysis of any point mutation in DNA. The amplification refractory mutation system (ARMS)". Nucleic Acids Research. **17** *(7): 2503–2516. doi:10.1093/*nar/17.7.2503. PMC 317639. PMID 2785681.

21. Stemmer WP, Crameri A, Ha KD, Brennan TM, Heyneker HL (1995). "Single-step assembly of a gene and entire plasmid from large numbers of oligodeoxyribonucleotides". Gene. **164** *(1): 49–53.* doi:10.1016/0378-1119(95)00511-4. PMID 7590320.

22. Innis MA, Myambo KB, Gelfand DH, Brow MA (1988). "DNA sequencing with Thermus aquaticus DNA polymerase and direct sequencing of polymerase chain reaction-amplified DNA". Proc Natl Acad Sci USA. **85** *(24): 9436–4940. doi:10.1073/pnas.85.24.9436. PMC 282767. PMID 3200828.*

23. Pierce KE & Wangh LJ (2007). "Linear-after-the-exponential polymerase chain reaction and allied technologies Real-time detection strategies for rapid, reliable diagnosis from single cells". Methods Mol Med. Methods in Molecular Medicine™. **132**: 65–85. doi:10.1007/978-1-59745-298-4_7. ISBN 978-1-58829-578-1. PMID 17876077.

24. Schwartz JJ, Lee C, Shendure J (2012). "Accurate gene synthesis with tag-directed retrieval of sequence-verified DNA molecules". Nature Methods. **9** *(9): 913–915. doi:10.1038/nmeth.2137. PMC 3433648. PMID 22886093.*

25. Vincent M, Xu Y, Kong H (2004). "Helicase-dependent isothermal DNA amplification". EMBO Reports. **5** *(8): 795–800. doi:10.1038/sj.embor.7400200. PMC 1249482. PMID 15247927.*

26. Chou Q, Russell M, Birch DE, Raymond J, Bloch W (1992). "Prevention of pre-PCR mis-priming and primer dimerization improves low-copy-number amplifications". Nucleic Acids Research. **20** *(7): 1717–1723. doi:10.1093/* nar/20.7.1717. PMC 312262. PMID 1579465.

27. Kellogg, DE; Rybalkin, I; Chen, S; Mukhamedova, N; Vlasik, T; Siebert, PD; Chenchik, A (1994). "TaqStart Antibody: "hot start" PCR facilitated by a neutralizing monoclonal antibody directed against Taq DNA polymerase". BioTechniques. **16** *(6): 1134–7. PMID 8074881.*

28. San Millan RM, Martinez-Ballesteros I, Rementeria A, Garaizar J, Bikandi J (2013). "Online exercise for the design and simulation of PCR and PCR-RFLP experiments". BMC Research Notes. **6**: 513. doi:10.1186/1756-0500-6-513. PMID 24314313.

29. E. Zietkiewicz; A. Rafalski & D. Labuda (1994). "Genome fingerprinting by simple sequence repeat (SSR)-anchored polymerase chain reaction amplification". Genomics. **20** *(2): 176–83. doi:10.1006/geno.1994.1151. PMID 8020964.*

30. Ochman H, Gerber AS, Hartl DL (1988). "Genetic Applications of an Inverse Polymerase Chain Reaction". Genetics. **120** *(3): 621–623. PMC 1203539. PMID 2852134.*

31. Mueller PR, Wold B (1988). "In vivo footprinting of a muscle specific enhancer by ligation mediated PCR". Science. **246** *(4931): 780–786. doi:10.1126/*

science.2814500. PMID 2814500.

32. Herman JG, Graff JR, Myöhänen S, Nelkin BD, Baylin SB (1996). "Methylation-specific PCR: a novel PCR assay for methylation status of CpG islands". Proc Natl Acad Sci USA. **93** *(13): 9821–9826. doi:10.1073/pnas.93.18.9821. PMC 38513. PMID 8790415.*
33. Isenbarger TA, Finney M, Ríos-Velázquez C, Handelsman J, Ruvkun G (2008). "Miniprimer PCR, a New Lens for Viewing the Microbial World". Applied and Environmental Microbiology. **74** *(3): 840–9. doi:10.1128/AEM.01933-07. PMC 2227730. PMID 18083877.*
34. Cenchao Shen; Wenjuan Yang; Qiaoli Ji; Hisaji Maki; Anjie Dong; Zhizhou Zhang (2009). "NanoPCR observation: different levels of DNA replication fidelity in nanoparticle-enhanced polymerase chain reactions". Nanotechnology. **20**: 455103. doi:10.1088/0957-4484/20/45/455103.
35. Shen, Cenchao (2013). An Overview of Nanoparticle-Assisted Polymerase Chain Reaction Technology. US: Wiley-Blackwell Publishing Ltd. pp. 97–106.
36. Horton RM, Hunt HD, Ho SN, Pullen JK, Pease LR (1989). "Engineering hybrid genes without the use of restriction enzymes: gene splicing by overlap exten-sion". Gene. **77** *(1): 61–68.* doi:10.1016/0378-1119(89)90359-4. PMID 2744488.
37. Moller, Simon (2006). PCR (THE BASICS). US: Taylor & Francis Group. p. 144.
38. David F, Turlotte E (1998). "Une méthode d'amplification génique isotherme" [An Isothermal Amplification Method]. Comptes Rendus de l'Académie des Sciences - Series III - Sciences de la Vie. **321** *(11): 909–914.* doi:10.1016/S0764-4469(99)80005-5. ISSN 0764-4469.
39. Bing DH, Boles C, Rehman FN, Audeh M, Belmarsh M, Kelley B, Adams CP (1996). "Bridge amplification: a solid phase PCR system for the amplification and detection of allelic differences in single copy genes". Genetic Identity Conference Proceedings, Seventh International Symposium on Human Identification.
40. Khan Z, Poetter K, Park DJ (2008). "Enhanced solid phase PCR: mechanisms to increase priming by solid support primers". Analytical Biochemistry. **375** *(2): 391–393. doi:10.1016/j.ab.2008.01.021. PMID 18267099.*
41. Raoult, D; G Aboudharam; E Crubezy; G Larrouy; B Ludes; M Drancourt (2000-11-07). "Molecular identification by "suicide PCR" of Yersinia pestis as the agent of medieval black death". Proc. Natl. Acad. Sci. U.S.A. **97** *(23): 12800–12803. doi:10.1073/pnas.220225197. ISSN 0027-8424. PMC 18844. PMID 11058154.*
42. Y.G. Liu & R. F. Whittier (1995). "Thermal asymmetric interlaced PCR: automatable amplification and sequencing of insert end fragments from P1 and YAC clones for chromosome walking". Genomics. **25** *(3): 674–81.*

doi:10.1016/0888-7543(95)80010-J. PMID 7759102.

43. Don RH, Cox PT, Wainwright BJ, Baker K, Mattick JS (1991). "'Touchdown' PCR to circumvent spurious priming during gene amplification". Nucl Acids Res. **19** *(14): 4008. doi:10.1093/*nar/19.14.4008. PMC 328507. PMID 1861999.

44. Myrick KV, Gelbart WM (2002). "Universal Fast Walking for direct and versatile determination of flanking sequence". Gene. **284** *(1–2): 125–131.* doi:10.1016/S0378-1119(02)00384-0. PMID 11891053.

45. Park DJ (2005). "A new 5' terminal murine GAPDH exon identified using 5'RACE LaNe". Molecular Biotechnology. **29** *(1): 39–46.* doi:10.1385/MB:29:1:39. PMID 15668518.

46. Park DJ (2004). "3'RACE LaNe: a simple and rapid fully nested PCR method to determine 3'-terminal cDNA sequence". Biotechniques. **36** *(4): 586–588, 590. PMID 15088375.*

47. Kleppe K, Ohtsuka E, Kleppe R, Molineux I, Khorana HG (1971). "Studies on polynucleotides. XCVI. Repair replications of short synthetic DNA's as catalyzed by DNA polymerases". J. Mol. Biol. **56** *(2): 341–361.* doi:10.1016/0022-2836(71)90469-4. PMID 4927950.

48. Rabinow, Paul (1996). Making PCR: A Story of Biotechnology. Chicago: University of Chicago Press. ISBN 0-226-70146-8.

49. Mullis, Kary *(1998). Dancing Naked in the Mind Field. New York: Pantheon Books. ISBN 0-679-44255-3.*

50. Mullis, Kary (1990). "The unusual origin of the polymerase chain reaction". Scientific American. **262** *(4): 56–61, 64–5.* doi:10.1038/scientificamerican0490-56. PMID 2315679.

CHAPTER 6

Gene Regulation

Chapter Contents

Gene regulation is the process of controlling which genes in a cell's DNA are expressed (used to make a functional product such as a protein). Different cells in a multicellular organism may express very different sets of genes, even though they contain the same DNA.

INTRODUCTION

Regulation of gene expression includes a wide range of mechanisms that are used by cells to increase or decrease the production of specific gene products (protein or RNA), and is informally termed gene regulation. Sophisticated programs of gene expression are widely observed in biology, for example to trigger developmental pathways, respond to environmental stimuli, or adapt to new food sources. Virtually any step of gene expression can be modulated, from transcriptional initiation, to RNA processing, and to the post-translational modification of a protein. Often, one gene regulator controls another, and so on, in a gene regulatory network.

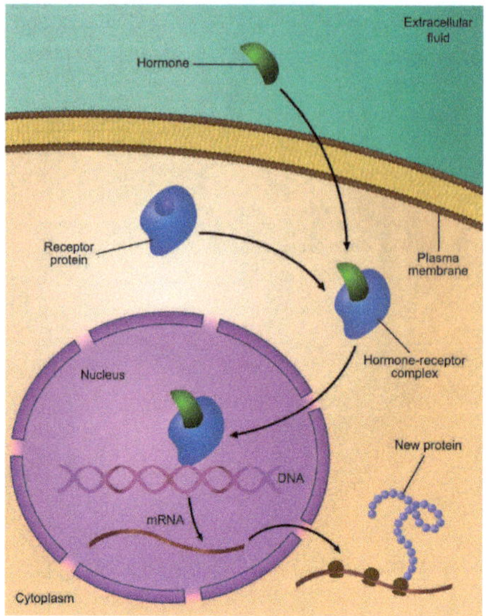

Regulation of gene expression by a hormone receptor

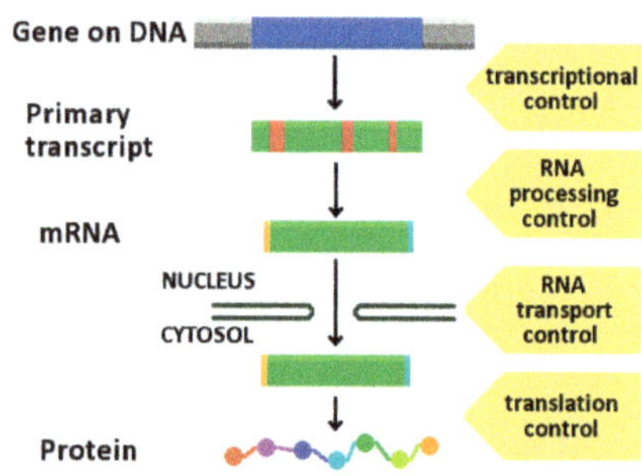

Diagram showing at which stages in the DNA-mRNA-protein pathway expression can be controlled

Gene regulation is essential for viruses, prokaryotes and eukaryotes as it increases the versatility and adaptability of an organism by allowing the cell to express protein when needed. Although as early as 1951, Barbara McClintock showed interaction between two genetic loci, Activator (Ac) and Dissociator (Ds), in the color formation of maize seeds, the first discovery of a gene regulation system is widely considered to be the identification in 1961 of the lac operon, discovered by Jacques Monod, in which some enzymes involved in lactose metabolism are expressed by *E. coli* only in the presence of lactose and absence of glucose.

In multicellular organisms, gene regulation drives cellular differentiation and morphogenesis in the embryo, leading to the creation of different cell types that possess different gene expression profiles from the same genome sequence. This explains how evolution actually works at a molecular level, and is central to the science of evolutionary developmental biology ("evo-devo").

The initiating event leading to a change in gene expression includes activation or deactivation of receptors.

MODIFICATION OF DNA

In eukaryotes, the accessibility of large regions of DNA can depend on its chromatin structure, which can be altered as a result of histone modifications directed by DNA methylation, ncRNA, or DNA-binding protein. Hence these modifications may up or down regulate the expression of a gene. Some of these modifications that regulate gene expression are inheritable and are referred to as epigenetic regulation.

Structural

Transcription of DNA is dictated by its structure. In general, the density of its packing is indicative of the frequency of transcription. Octameric protein complexes called nucleosomes are responsible for the amount of supercoiling of DNA, and these complexes can be temporarily modified by processes such as phosphorylation or more permanently modified by processes such as methylation. Such modifications are considered to be responsible for more or less permanent changes in gene expression levels.

Chemical

Methylation of DNA is a common method of gene silencing. DNA is typically methylated by methyltransferase enzymes on cytosine nucleotides in a CpG dinucleotide sequence (also called "CpG islands when densely clustered). Analysis of the pattern of methylation in a given region of DNA (which can be a promoter) can be achieved through a method called bisulfite mapping. Methylated cytosine residues are unchanged by the treatment, whereas unmethylated ones are changed to uracil. The differences are analyzed by DNA sequencing or by methods developed to quantify SNPs, such as Pyrosequencing (Biotage) or MassArray (Sequenom), measuring the relative amounts of C/T at the CG dinucleotide. Abnormal methylation patterns are thought to be involved in oncogenesis.

Histone acetylation is also an important process in transcription. Histone acetyltransferase enzymes (HATs) such as CREB-binding protein also dissociate the DNA from the histone complex, allowing transcription to proceed. Often, DNA methylation and histone deacetylation work together in gene silencing. The combination of the two seems to be a signal for DNA to be packed more densely, lowering gene expression.

REGULATION OF TRANSCRIPTION

Regulation of transcription thus controls when transcription occurs and how much RNA is created. Transcription of a gene by RNA polymerase can be regulated by several mechanisms. Specificity factors alter the specificity of RNA polymerase for a given promoter or set of promoters, making it more or less likely to bind to them (i.e., sigma factors used in prokaryotic transcription). Repressors bind to the Operator, coding sequences on the DNA strand that are close to or overlapping the promoter region, impeding RNA polymerase's progress along the strand, thus impeding the expression of the gene.The image to the right demonstrates regulation by a repressor in the lac operon. General transcription factors position RNA polymerase at the start of a protein-coding sequence and then release the polymerase to transcribe the mRNA. Activators enhance the interaction between RNA polymerase and a particular promoter, encouraging the expression of the gene. Activators do this by increasing the attraction of RNA polymerase for the promoter, through interactions with subunits of the RNA polymerase or indirectly by changing the structure of the DNA. Enhancers are sites on the DNA helix that are bound by activators in order to loop the DNA bringing a specific promoter to the initiation complex. Enhancers are much more common in eukaryotes than prokaryotes, where only a few examples exist (to date). Silencers are regions of DNA sequences that, when bound by particular transcription factors, can silence expression of the gene.

REGULATION OF TRANSCRIPTION IN CANCER

In vertebrates, the majority of gene promoters contain a CpG island with numerous CpG sites. When many of a gene's promoter CpG sites are methylated the gene becomes silenced. Colorectal cancers typically have 3 to 6 driver mutations and 33 to 66 hitchhiker or passenger mutations. However, transcriptional silencing may be of more importance than mutation in causing progression to cancer. For example, in colorectal cancers about 600 to 800 genes are transcriptionally silenced by CpG island methylation (see regulation of transcription in cancer). Transcriptional repression in cancer can also occur by other epigenetic mechanisms, such as altered expression of microRNAs. In breast cancer, transcriptional repression of BRCA1 may occur more frequently by over-expressed microRNA-182 than by hypermethylation of the BRCA1 promoter (see Low expression of BRCA1 in breast and ovarian cancers).

POST-TRANSCRIPTIONAL REGULATION

Post-transcriptional regulation is the control of gene expression at the RNA level, therefore between the transcription and the translation of the gene.

Mechanism

After being produced, the stability and distribution of the different transcripts is regulated (post-transcriptional regulation) by means of RNA binding protein (RBP) that control the various steps and rates cripts: events such as alternative splicing, nuclear degradation (exosome), processing, nuclear export (three alternative pathways), sequestration in P-bodies for storage or degradation and ultimately translation. These proteins achieve these events thanks to a RNA recognition motif (RRM) that binds a specific sequence or secondary structure of the transcripts, typically at the 5' and 3' UTR of the transcript.

Modulating the capping, splicing, addition of a Poly(A) tail, the sequence-specific nuclear export rates and in several contexts sequestration of the RNA transcript occurs in eukaryotes but not in prokaryotes. This modulation is a result of a protein or transcript which in turn is regulated and may have an affinity for certain sequences.

- **Capping** changes the five prime end of the mRNA to a three prime end by 5'-5' linkage, which protects the mRNA from 5' exonuclease, which degrades foreign RNA. The cap also helps in ribosomal binding.
- **Splicing** removes the introns, noncoding regions that are transcribed into RNA, in order to make the mRNA able to create proteins. Cells do this by spliceosomes binding on either side of an intron, looping the intron into a circle and then cleaving it off. The two ends of the exons are then joined together.
- **Addition of poly(A) tail** otherwise known as polyadenylation. That is, a stretch of RNA that is made solely of adenine bases is added to the 3' end, and acts as a buffer to the 3' exonuclease in order to increase the half life of mRNA. In addition, a long poly(A) tail can increase translation. Poly(A)-binding protein (PABP) binds to a long poly(A) tail and mediates the interaction between EIF4E and EIF4G which encourages the initiation of translation.
- **RNA editing** is a process which results in sequence variation in the RNA molecule, and is catalyzed by enzymes. These enzymes include the Adenosine Deaminase Acting on RNA (ADAR) enzymes, which convert specific adenosine residues to inosine in an mRNA molecule by hydrolytic deamination. Three ADAR enzymes have been cloned, ADAR1, ADAR2 and ADAR3, although only the first two subtypes have been shown to have RNA editing activity. Many mRNAs are vulnerable to the effects of RNA editing, including the glutamate receptor subunits GluR2, GluR3, GluR4, GluR5 and GluR6 (which are components of the AMPA and kainate receptors), the serotonin2C receptor, the GABA-alpha3 receptor subunit, the tryptophan hydroxlase enzyme TPH2, the hepatitis delta virus and more than 16% of microRNAs. In addition to ADAR enzymes, CDAR enzymes exist and these convert cytosines in specific RNA molecules, to uracil. These enzymes are termed ‹APOBEC' and have genetic loci at 22q13, a region close to the chromosomal deletion which occurs in velocardiofacial syndrome (22q11) and which is linked to psychosis. RNA editing is extensively studied in relation to infectious diseases, because the editing process alters viral function.
- **mRNA Stability** can be manipulated in order to control its half-life, and the poly(A) tail has some effect on this stability, as previously stated. Stable mRNA can have a half life of up to a day or more which allows for the production of more protein product; unstable mRNA is used in regulation that must occur quickly.

microRNA Mediated Regulation

MicroRNAs (miRNAs) appear to regulate the expression of more than 60% of protein coding genes of the human genome. If an miRNA is abundant it can behave as a "switch," turning some genes on or off. However, altered expression of many miRNAs only leads to a modest 1.5- to 4-fold change in protein expression of their target genes. Individual miRNAs often repress several hundred target genes. Repression usually occurs either through translational silencing of the mRNA or through degradation of the mRNA, via complementary binding, mostly to specific sequences in the 3' untranslated region of the target gene's mRNA.The mechanism of translational silencing or degradation of mRNA is implemented through the RNA-induced silencing complex (RISC).

Feedback in the Regulation of RNA Binding Proteins

In metazoans and bacteria, many genes involved in post-post transcriptional regulation are regulated post transcriptionally. For Drosophila RBPs associated with splicing or nonsense mediated decay, analyses of protein-protein and protein-RNA interaction profiles have revealed ubiquitous interactions with RNA and protein products of the same gene. It remains unclear whether these observations are driven by ribosome proximal or ribosome mediated contacts, or if some protein complexes, particularly RNPs, undergo co-translational assembly.

Significance

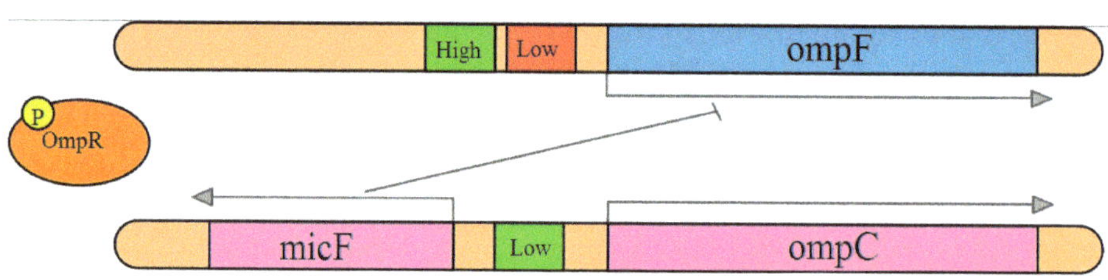

A prokaryotic example: Salmonella enterica (a pathogenic γ-proteobacterium) can express two alternative porins depending on the external environment (gut or murky water), this system involves EnvZ (osomotic sensor) which activates OmpR (transcription factor) which can bind to a high affinity promoter even at low concentrations and the low affinity promoter only at high concentrations (by definition): when the concentration of this transcription factor is high it activates OmpC and micF and inhibits OmpF, OmpF is further inhibited post-transcriptionally by micF RNA which binds to the OmpF transcript

This area of study has recently gained more importance due to the increasing evidence that post-transcriptional regulation plays a larger role than previously expected. Even though protein with DNA binding domains are more abundant than protein with RNA binding domains, a recent study by Cheadle et al. (2005) showed that during T-cell activation 55% of significant changes at the steady-state level had no corresponding changes at the transcriptional level, meaning they were a result of stability regulation alone.

Furthermore, RNA found in the nucleus is more complex than that found in the cytoplasm: more than 95% (bases) of the RNA synthesized by RNA polymerase II never reaches the cytoplasm. The main reason for this is due to the removal of introns which account for 80% of the total bases. Some studies have shown that even after processing the levels of mRNA between the cytoplasm and the nucleus differ greatly.

Developmental biology is a good source of models of regulation, but due to the technical difficulties it was easier to determine the transcription factor cascades than regulation at the RNA level. In fact several key genes such as nanos are known to bind RNA but often their targets are unknown. Although RNA binding proteins may regulate post transcriptionally large amount of the transcriptome, the targeting of a single gene is of interest to the scientific community for medical reasons, this is RNA interference and microRNAs which are both examples of posttranscriptional regulation, which regulate the destruction of RNA and change the chromatin structure. To study post-transcriptional regulation several techniques are used, such as RIP-Chip (RNA immunoprecipitation on chip).

microRNA Role in Cancer

Deficiency of expression of a DNA repair gene occurs in many cancers (see DNA repair defect and cancer risk and microRNA and DNA repair). Altered microRNA (miRNA) expression that either decreases accurate DNA repair or increases inaccurate microhomology-mediated end joining (MMEJ) DNA repair is often observed in cancers. Deficiency of accurate DNA repair may be a major source of the high frequency of mutations in cancer (see mutation frequencies in cancers). Repression of DNA repair genes in cancers by changes in the levels of microRNAs may be a more frequent cause of repression than mutation or epigenetic methylation of DNA repair genes.

For instance, BRCA1 is employed in the accurate homologous recombinational repair (HR) pathway. Deficiency of BRCA1 can cause breast cancer. Down-regulation of BRCA1 due to mutation occurs in about 3% of breast cancers. Down-regulation of BRCA1 due to methylation of its promoter occurs in about 14% of breast cancers. However, increased expression of miR-182 down-regulates BRCA1 mRNA and protein expression, and increased miR-182 is found in 80% of breast cancers.

In another example, a mutated constitutively (persistently) expressed version of the oncogene c-Myc is found in many cancers. Among many functions, c-Myc negatively regulates microRNAs miR-150 and miR-22. These microRNAs normally repress expression of two genes essential for MMEJ, Lig3 and Parp1, thereby inhibiting this inaccurate, mutagenic DNA repair pathway. Muvarak et al. showed, in leukemias, that constitutive expression of c-Myc, leading to down-regulation of miR-150 and miR-22, allowed increased expression of Lig3 and Parp1. This generates genomic instability through increased inaccurate MMEJ DNA repair, and likely contributes to progression to leukemia.

To show the frequent ability of microRNAs to alter DNA repair expression, Hatano et al. performed a large screening study, in which 810 microRNAs were transfected into cells that were then subjected to ionizing radiation (IR). For 324 of these microRNAs, DNA repair was reduced (cells were killed more efficiently by IR) after transfection. For a further 75 microRNAs, DNA repair was increased, with less cell death after IR. This indicates that alterations in microRNAs may often down-regulate DNA repair, a likely important early step in progression to cancer.

THREE PRIME UNTRANSLATED REGIONS AND MICRORNAS

Three prime untranslated regions (3'-UTRs) of messenger RNAs (mRNAs) often contain regulatory sequences that post-transcriptionally influence gene expression. Such 3'-UTRs often contain both binding sites for microRNAs (miRNAs) as well as for regulatory proteins. By binding to specific sites within the 3'-UTR, miRNAs can decrease gene expression of various mRNAs by either inhibiting translation or directly causing degradation of the transcript. The 3'-UTR also may have silencer regions that bind repressor proteins that inhibit the expression of a mRNA.

The 3'-UTR often contains miRNA response elements (MREs). MREs are sequences to which miRNAs bind. These are prevalent motifs within 3'-UTRs. Among all regulatory motifs within the 3'-UTRs (e.g. including silencer regions), MREs make up about half of the motifs.

As of 2014, the miRBase web site, an archive of miRNA sequences and annotations, listed 28,645 entries in 233 biologic species. Of these, 1,881 miRNAs were in annotated human miRNA loci. miRNAs were predicted to have an average of about four hundred target mRNAs (affecting expression of several hundred genes).Freidman et al. estimate that >45,000 miRNA target sites within human mRNA 3'-UTRs are conserved above background levels, and >60% of human protein-coding genes have been under selective pressure to maintain pairing to miRNAs.

Direct experiments show that a single miRNA can reduce the stability of hundreds of unique mRNAs. Other experiments show that a single miRNA may repress the production of hundreds of proteins, but that this repression often is relatively mild (less than 2-fold).

The effects of miRNA dysregulation of gene expression seem to be important in cancer. For instance, in gastrointestinal cancers, a 2015 paper identified nine miRNAs as epigenetically altered and effective in down-regulating DNA repair enzymes.

The effects of miRNA dysregulation of gene expression also seem to be important in neuropsychiatric disorders, such as schizophrenia, bipolar disorder, major depressive disorder, Parkinson's disease, Alzheimer's disease and autism spectrum disorders.

REGULATION OF TRANSLATION

The translation of mRNA can also be controlled by a number of mechanisms, mostly at the level of initiation. Recruitment of the small ribosomal subunit can indeed be modulated by mRNA secondary structure, antisense RNA binding, or protein binding. In both prokaryotes and eukaryotes, a large number of RNA binding proteins exist, which often are directed to their target sequence by the secondary structure of the transcript, which may change depending on certain conditions, such as temperature or presence of a ligand (aptamer). Some transcripts act as ribozymes and self-regulate their expression.

EXAMPLES OF GENE REGULATION

- Enzyme induction is a process in which a molecule (e.g., a drug) induces (i.e., initiates or enhances) the expression of an enzyme.
- The induction of heat shock proteins in the fruit fly Drosophila melanogaster.
- The Lac operon is an interesting example of how gene expression can be regulated.
- Viruses, despite having only a few genes, possess mechanisms to regulate their gene expression, typically into an early and late phase, using collinear systems regulated by anti-terminators (lambda phage) or splicing modulators (HIV).
- GAL4 is a transcriptional activator that controls the expression of GAL1, GAL7, and GAL10 (all of which code for the metabolic of galactose in yeast). The GAL4/UAS system has been used in a variety of organisms across various phyla to study gene expression.

Developmental Biology

A large number of studied regulatory systems come from developmental biology. Examples include:

- The colinearity of the Hox gene cluster with their nested antero-posterior patterning
- It has been speculated that pattern generation of the hand (digits - interdigits) The gradient of Sonic hedgehog (secreted inducing factor) from the zone of polarizing activity in the limb, which creates a gradient of active Gli3, which activates Gremlin, which inhibits BMPs also secreted in the limb, resulting in the formation of an alternating pattern of activity as a result of this reaction-diffusion system.
- Somitogenesis is the creation of segments (somites) from a uniform tissue (Pre-somitic Mesoderm). They are formed sequentially from anterior to posterior. This is achieved in amniotes possibly by means of two opposing gradients, Retinoic acid in the anterior (wavefront) and Wnt and Fgf in the posterior, coupled to an oscillating pattern (segmentation clock) composed of FGF + Notch and Wnt in antiphase.
- Sex determination in the soma of a Drosophila requires the sensing of the ratio of autosomal genes to sex chromosome-encoded genes, which results in the production of sexless splicing factor in females, resulting in the female isoform of doublesex.

CIRCUITRY

Up-Regulation and down-Regulation

Up-regulation is a process that occurs within a cell triggered by a signal (originating internal or external to the cell), which results in increased expression of one or more genes and as a result the protein(s) encoded by those genes. Conversely, down-regulation is a process resulting in decreased gene and corresponding protein expression.

- Up-regulation occurs, for example, when a cell is deficient in some kind of receptor. In this case, more receptor protein is synthesized and transported to the membrane of the cell and, thus, the sensitivity of the cell is brought back to normal, reestablishing homeostasis.
- Down-regulation occurs, for example, when a cell is overstimulated by a neurotransmitter, hormone, or drug for a prolonged period of time, and the expression of the receptor protein is decreased in order to protect the cell (see also tachyphylaxis).

Inducible *vs.* Repressible Systems

Gene Regulation can be summarized by the response of the respective system:

- Inducible systems - An inducible system is off unless there is the presence of some molecule (called an inducer) that allows for gene expression. The molecule is said to "induce expression". The manner by which this happens is dependent on the control mechanisms as well as differences between prokaryotic and eukaryotic cells.
- Repressible systems - A repressible system is on except in the presence of some molecule (called a corepressor) that suppresses gene expression. The molecule is said to "repress expression". The manner by which this happens is dependent on the control mechanisms as well as differences between prokaryotic and eukaryotic cells.

The GAL4/UAS system is an example of both an inducible and repressible system. GAL4 binds an upstream activation sequence (UAS) to activate the transcription of the GAL1/GAL7/GAL10 cassette. On the other hand, a MIG1 response to the presence of glucose can inhibit GAL4 and therefore stop the expression of the GAL1/GAL7/GAL10 cassette.

Theoretical circuits

- Repressor/Inducer: an activation of a sensor results in the change of expression of a gene
- negative feedback: the gene product downregulates its own production directly or indirectly, which can result in
  - keeping transcript levels constant/proportional to a factor
  - inhibition of run-away reactions when coupled with a positive feedback loop
  - creating an oscillator by taking advantage in the time delay of transcription and translation, given that the mRNA and protein half-life is shorter
- positive feedback: the gene product upregulates its own production directly or indirectly, which can result in

- signal amplification
- bistable switches when two genes inhibit each other and both have positive feedback
- pattern generation

EPIGENETICS

Epigenetics studies genetic effects not entirely encoded in the DNA sequence of an organism, hence the prefix epi- (Greek: επί- over, outside of, around). Such effects on cellular and physiological phenotypic traits may result from external or environmental factors that switch genes on and off and affect how cells express gene products.These alterations may or may not be heritable.

The term also refers to the changes themselves: functionally relevant changes to the genome that do not involve a change in the nucleotide sequence. Examples of mechanisms that produce such changes are DNA methylation and histone modification, each of which alters how genes are expressed without altering the underlying DNA sequence. Gene expression can be controlled through the action of repressor proteins that attach to silencer regions of the DNA. These epigenetic changes may last through cell divisions for the duration of the cell's life, and may also last for multiple generations even though they do not involve changes in the underlying DNA sequence of the organism; instead, non-genetic factors cause the organism's genes to behave (or express themselves) differently.

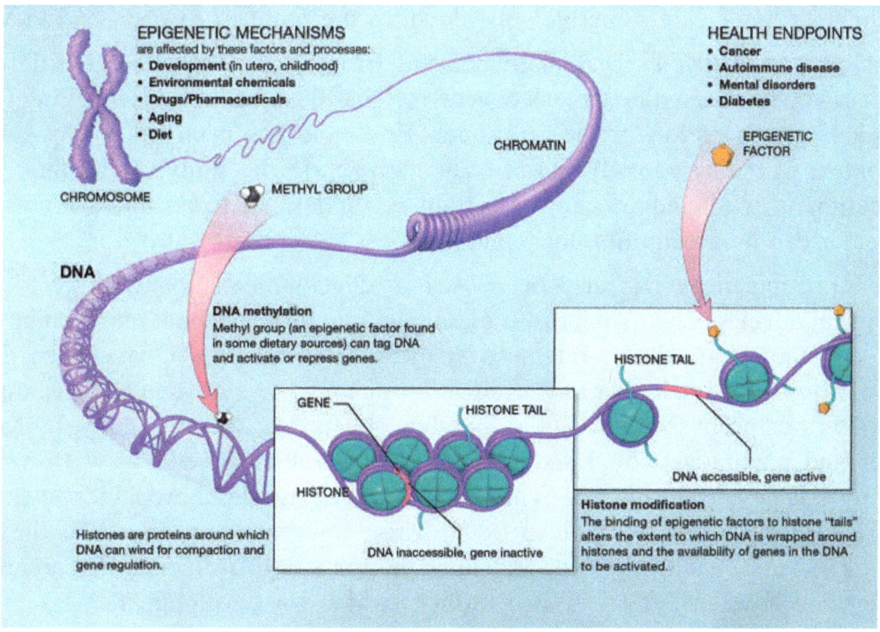

Epigenetic mechanisms

One example of an epigenetic change in eukaryotic biology is the process of cellular differentiation. During morphogenesis, totipotent stem cells become the various pluripotent cell lines of the embryo, which in turn become fully differentiated cells. In other words, as a single fertilized egg cell – the zygote – continues to divide, the resulting daughter cells change into all the different cell types in an organism, including neurons, muscle cells, epithelium, endothelium of blood vessels, etc., by activating some genes while inhibiting the expression of others.

The term epigenetics in its contemporary usage emerged in the 1990s, but for some years has been used in somewhat variable meanings. A consensus definition of the concept of epigenetic trait as "stably heritable phenotype resulting from changes in a chromosome without alterations in the DNA sequence" was formulated at a Cold Spring Harbor meeting in 2008.

The term epigenesis has a generic meaning "extra growth", taken directly from Koine Greek, used in English since the 17th century.

From this, and the associated adjective epigenetic, the term epigenetics was coined by C. H. Waddington in 1942 as pertaining to epigenesis in parallel to Valentin Haecker's 'phenogenetics' (Pänogenetik). Epigenesis in the context of biology refers to the differentiation of cells from their initial totipotent state in embryonic development.

When Waddington coined the term the physical nature of genes and their role in heredity was not known; he used it as a conceptual model of how genes might interact with their surroundings to produce a phenotype; he used the phrase epigenetic landscape as a metaphor for biological development. Waddington held that cell fates were established in development much like a marble rolls down to the point of lowest local elevation.

Waddington suggested visualising increasing irreversibility of cell type differentiation as ridges rising between the valleys where the marbles (cells) are travelling.In recent times Waddington's notion of the epigenetic landscape has been rigorously formalized in the context of the systems dynamics state approach to the study of cell-fate. Cell-fate determination is predicted to exhibit certain dynamics, such as attractor-convergence (the attractor can be an equilibrium point, limit cycle or strange attractor) or oscillatory.

The term epigenetic has also been used in developmental psychology to describe psychological development as the result of an ongoing, bi-directional interchange between heredity and the environment. Interactivist ideas of development have been discussed in various forms and under various names throughout the 19th and 20th centuries. An early version was proposed, among the founding statements in embryology, by Karl Ernst von Baer and popularized by Ernst Haeckel. A radical epigenetic view (physiological epigenesis) was developed by Paul Wintrebert. Another variation, probabilistic epigenesis, was presented by Gilbert Gottlieb in 2003. This view encompasses all of the possible developing factors on an organism and how they not only influence the organism and each other, but how the organism also influences its own development.

The developmental psychologist Erik Erikson used the term epigenetic principle in his book Identity: Youth and Crisis (1968), and used it to encompass the notion that we develop through an unfolding of our personality in predetermined stages, and that our environment and surrounding culture influence how we progress through these stages. This biological unfolding in relation to our socio-cultural settings is done in stages of psychosocial development, where "progress through each stage is in part determined by our success, or lack of success, in all the previous stages."

Contemporary

Robin Holliday defined epigenetics as "the study of the mechanisms of temporal and spatial control of gene activity during the development of complex organisms." Thus epigenetic can be used to describe anything other than DNA sequence that influences the development of an organism.

The more recent usage of the word in science has a stricter definition. It is, as defined by Arthur Riggs and colleagues, the study of mitotically and/or meiotically heritable changes in gene function that cannot be explained by changes in DNA sequence. The Greek prefix epi- in epigenetics implies features that are "on top of" or "in addition to" genetics; thus epigenetic traits exist on top of or in addition to the traditional molecular basis for inheritance.

The term epigenetics, however, has been used to describe processes which have not been demonstrated to be heritable such as histone modification; there are therefore attempts to redefine it in broader terms that would avoid the constraints of requiring heritability. For example, Sir Adrian Bird defined epigenetics as "the structural adaptation of chromosomal regions so as to register, signal or perpetuate altered activity states." This definition would be inclusive of transient modifications associated with DNA repair or cell-cycle phases as well as stable changes maintained across multiple cell generations, but exclude others such as templating of membrane architecture and prions unless they impinge on chromosome function. Such redefinitions however are not universally accepted and are still subject to dispute. The NIH "Roadmap Epigenomics Project," ongoing as of 2016, uses the following definition: ...For purposes of this program, epigenetics refers to both heritable changes in gene activity and expression (in the progeny of cells or of individuals) and also stable, long-term alterations in the transcriptional potential of a cell that are not necessarily heritable.

In 2008, a consensus definition of the epigenetic trait, "stably heritable phenotype resulting from changes in a chromosome without alterations in the DNA sequence", was made at a Cold Spring Harbor meeting.

The similarity of the word to "genetics" has generated many parallel usages. The "epigenome" is a parallel to the word "genome", referring to the overall epigenetic state of a cell, and epigenomics refers to more global analyses of epigenetic changes across the entire genome. The phrase "genetic code" has also been adapted—the "epigenetic code" has been used to describe the set of epigenetic features that create different phenotypes in different cells. Taken to its extreme, the "epigenetic code" could represent the total state of the cell, with the position of each molecule accounted for in an epigenomic map, a diagrammatic representation of the gene expression, DNA methylation and histone modification status of a particular genomic region. More typically, the term is used in reference to systematic efforts to measure specific, relevant forms of epigenetic information such as the histone code or DNA methylation patterns.

Molecular Basis

Epigenetic changes modify the activation of certain genes, but not the genetic code sequence of DNA. The microstructure (not code) of DNA itself or the associated chromatin proteins may be modified, causing activation or silencing. This mechanism enables differentiated cells in a multicellular organism to express only the genes that are necessary for their own activity. Epigenetic changes are preserved when cells divide. Most epigenetic changes only occur within the course of one individual organism's lifetime; however, if gene inactivation occurs in a sperm or egg cell that results in fertilization, then some epigenetic changes can be transferred to the next generation. This raises the question of whether or not epigenetic changes in an organism can alter the basic structure of its DNA (see Evolution, below), a form of Lamarckism.

Specific epigenetic processes include paramutation, bookmarking, imprinting, gene silencing, X chromosome inactivation, position effect, reprogramming, transvection, maternal effects, the progress of carcinogenesis, many effects of teratogens, regulation of histone modifications and heterochromatin, and technical limitations affecting parthenogenesis and cloning.

DNA damage can also cause epigenetic changes. DNA damage is very frequent, occurring on average about 60,000 times a day per cell of the human body. These damages are largely repaired, but at the site of a DNA repair, epigenetic changes can remain. In particular, a double strand break in DNA can initiate unprogrammed epigenetic gene silencing both by causing DNA methylation as well as by promoting silencing types of histone modifications (chromatin remodeling - see next section). In addition, the enzyme Parp1 (poly(ADP)-ribose polymerase) and its product poly(ADP)-ribose (PAR) accumulate at sites of DNA damage as part of a repair process. This accumulation, in turn, directs recruitment and activation of the chromatin remodeling protein ALC1 that can cause nucleosome remodeling. Nucleosome remodeling has been found to cause, for instance, epigenetic silencing of DNA repair gene MLH1. DNA damaging chemicals, such as benzene, hydroquinone, styrene, carbon tetrachloride and trichloroethylene, cause considerable hypomethylation of DNA, some through the activation of oxidative stress pathways.

Foods are known to alter the epigenetics of rats on different diets. Some food components epigenetically increase the levels of DNA repair enzymes such as MGMT and MLH1 and p53. Other food components can reduce DNA damage, such as soy isoflavones and bilberry anthocyanins.

Epigenetic research uses a wide range of molecular biologic techniques to further our understanding of epigenetic phenomena, including chromatin immunoprecipitation (together with its large-scale variants ChIP-on-chip and ChIP-Seq), fluorescent in situ hybridization, methylation-sensitive restriction enzymes, DNA adenine methyltransferase identification (DamID) and bisulfite sequencing. Furthermore, the use of bioinformatic methods is playing an increasing role (computational epigenetics).

Computer simulations and molecular dynamics approaches revealed the atomistic motions associated with the molecular recognition of the histone tail through an allosteric mechanism.

Mechanisms

Several types of epigenetic inheritance systems may play a role in what has become known as cell memory, note however that not all of these are universally accepted to be examples of epigenetics.

Covalent Modifications

Covalent modifications of either DNA (e.g. cytosine methylation and hydroxymethylation) or of histone proteins (e.g. lysine acetylation, lysine and arginine methylation, serine and threonine phosphorylation, and lysine ubiquitination and sumoylation) play central roles in many types of epigenetic inheritance. Therefore, the word "epigenetics" is sometimes used as a synonym for these processes. However, this can be misleading. Chromatin remodeling is not always inherited, and not all epigenetic inheritance involves chromatin remodeling.

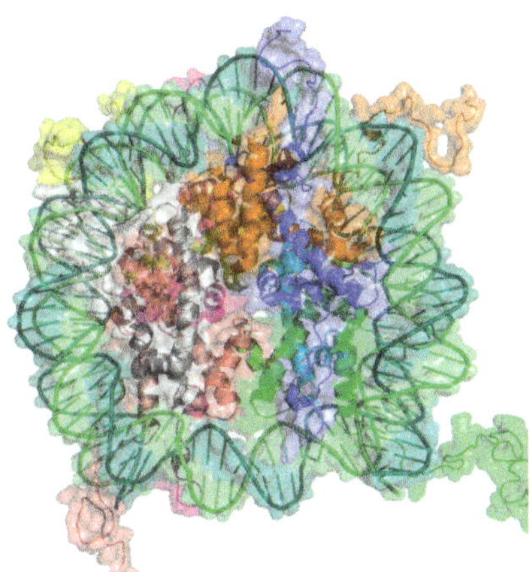

DNA associates with histone proteins to form chromatin.

Because the phenotype of a cell or individual is affected by which of its genes are transcribed, heritable transcription states can give rise to epigenetic effects. There are several layers of regulation of gene expression. One way that genes are regulated is through the remodeling of chromatin. Chromatin is the complex of DNA and the histone proteins with which it associates. If the way that DNA is wrapped around the histones changes, gene expression can change as well. Chromatin remodeling is accomplished through two main mechanisms:

- The first way is post translational modification of the amino acids that make up histone proteins. Histone proteins are made up of long chains of amino acids. If the amino acids that are in the chain are changed, the shape of the histone might be modified. DNA is not completely unwound during replication. It is possible, then, that the modified histones may be carried into each new copy of the DNA. Once there, these histones may act as templates, initiating the surrounding new histones to be shaped in the new manner. By altering the shape of the histones around them, these modified histones would ensure that a lineage-specific transcription program is maintained after cell division.

- The second way is the addition of methyl groups to the DNA, mostly at CpG sites, to convert cytosine to 5-methylcytosine. 5-Methylcytosine performs much like a regular cytosine, pairing with a guanine in double-stranded DNA. However, some areas of the genome are methylated more heavily than others, and highly methylated areas tend to be less transcriptionally active, through a mechanism not fully understood. Methylation of cytosines can also persist from the germ line of one of the parents into the zygote, marking the chromosome as being inherited from one parent or the other (genetic imprinting).

Mechanisms of heritability of histone state are not well understood; however, much is known about the mechanism of heritability of DNA methylation state during cell division and differentiation. Heritability of methylation state depends on certain enzymes (such as DNMT1) that have a higher affinity for 5-methylcytosine than for cytosine. If this enzyme reaches a "hemimethylated" portion of DNA (where 5-methylcytosine is in only one of the two DNA strands) the enzyme will methylate the other half.

Although histone modifications occur throughout the entire sequence, the unstructured N-termini of histones (called histone tails) are particularly highly modified. These modifications include acetylation, methylation, ubiquitylation, phosphorylation, sumoylation, ribosylation and citrullination. Acetylation is the most highly studied of these modifications. For example, acetylation of the K14 and K9 lysines of the tail of histone H3 by histone acetyltransferase enzymes (HATs) is generally related to transcriptional competence.

One mode of thinking is that this tendency of acetylation to be associated with "active" transcription is biophysical in nature. Because it normally has a positively charged nitrogen at its end, lysine can bind the negatively charged phosphates of the DNA backbone. The acetylation event converts the positively charged amine group on the side chain into a neutral amide linkage. This removes the positive charge, thus loosening the DNA from the histone. When this occurs, complexes like SWI/SNF and other transcriptional factors can bind to the DNA and allow transcription to occur. This is the "cis" model of epigenetic function. In other words, changes to the histone tails have a direct effect on the DNA itself.

Another model of epigenetic function is the "trans" model. In this model, changes to the histone tails act indirectly on the DNA. For example, lysine acetylation may create a binding site for chromatin-modifying enzymes (or transcription machinery as well). This chromatin remodeler can then cause changes to the state of the chromatin. Indeed, a bromodomain — a protein domain that specifically binds acetyl-lysine — is found in many enzymes that help activate transcription, including the SWI/SNF complex. It may be that acetylation acts in this and the previous way to aid in transcriptional activation.

The idea that modifications act as docking modules for related factors is borne out by histone methylation as well. Methylation of lysine 9 of histone H3 has long been associated with constitutively transcriptionally silent chromatin (constitutive heterochromatin). It has been determined that a chromodomain (a domain that specifically binds methyl-lysine) in the transcriptionally repressive protein HP1 recruits HP1 to K9 methylated regions. One example that seems to refute this biophysical model for methylation is that tri-methylation of histone H3 at lysine 4 is strongly associated with (and required for full) transcriptional activation. Tri-methylation in this case would introduce a fixed positive charge on the tail.

It has been shown that the histone lysine methyltransferase (KMT) is responsible for this methylation activity in the pattern of histones H3 & H4. This enzyme utilizes a catalytically active site called the SET domain (Suppressor of variegation, Enhancer of zeste, Trithorax). The SET domain is a 130-amino acid sequence involved in modulating gene activities. This domain has been demonstrated to bind to the histone tail and causes the methylation of the histone.

Differing histone modifications are likely to function in differing ways; acetylation at one position is likely to function differently from acetylation at another position. Also, multiple modifications may occur at the same time, and these modifications may work together to change the behavior of the nucleosome. The idea that multiple dynamic modifications regulate gene transcription in a systematic and reproducible way is called the histone code, although the idea that histone state can be read linearly as a digital information carrier has been largely debunked. One of the best-understood systems that orchestrates chromatin-based silencing is the SIR protein based silencing of the yeast hidden mating type loci HML and HMR.

DNA methylation frequently occurs in repeated sequences, and helps to suppress the expression and mobility of 'transposable elements': Because 5-methylcytosine can be spontaneously deaminated (replacing nitrogen by oxygen) to thymidine, CpG sites are frequently mutated and become rare in the genome, except at CpG islands where they remain unmethylated. Epigenetic changes of this type thus have the potential to direct increased frequencies of permanent genetic mutation. DNA methylation patterns are known to be established and modified in response to environmental factors by a complex interplay of at least three independent DNA methyltransferases, DNMT1, DNMT3A, and DNMT3B, the loss of any of which is lethal in mice. DNMT1 is the most abundant methyltransferase in somatic cells, localizes to replication foci, has a 10–40-fold preference for hemimethylated DNA and interacts with the proliferating cell nuclear antigen (PCNA).

By preferentially modifying hemimethylated DNA, DNMT1 transfers patterns of methylation to a newly synthesized strand after DNA replication, and therefore is often referred to as the 'maintenance' methyltransferase. DNMT1 is essential for proper embryonic development, imprinting and X-inactivation. To emphasize the difference of this molecular mechanism of inheritance from the canonical Watson-Crick base-pairing mechanism of transmission of genetic information, the term 'Epigenetic templating' was introduced. Furthermore, in addition to the maintenance and transmission of methylated DNA states, the same principle could work in the maintenance and transmission of histone modifications and even cytoplasmic (structural) heritable states.

Histones H3 and H4 can also be manipulated through demethylation using histone lysine demethylase (KDM). This recently identified enzyme has a catalytically active site called the Jumonji domain (JmjC). The demethylation occurs when JmjC utilizes multiple cofactors to hydroxylate the methyl group, thereby removing it. JmjC is capable of demethylating mono-, di-, and tri-methylated substrates.

Chromosomal regions can adopt stable and heritable alternative states resulting in bistable gene expression without changes to the DNA sequence. Epigenetic control is often associated with alternative covalent modifications of histones. The stability and heritability of states of larger chromosomal regions are suggested to involve positive feedback where modified nucleosomes recruit enzymes that similarly modify nearby nucleosomes. A simplified stochastic model for this type of epigenetics is found here.

It has been suggested that chromatin-based transcriptional regulation could be mediated by the effect of small RNAs. Small interfering RNAs can modulate transcriptional gene expression via epigenetic modulation of targeted promoters.

RNA Transcripts

Sometimes a gene, after being turned on, transcribes a product that (directly or indirectly) maintains the activity of that gene. For example, Hnf4 and MyoD enhance the transcription of many liver- and muscle-specific genes, respectively, including their own, through the transcription factor activity of the proteins they encode. RNA signalling includes differential recruitment of a hierarchy of generic chromatin modifying complexes and DNA methyltransferases to specific loci by RNAs during differentiation and development. Other epigenetic changes are mediated by the production of different splice forms of RNA, or by formation of double-stranded RNA (RNAi). Descendants of the cell in which the gene was turned on will inherit this activity, even if the original stimulus for gene-activation is no longer present. These genes are often turned on or off by signal transduction, although in some systems where syncytia or gap junctions are important, RNA may spread directly to other cells or nuclei by diffusion. A large amount of RNA and protein is contributed to the zygote by the mother during oogenesis or via nurse cells, resulting in maternal effect phenotypes. A smaller quantity of sperm RNA is transmitted from the father, but there is recent evidence that this epigenetic information can lead to visible changes in several generations of offspring.

MicroRNAs

MicroRNAs (miRNAs) are members of non-coding RNAs that range in size from 17 to 25 nucleotides. miRNAs regulate a large variety of biological functions in plants and animals. So far, in 2013, about 2000 miRNAs have been discovered in humans and these can be found online in a miRNA database. Each miRNA expressed in a cell may target about 100 to 200 messenger RNAs that it downregulates. Most of the downregulation of mRNAs occurs by causing the decay of the targeted mRNA, while some downregulation occurs at the level of translation into protein.

It appears that about 60% of human protein coding genes are regulated by miRNAs. Many miRNAs are epigenetically regulated. About 50% of miRNA genes are associated with CpG islands, that may be repressed by epigenetic methylation. Transcription from methylated CpG islands is strongly and heritably repressed.Other miRNAs are epigenetically regulated by either histone modifications or by combined DNA methylation and histone modification.

mRNA

In 2011, it was demonstrated that the methylation of mRNA plays a critical role in human energy homeostasis. The obesity-associated FTO gene is shown to be able to demethylate N6-methyladenosine in RNA.

sRNAs

sRNAs are small (50–250 nucleotides), highly structured, non-coding RNA fragments found in bacteria. They control gene expression including virulence genes in pathogens and are viewed as new targets in the fight against drug-resistant bacteria. They play an important role in many biological processes, binding to mRNA and protein targets in prokaryotes. Their phylogenetic analyses, for example through sRNA–mRNA target interactions or protein binding properties, are used to build comprehensive databases. sRNA-gene maps based on their targets in microbial genomes are also constructed.

Prions

Prions are infectious forms of proteins. In general, proteins fold into discrete units that perform distinct cellular functions, but some proteins are also capable of forming an infectious conformational state known as a prion. Although often viewed in the context of infectious disease, prions are more loosely defined by their ability to catalytically convert other native state versions of the same protein to an infectious conformational state. It is in this latter sense that they can be viewed as epigenetic agents capable of inducing a phenotypic change without a modification of the genome.

Fungal prions are considered by some to be epigenetic because the infectious phenotype caused by the prion can be inherited without modification of the genome. PSI+ and URE3, discovered in yeast in 1965 and 1971, are the two best studied of this type of prion. Prions can have a phenotypic effect through the sequestration of protein in aggregates, thereby reducing that protein's activity. In PSI+ cells, the loss of the Sup35 protein (which is involved in termination of translation) causes ribosomes to have a higher rate of read-through of stop codons, an effect that results in suppression of nonsense mutations in other genes.The ability of Sup35 to form prions may be a conserved trait. It could confer an adaptive advantage by giving cells the ability to switch into a PSI+ state and express dormant genetic features normally terminated by stop codon mutations.

Structural Inheritance

In ciliates such as Tetrahymena and Paramecium, genetically identical cells show heritable differences in the patterns of ciliary rows on their cell surface. Experimentally altered patterns can be transmitted to daughter cells. It seems existing structures act as templates for new structures. The mechanisms of such inheritance are unclear, but reasons exist to assume that multicellular organisms also use existing cell structures to assemble new ones.

Nucleosome Positioning

Eukaryotic genomes have numerous nucleosomes. Nucleosome position is not random, and determine the accessibility of DNA to regulatory proteins. This determines differences in gene expression and cell differentiation. It has been shown that at least some nucleosomes are retained in sperm cells (where most but not all histones are replaced by protamines). Thus nucleosome positioning is to some degree inheritable. Recent studies have uncovered connections between nucleosome positioning and other epigenetic factors, such as DNA methylation and hydroxymethylation

Functions and Consequences

Development

Developmental epigenetics can be divided into predetermined and probabilistic epigenesis. Predetermined epigenesis is a unidirectional movement from structural development in DNA to the functional maturation of the protein. "Predetermined" here means that development is scripted and predictable. Probabilistic epigenesis on the other hand is a bidirectional structure-function development with experiences and external molding development.

Somatic epigenetic inheritance, particularly through DNA and histone covalent modifications and nucleosome repositioning, is very important in the development of multicellular eukaryotic organisms. The genome sequence is static (with some notable exceptions), but cells differentiate into many different types, which perform different functions, and respond differently to the environment and intercellular signalling. Thus, as individuals develop, morphogens activate or silence genes in an epigenetically heritable fashion, giving cells a memory. In mammals, most cells terminally differentiate, with only stem cells retaining the ability to differentiate into several cell types ("totipotency" and "multipotency"). In mammals, some stem cells continue producing new differentiated cells throughout life, such as in neurogenesis, but mammals are not able to respond to loss of some tissues, for example, the inability to regenerate limbs, which some other animals are capable of. Epigenetic modifications regulate the transition from neural stem cells to glial progenitor cells (for example, differentiation into oligodendrocytes is regulated by the deacetylation and methylation of histones. Unlike animals, plant cells do not terminally differentiate, remaining totipotent with the ability to give rise to a new individual plant. While plants do utilise many of the same epigenetic mechanisms as animals, such as chromatin remodeling, it has been hypothesised that some kinds of plant cells do not use or require "cellular memories", resetting their gene expression patterns using positional information from the environment and surrounding cells to determine their fate.

Epigenetic changes can occur in response to environmental exposure—for example, mice given some dietary supplements have epigenetic changes affecting expression of the agouti gene, which affects their fur color, weight, and propensity to develop cancer.

Controversial results from one study suggested that traumatic experiences might produce an epigenetic signal that is capable of being passed to future generations. Mice were trained, using foot shocks, to fear a cherry blossom odor. The investigators reported that the mouse offspring had an increased aversion to this specific odor. They suggested epigenetic changes that increase gene expression, rather than in DNA itself, in a gene, M71, that governs the functioning of an odor receptor in the nose that responds specifically to this cherry blossom smell. There were physical changes that correlated with olfactory (smell) function in the brains of the trained mice and their descendants. Several criticisms were reported, including the study's low statistical power as evidence of some irregularity such as bias in reporting results. Due to limits of sample size, there is a probability that an effect will not be demonstrated to within statistical significance even if it exists. The criticism suggested that the probability that all the experiments reported would show positive results if an identical protocol was followed, assuming the claimed effects exist, is merely 0.4%. The authors also did not indicate which mice were siblings, and treated all of the mice as statistically independent. The original researchers pointed out negative results in the paper's appendix that the criticism omitted in its calculations, and undertook to track which mice were siblings in the future.

Transgenerational

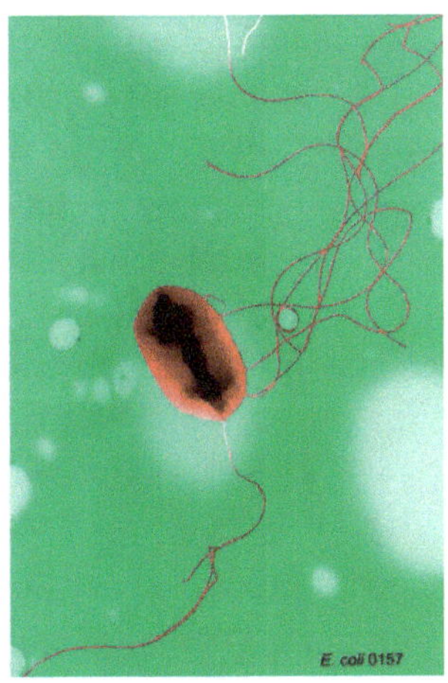

Escherichia coli bacteria

Epigenetics can affect evolution when epigenetic changes are heritable. A sequestered germ line or Weismann barrier is specific to animals, and epigenetic inheritance is more common in plants and microbes. Eva Jablonka, Marion J. Lamb and Étienne Danchin have argued that these effects may require enhancements to the standard conceptual framework of the modern synthesis and have called for an extended evolutionary synthesis. Other evolutionary biologists have incorporated epigenetic inheritance into population genetics models and are openly skeptical, stating that epigenetic mechanisms such as DNA methylation and histone modification are genetically inherited under the control of natural selection.

Two important ways in which epigenetic inheritance can be different from traditional genetic inheritance, with important consequences for evolution, are that rates of epimutation can be much faster than rates of mutation and the epimutations are more easily reversible. In plants heritable DNA methylation mutations are 100.000 times more likely to occur compared to DNA mutations.An epigenetically inherited element such as the PSI+ system can act as a "stop-gap", good enough for short-term adaptation that allows the lineage to survive for long enough for mutation and/or recombination to genetically assimilate the adaptive phenotypic change.The existence of this possibility increases the evolvability of a species.

More than 100 cases of transgenerational epigenetic inheritance phenomena have been reported in a wide range of organisms, including prokaryotes, plants, and animals. For instance, Mourning Cloak butterflies will change color through hormone changes in response to experimentation of varying temperatures.

The filamentous fungus Neurospora crassa is a prominent model system for understanding the control and function of cytosine methylation. In this organisms, DNA methylation is associated with relics of a genome defense system called RIP (repeat-induced point mutation) and silences gene expression by inhibiting transcription elongation.

The yeast prion PSI is generated by a conformational change of a translation termination factor, which is then inherited by daughter cells. This can provide a survival advantage under adverse conditions. This is an example of epigenetic regulation enabling unicellular organisms to respond rapidly to environmental stress. Prions can be viewed as epigenetic agents capable of inducing a phenotypic change without modification of the genome.

Direct detection of epigenetic marks in microorganisms is possible with single molecule real time sequencing, in which polymerase sensitivity allows for measuring methylation and other modifications as a DNA molecule is being sequenced. Several projects have demonstrated the ability to collect genome-wide epigenetic data in bacteria.

Epigenetics in Bacteria

While epigenetics is of fundamental importance in eukaryotes, especially metazoans, it plays a different role in bacteria. Most importantly, eukaryotes use epigenetic mechanisms primarily to regulate gene expression which bacteria rarely do. However, bacteria make widespread use of postreplicative DNA methylation for the epigenetic control of DNA-protein interactions. Bacteria also use DNA adenine methylation (rather than DNA cytosine methylation) as an epigenetic signal. DNA adenine methylation is important in bacteria virulence in organisms such as Escherichia coli, Salmonella, Vibrio, Yersinia, Haemophilus, and Brucella. In Alphaproteobacteria, methylation of adenine regulates the cell cycle and couples gene transcription to DNA replication. In Gammaproteobacteria, adenine methylation provides signals for DNA replication, chromosome segregation, mismatch repair, packaging of bacteriophage, transposase activity and regulation of gene expression. There exists a genetic switch controlling Streptococcus pneumoniae (the pneumococcus) that allows the bacterium to randomly change its characteristics into six alternative states that could pave the way to improved vaccines. Each form is randomly generated by a phase variable methylation system. The ability of the pneumococcus to cause deadly infections is different in each of these six states. Similar systems exist in other bacterial genera.

Medicine

Epigenetics has many and varied potential medical applications. In 2008, the National Institutes of Health announced that $190 million had been earmarked for epigenetics research over the next five years. In announcing the funding, government officials noted that epigenetics has the potential to explain mechanisms of aging, human development, and the origins of cancer, heart disease, mental illness, as well as several other conditions. Some investigators, like Randy Jirtle, PhD, of Duke University Medical Center, think epigenetics may ultimately turn out to have a greater role in disease than genetics.

Twins

Direct comparisons of identical twins constitute an optimal model for interrogating environmental epigenetics. In the case of humans with different environmental exposures, monozygotic (identical) twins were epigenetically indistinguishable during their early years, while older twins had remarkable differences in the overall content and genomic distribution of 5-methylcytosine DNA and histone acetylation. The twin pairs who had spent less of their lifetime together and/or had greater differences in their medical histories were those who showed the largest differences in their levels of 5-methylcytosine DNA and acetylation of histones H3 and H4.

Dizygotic (fraternal) and monozygotic (identical) twins show evidence of epigenetic influence in humans. DNA sequence differences that would be abundant in a singleton-based study do not interfere with the analysis. Environmental differences can produce long-term epigenetic effects, and different developmental monozygotic twin subtypes may be different with respect to their susceptibility to be discordant from an epigenetic point of view.

A high-throughput study, which denotes technology that looks at extensive genetic markers, focused on epigenetic differences between monozygotic twins to compare global and locus-specific changes in DNA methylation and histone modifications in a sample of 40 monozygotic twin pairs. In this case, only healthy twin pairs were studied, but a wide range of ages was represented, between 3 and 74 years. One of the major conclusions from this study was that there is an age-dependent accumulation of epigenetic differences between the two siblings of twin pairs. This accumulation suggests the existence of epigenetic "drift". Epigenetic drift is the term given to epigenetic modifications as they occur as a direct function with age. While age is a known risk factor for many diseases, age-related methylation has been found to occur differentially at specific sites along the genome. Over time, this can result in measurable differences between biological and chronological age. Epigenetic changes have been found to be reflective of lifestyle and may act as functional biomarkers of disease before clinical threshold is reached.

A more recent study, where 114 monozygotic twins and 80 dizygotic twins were analyzed for the DNA methylation status of around 6000 unique genomic regions, concluded that epigenetic similarity at the time of blastocyst splitting may also contribute to phenotypic similarities in monozygotic co-twins. This supports the notion that microenvironment at early stages of embryonic development can be quite important for the establishment of epigenetic marks. Congenital genetic disease is well understood and it is clear that epigenetics can play a role, for example, in the case of Angelman syndrome and Prader-Willi syndrome. These are normal genetic diseases caused by gene deletions or inactivation of the genes, but are unusually common because individuals are essentially hemizygous because of genomic imprinting, and therefore a single gene knock out is sufficient to cause the disease, where most cases would require both copies to be knocked out.

Genomic Imprinting

Some human disorders are associated with genomic imprinting, a phenomenon in mammals where the father and mother contribute different epigenetic patterns for specific genomic loci in their germ cells. The best-known case of imprinting in human disorders is that of Angelman syndrome and Prader-Willi syndrome—both can be produced by the same genetic mutation, chromosome 15q partial deletion, and the particular syndrome that will develop depends on whether the mutation is inherited from the child's mother or from their father. This is due to the presence of genomic imprinting in the region. Beckwith-Wiedemann syndrome is also associated with genomic imprinting, often caused by abnormalities in maternal genomic imprinting of a region on chromosome 11.

Rett syndrome is underlain by mutations in the MECP2 gene despite no large-scale changes in expression of MeCP2 being found in microarray analyses. BDNF is downregulated in the MECP2 mutant resulting in Rett syndrome.

In the Överkalix study, paternal (but not maternal) grandsons of Swedish men who were exposed during preadolescence to famine in the 19th century were less likely to die of cardiovascular disease. If food was plentiful, then diabetes mortality in the grandchildren increased, suggesting that this was a transgenerational epigenetic inheritance. The opposite effect was observed for females—the paternal (but not maternal) granddaughters of women who experienced famine while in the womb (and therefore while their eggs were being formed) lived shorter lives on average.

Cancer

A variety of epigenetic mechanisms can be perturbed in different types of cancer. Epigenetic alterations of DNA repair genes or cell cycle control genes are very frequent in sporadic (non-germ line) cancers, being significantly more common than germ line (familial) mutations in these sporadic cancers. Epigenetic alterations are important in cellular transformation to cancer, and their manipulation holds great promise for cancer prevention, detection, and therapy.Several medications which have epigenetic impact are used in several of these diseases. These aspects of epigenetics are addressed in cancer epigenetics.

Addiction

Addiction is a disorder of the brain's reward system which arises through transcriptional and neuroepigenetic mechanisms and occurs over time from chronically high levels of exposure to an addictive stimulus (e.g., morphine, cocaine, sexual intercourse, gambling, etc.). Transgenerational epigenetic inheritance of addictive phenotypes has been noted to occur in preclinical studies.

Anxiety

Transgenerational epigenetic inheritance of anxiety-related phenotypes has been reported in a preclinical study using mice. In this investigation, transmission of paternal stress-induced traits across generations involved small non-coding RNA signals transmitted via the male germline.

Depression

Epigenetic inheritance of depression-related phenotypes has also been reported in a preclinically. Inheritance of paternal stress-induced traits across generations involved small non-coding RNA signals transmitted via the paternal germline.

Fear Conditioning

Studies on mice have shown that certain conditional fears can be inherited from either parent. In one example, mice were conditioned to fear a strong scent, acetophenone, by accompanying the smell with an electric shock. Consequently, the mice learned to fear the scent of acetophenone alone. It was discovered that this fear could be passed down to the mice offspring. Despite the offspring never experiencing the electric shock themselves the mice still display a fear of the acetophenone scent, because they inherited the fear epigenetically by site-specific DNA methylation. These epigenetic changes lasted up to two generations without reintroducing the shock.

Research

The two forms of heritable information, namely genetic and epigenetic, are collectively denoted as dual inheritance. Members of the APOBEC/AID family of cytosine deaminases may concurrently influence genetic and epigenetic inheritance using similar molecular mechanisms, and may be a point of crosstalk between these conceptually compartmentalized processes.

Fluoroquinolone antibiotics induce epigenetic changes in mammalian cells through iron chelation. This leads to epigenetic effects through inhibition of α-ketoglutarate-dependent dioxygenases that require iron as a co-factor.

Various pharmacological agents are applied for the production of induced pluripotent stem cells (iPSC) or maintain the embryonic stem cell (ESC) phenotypic via epigenetic approach. Adult stem cells like bone marrow stem cells have also shown a potential to differentiate into cardiac competent cells when treated with G9a histone methyltransferase inhibitor BIX01294.

Caution

Due to the early stages of epigenetics as a science and to the sensationalism surrounding it, surgical oncologist David Gorski and geneticist Adam Rutherford caution against the drawing and proliferation of false and pseudoscientific conclusions from new age authors such as Deepak Chopra and Bruce Lipton.

In popular culture

In Neal Stephenson's 2015 novel Seveneves, survivors of a worldwide holocaust are tasked with seeding new life on a dormant Earth. Rather than create specific breeds of animals to be hunters, scavengers, or prey, species like "canids" are developed with mutable epigenetic traits, with the intention that the animals would quickly transform into the necessary roles that would be required for an ecosystem to rapidly evolve. Additionally, a race of humans, "Moirans," are created to survive in space, with the hope that this subspecies of human would be able to adapt to unforeseeable dangers and circumstances, via an epigenetic process called "going epi".

Study Methods

In general, most experiments investigating differential expression used whole cell extracts of RNA, called steady-state levels, to determine which genes changed and by how much they did. These are, however, not informative of where the regulation has occurred and may actually mask conflicting regulatory processes (see post-transcriptional regulation), but it is still the most commonly analysed (quantitative PCR and DNA microarray).

When studying gene expression, there are several methods to look at the various stages. In eukaryotes these include:

- The local chromatin environment of the region can be determined by ChIP-chip analysis by pulling down RNA Polymerase II, Histone 3 modifications, Trithorax-group protein, Polycomb-group protein, or any other DNA-binding element to which a good antibody is available.
- Epistatic interactions can be investigated by synthetic genetic array analysis
- Due to post-transcriptional regulation, transcription rates and total RNA levels differ significantly. To measure the transcription rates nuclear run-on assays can be done and newer high-throughput methods are being developed, using thiol labelling instead of radioactivity.
- Only 5% of the RNA polymerised in the nucleus actually exits, and not only introns, abortive products, and non-sense transcripts are degradated. Therefore, the differences in nuclear and cytoplasmic levels can be see by separating the two fractions by gentle lysis.
- Alternative splicing can be analysed with a splicing array or with a tiling array (see DNA microarray).
- All in vivo RNA is complexed as RNPs. The quantity of transcripts bound to

specific protein can be also analysed by RIP-Chip. For example, DCP2 will give an indication of sequestered protein; ribosome-bound gives and indication of transcripts active in transcription (although it should be noted that a more dated method, called polysome fractionation, is still popular in some labs)

- Protein levels can be analysed by Mass spectrometry, which can be compared only to quantitative PCR data, as microarray data is relative and not absolute.
- RNA and protein degradation rates are measured by means of transcription inhibitors (actinomycin D or α-amanitin) or translation inhibitors (Cycloheximide), respectively.

Genetic Methods

Experimental analysis of proteins typically requires expression and purification of proteins. Expression is achieved by manipulating DNA that encodes the protein(s) of interest. Hence, protein analysis usually requires DNA methods, especially cloning. Some examples of genetic methods include conceptual translation, Site-directed mutagenesis, using a fusion protein, and matching allele with disease states. Some proteins have never been directly sequenced, however by translating codons from known mRNA sequences into amino acids by a method known as conceptual translation. (See genetic code.) Site-directed mutagenesis selectively introduces mutations that change the structure of a protein. The function of parts of proteins can be better understood by studying the change in phenotype as a result of this change. Fusion proteins are made by inserting protein tags, such as the His-tag, to produce a modified protein that is easier to track. An example of this would be GFP-Snf2H which consists of a protein bound to a green fluorescent protein to form a hybrid protein. By analyzing DNA alleles can be identified as being associated with disease states, such as in calculation of LOD scores.

Protein Extraction from Tissues

Protein extraction from tissues with tough extracellular matrices (e.g., biopsy samples, venous tissues, cartilage, skin) is often achieved in a laboratory setting by impact pulverization in liquid nitrogen. Samples are frozen in liquid nitrogen and subsequently subjected to impact or mechanical grinding. As water in the samples becomes very brittle at these temperature, the samples are often reduced to a collection of fine fragments, which can then be dissolved for protein extraction. Stainless steel devices known as tissue pulverizers are sometimes used for this purpose.

Advantages of these devices include high levels of protein extraction from small, valuable samples, disadvantages include low-level cross-over contamination.

Protein Purification

- Protein isolation
    - Chromatography methods: ion exchange, size-exclusion chromatography (or gel filtration), affinity chromatography

- Protein extraction and solubilization
- Concentrating protein solutions
- Gel electrophoresis
    - Gel electrophoresis under denaturing conditions
    - Gel electrophoresis under non-denaturing conditions
    - 2D gel electrophoresis
- Electrofocusing

Detecting Proteins

Non-specific methods that detect total protein only

- Absorbance: Read at 280 or 205 nm. Can be very inaccurate. Detection in the range of 100 μg/mL to 1 mg/mL. Ratio of absorbance readings taken at 260/280 can indicate purity/contamination of the sample (pure samples have a ratio <0.8)
- Bradford protein assay: Detection in the range of ~1 mg/mL
- Biuret Test Derived Assays:
    - Bicinchoninic acid assay (BCA assay): Detection down to 0.5 μg/mL
    - Lowry Protein assay: Detection in the range of 0.01–1.0 mg/mL
- Fluorescamine: Quantifies proteins and peptides in solution if primary amine are present in the amino acids
- Amido black: Detection in the range of 1-12 μg/mL
- Colloidal gold: Detection in the range of 20 - 640 ng/mL
- Nitrogen detection:
    - Kjeldahl method: used primarily for food and requires oxidation of material
    - Dumas method: used primarily for food and requires combustion of material

Specific methods which can detect amount of a single protein

- Spectrometry methods:
    - High-performance liquid chromatography (HPLC): Chromatography method to detect proteins or peptides
    - Liquid chromatography–mass spectrometry (LC/MS): Can detect proteins at low concentrations (ng/mL to pg/mL) in blood and body fluids, such as for Pharmacokinetics.
- Antibody dependent methods:
    - Enzyme-linked immunosorbent assay (ELISA): Specifically can detect protein down to pg/mL.
    - Protein immunoprecipitation: technique of precipitating a protein antigen out of solution using an antibody that specifically binds to that particular

protein.

- Immunoelectrophoresis: separation and characterization of proteins based on electrophoresis and reaction with antibodies.
- Western blot: couples gel electrophoresis and incubation with antibodies to detect specific proteins in a sample of tissue homogenate or extract (a type of Immunoelectrophoresis technique).
- Protein immunostaining

POST-TRANSLATIONAL MODIFICATION

Post-translational modification of insulin. At the top, the ribosome translates a mRNA sequence into a protein, insulin, and passes the protein through the endoplasmic reticulum, where it is cut, folded and held in shape by disulfide (-S-S-) bonds. Then the protein passes through the golgi apparatus, where it is packaged into a vesicle. In the vesicle, more parts are cut off, and it turns into mature insulin.

Post-translational modifications can occur on the amino acid side chains or at the protein's C- or N- termini. They can extend the chemical repertoire of the 20 standard amino acids by introducing new functional groups such as phosphate, acetate, amide groups, or methyl groups. Phosphorylation is a very common mechanism for regulating the activity of enzymes and is the most common post-translational modification.Many eukaryotic proteins also have carbohydrate molecules attached to them in a process called glycosylation, which can promote protein folding and improve stability as well as serving regulatory functions. Attachment of lipid molecules, known as lipidation, often targets a protein or part of a protein attached to the cell membrane.

Other forms of post-translational modification consist of cleaving peptide bonds, as in processing a propeptide to a mature form or removing the initiator methionine residue. The formation of disulfide bonds from cysteine residues may also be referred to as a post-translational modification. For instance, the peptide hormone insulin is cut twice after disulfide bonds are formed, and a propeptide is removed from the middle of the chain; the resulting protein consists of two polypeptide chains connected by disulfide bonds.

Some types of post-translational modification are consequences of oxidative stress. Carbonylation is one example that targets the modified protein for degradation and can result in the formation of protein aggregates. Specific amino acid modifications can be used as biomarkers indicating oxidative damage.

Sites that often undergo post-translational modification are those that have a functional group that can serve as a nucleophile in the reaction: the hydroxyl groups of serine, threonine, and tyrosine; the amine forms of lysine, arginine, and histidine; the thiolate anion of cysteine; the carboxylates of aspartate and glutamate; and the N- and C-termini. In addition, although the amides of asparagine and glutamine are weak nucleophiles, both can serve as attachment points for glycans. Rarer modifications can occur at oxidized methionines and at some methylenes in side chains.

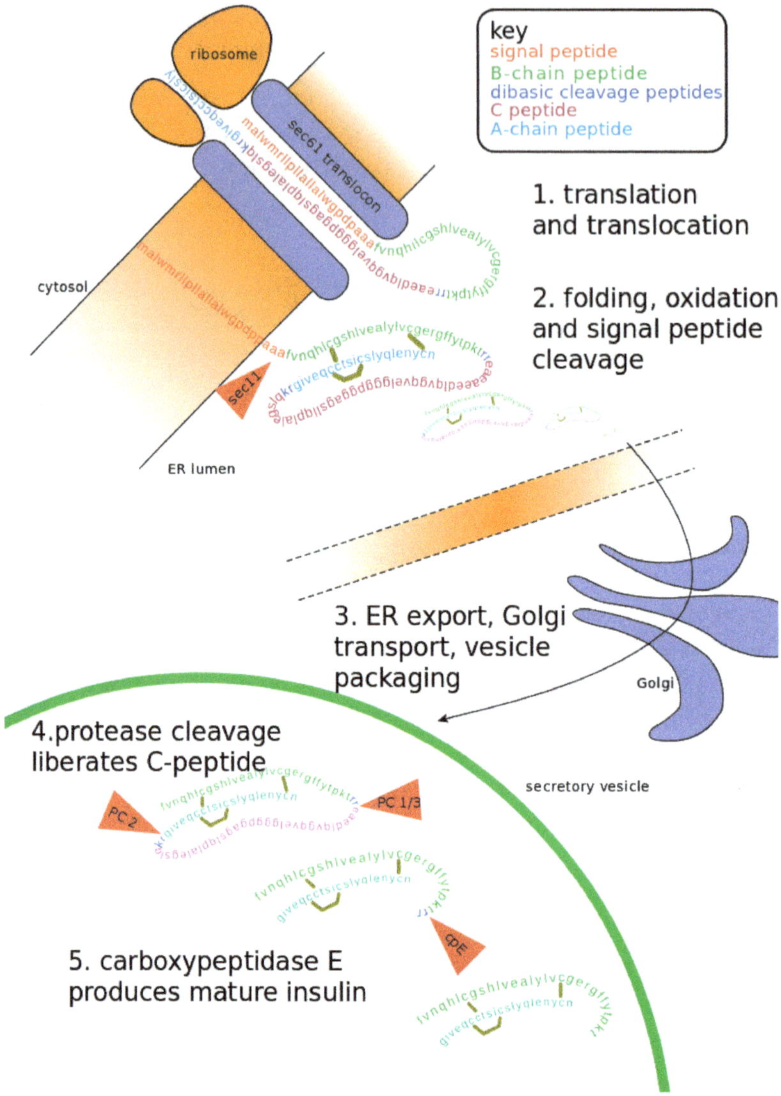

Post-translational modification (**PTM**) refers to the covalent and generally enzymatic modification of proteins during or after protein biosynthesis. Proteins are synthesized by ribosomes translating mRNA into polypeptide chains, which may then undergo PTM to form the mature protein product. PTMs are important components in cell signaling.

Post-translational modification of proteins can be experimentally detected by a variety of techniques, including mass spectrometry, Eastern blotting, and Western blotting.

PTMs Involving Addition of Functional Groups

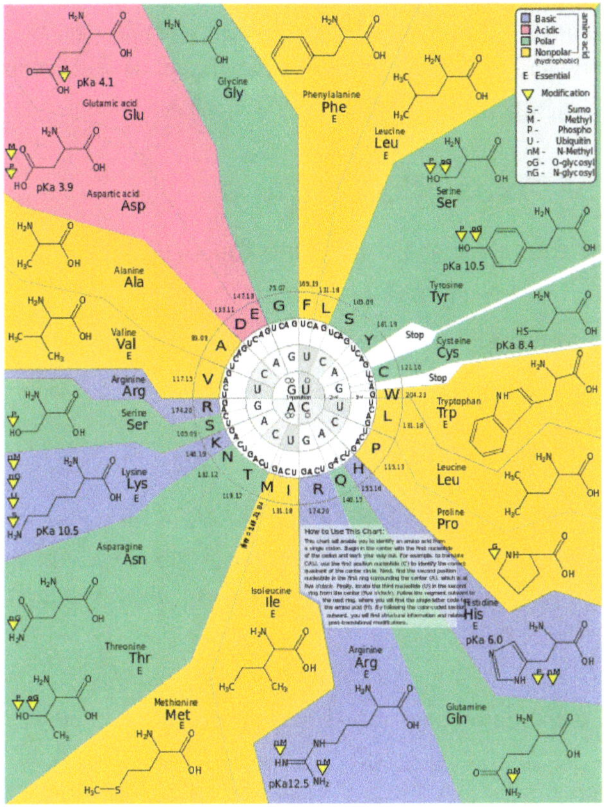

The genetic code diagram showing the amino acid residues as target of modification.

Addition by an enzyme in vivo

Hydrophobic Groups for Membrane Localization

- Myristoylation, attachment of myristate, a C_{14} saturated acid
- Palmitoylation, attachment of palmitate, a C_{16} saturated acid
- Isoprenylation or prenylation, the addition of an isoprenoid group (e.g. farnesol and geranylgeraniol)
    - Farnesylation
    - Geranylgeranylation
- Glypiation, glycosylphosphatidylinositol (GPI) anchor formation via an amide bond to C-terminal tail

Cofactors for Enhanced Enzymatic Activity

- Lipoylation, attachment of a lipoate (C_8) functional group

- Flavin moiety (FMN or FAD) may be covalently attached
- Heme C attachment via thioether bonds with cysteins
- Phosphopantetheinylation, the addition of a 4'-phosphopantetheinyl moiety from coenzyme A, as in fatty acid, polyketide, non-ribosomal peptide and leucine biosynthesis
- Retinylidene Schiff base formation

Modifications of Translation Factors

- Diphthamide formation (on a histidine found in eEF2)
- Ethanolamine phosphoglycerol attachment (on glutamate found in eEF1α)
- Hypusine formation (on conserved lysine of eIF5A (eukaryotic) and aIF5A (archaeal))

Smaller Chemical Groups

- Acylation, e.g. *O*-acylation (esters), *N*-acylation (amides), *S*-acylation (thioesters)
    - Acetylation, the addition of an acetyl group, either at the N-terminus of the protein or at lysine residues. *See also histone acetylation.* The reverse is called deacetylation.
    - Formylation
- Alkylation, the addition of an alkyl group, e.g. methyl, ethyl
    - Methylation the addition of a methyl group, usually at lysine or arginine residues. The reverse is called demethylation.
- Amide bond formation
    - Amidation at C-terminus
    - Amino acid addition
        - Arginylation, a tRNA-mediation addition
        - Polyglutamylation, covalent linkage of glutamic acid residues to the N-terminus of tubulin and some other proteins. (See tubulin polyglutamylase)
        - Polyglycylation, covalent linkage of one to more than 40 glycine residues to the tubulin C-terminal tail
- Butyrylation
- gamma-carboxylation dependent on Vitamin K
- glycosylation, the addition of a glycosyl group to either arginine, asparagine, cysteine, hydroxylysine, serine, threonine, tyrosine, or tryptophan resulting in a glycoprotein. Distinct from glycation, which is regarded as a nonenzymatic attachment of sugars.
    - polysialylation, addition of polysialic acid, PSA, to NCAM
- malonylation
- hydroxylation

- iodination (e.g. of thyroglobulin)
- nucleotide addition such as ADP-ribosylation
- oxidation
- phosphate ester (*O*-linked) or phosphoramidate (*N*-linked) formation
    - phosphorylation, the addition of a phosphate group, usually to serine, threonine, and tyrosine (*O*-linked), or histidine (*N*-linked)
    - adenylylation, the addition of an adenylyl moiety, usually to tyrosine (*O*-linked), or histidine and lysine (*N*-linked)
- propionylation
- pyroglutamate formation
- *S*-glutathionylation
- *S*-nitrosylation
- *S*-sulfenylation (*aka S*-sulphenylation), reversible covalent attachment of hydroxide to the thiol group of cysteine residues
- succinylation addition of a succinyl group to lysine
- sulfation, the addition of a sulfate group to a tyrosine.

Non-enzymatic additions in vivo

- glycation, the addition of a sugar molecule to a protein without the controlling action of an enzyme.
- carbamylation the addition of Isocyanic acid to an N-terminus of either lysine, histidine, taurine, arginine, or cysteine.
- carbonylation the addition of carbon monoxide to other organic/inorganic compounds.

Non-enzymatic Additions in Vitro

- biotinylation, acylation of conserved lysine residues with a biotin appendage
- pegylation

Other Proteins or Peptides

- ISGylation, the covalent linkage to the ISG15 protein (Interferon-Stimulated Gene 15)
- SUMOylation, the covalent linkage to the SUMO protein (Small Ubiquitin-related MOdifier)
- ubiquitination, the covalent linkage to the protein ubiquitin.
- Neddylation, the covalent linkage to Nedd
- Pupylation, the covalent linkage to the Prokaryotic ubiquitin-like protein

Chemical Modification of Amino Acids

- citrullination, or **deimination**, the conversion of arginine to citrulline
- deamidation, the conversion of glutamine to glutamic acid or asparagine to aspartic acid
- eliminylation, the conversion to an alkene by beta-elimination of phosphothreonine and phosphoserine, or dehydration of threonine and serine, as well as by decarboxylation of cysteine
- carbamylation of the N-terminus (an α-amine) and the side-chain amino groups of lysine and, usually to a lesser extent, arginine

Structural Changes

- disulfide bridges, the covalent linkage of two cysteine amino acids
- proteolytic cleavage, cleavage of a protein at a peptide bond
- racemization
    - of proline by prolyl isomerase
    - of serine by protein-serine epimerase
    - of alanine in dermorphin, a frog opioid peptide
    - of methionine in deltorphin, also a frog opioid peptide
- protein splicing, self-catalytic removal of inteins analogous to mRNA processing

Statistics

In 2011, statistics of each post-translational modification experimentally and putatively detected have been compiled using proteome-wide information from the Swiss-Prot database. The 10 most common experimentally found modifications were as follows:

Frequency	Modification
58383	Phosphorylation
6751	Acetylation
5526	N-linked glycosylation
2844	Amidation
1619	Hydroxylation
1523	Methylation
1133	O-linked glycosylation
878	Ubiquitylation
826	Pyrrolidone Carboxylic Acid
504	Sulfation

REFERENCES

1. Spector, Tim (2012). Identically Different: Why You Can Change Your Genes. London: Weidenfeld & Nicolson. p. 8.
2. Carey N. (2011): Epigenetics revolution: How modern biology is rewriting our understanding of genetics, disease and inheritance. Icon Books, London, ISBN 978-1-84831-315-6; ISBN 978-1-84831-316-3.
3. Moore, David S. (2015). The Developing Genome: An Introduction to Behavioral Epigenetics (1st ed.). Oxford University Press. ISBN 978-0199922345.
4. Epigenetics. Icahn School of Medicine at Mount Sinai.
5. Ledford H (2008). Disputed definitions. Nature. 455 (7216): 1023–8. doi:10.1038/4551023a. PMID 18948925.
6. Bird A (May 2007). Perceptions of epigenetics. Nature. 447 (7143): 396–8. Bibcode:2007Natur.447..396B. doi:10.1038/nature05913. PMID 17522671.
7. Reik W (May 2007). Stability and flexibility of epigenetic gene regulation in mammalian development. Nature. 447 (7143): 425–32. Bibcode:2007Natur.447..425R. doi:10.1038/nature05918. PMID 17522676.
8. C. H. Waddington (1953). The Epigenetics of Birds. Cambridge University Press. pp. 1–. ISBN 978-1-107-44047-0. (2014 edition)
9. Hall BK (15 January 2004). In search of evolutionary developmental mechanisms: the 30-year gap between 1944 and 1974. 302. 302 (1): 5–18. doi:10.1002/jez.b.20002. PMID 14760651.
10. Alvarez-Buylla ER, Chaos A, Aldana M, Benítez M, Cortes-Poza Y, Espinosa-Soto C, Hartasánchez DA, Lotto RB, Malkin D, Escalera Santos GJ, Padilla-Longoria P (November 3, 2008). Floral Morphogenesis: Stochastic Explorations of a Gene Network Epigenetic Landscape. PLoS ONE. 3 (11): e3626. Bibcode:2008PLoSO...3.3626A. doi:10.1371/journal.pone.0003626. PMC 2572848. PMID 18978941.
11. Rabajante JF, Babierra AL (January 30, 2015). Branching and oscillations in the epigenetic landscape of cell-fate determination. Progress in Biophysics and Molecular Biology. 117 (2–3): 240–9. doi:10.1016/j.pbiomolbio.2015.01.006. PMID 25641423.
12. Gottlieb G (1991). Epigenetic systems view of human development. Developmental Psychology. 27 (1): 33–34. doi:10.1037/0012-1649.27.1.33.
13. Gilbert Gottlieb. Probabilistic epigenesis, Developmental Science 10:1 (2007), 1–11
14. Boeree, C. George, (1997/2006), Personality Theories, Erik Erikson
15. Erikson, Erik (1968). Identity: Youth and Crisis. Chapter 3: W.W. Norton and Company, Inc. p. 92.
16. Holliday R (Jan 30, 1990). DNA Methylation and Epigenetic Inheritance. Philosophical Transactions of the Royal Society of London. Series B, Biological Sciences. 326 (1235): 329–338. Bibcode:1990RSPTB.326..329H. doi:10.1098/

rstb.1990.0015.

17. Riggs AD, Russo VE, Martienssen RA (1996). Epigenetic mechanisms of gene regulation. Plainview, N.Y.: Cold Spring Harbor Laboratory Press. ISBN 0-87969-490-4.
18. Beware the pseudo gene genies Beware the pseudo gene genies The Guardian
19. Berger SL, Kouzarides T, Shiekhattar R, Shilatifard A (2009). An operational definition of epigenetics. Genes Dev. 23 (7): 781–3. doi:10.1101/gad.1787609. PMC 3959995. PMID 19339683.
20. Chandler VL (February 2007). Paramutation: from maize to mice. Cell. 128 (4): 641–5. doi:10.1016/j.cell.2007.02.007. PMID 17320501.
21. Kovalchuk O, Baulch JE (January 2008). Epigenetic changes and nontargeted radiation effects—is there a link?. Environ. Mol. Mutagen. 49 (1): 16–25. doi:10.1002/em.20361. PMID 18172877.
22. Ilnytskyy Y, Kovalchuk O (September 2011). Non-targeted radiation effects-an epigenetic connection. Mutat. Res. 714 (1–2): 113–25. doi:10.1016/j.mrfmmm.2011.06.014. PMID 21784089.
23. Friedl AA, Mazurek B, Seiler DM (2012). Radiation-induced alterations in histone modification patterns and their potential impact on short-term radiation effects. Front Oncol. 2: 117. doi:10.3389/fonc.2012.00117. PMC 3445916. PMID 23050241.
24. Cuozzo C, Porcellini A, Angrisano T, Morano A, Lee B, Di Pardo A, Messina S, Iuliano R, Fusco A, Santillo MR, Muller MT, Chiariotti L, Gottesman ME, Avvedimento EV (July 2007). DNA damage, homology-directed repair, and DNA methylation. PLoS Genet. 3 (7): e110. doi:10.1371/journal.pgen.0030110. PMC 1913100. PMID 17616978.
25. O'Hagan HM, Mohammad HP, Baylin SB (2008). Lee JT, ed. Double strand breaks can initiate gene silencing and SIRT1-dependent onset of DNA methylation in an exogenous promoter CpG island. PLoS Genet. 4 (8): e1000155. doi:10.1371/journal.pgen.1000155. PMC 2491723. PMID 18704159.
26. Malanga M, Althaus FR (2005). The role of poly(ADP-ribose) in the DNA damage signaling network. Biochem Cell Biol. 83 (3): 354-364. doi:10.1139/o05-038. PMID 15959561.
27. Gottschalk AJ, Timinszky G, Kong SE, Jin J, Cai Y, Swanson SK, Washburn MP, Florens L, Ladurner AG, Conaway JW, Conaway RC (August 2009). Poly(ADP-ribosyl)ation directs recruitment and activation of an ATP-dependent chromatin remodeler. Proc. Natl. Acad. Sci. U.S.A. 106 (33): 13770–4. Bibcode:2009PNAS..10613770G. doi:10.1073/pnas.0906920106. PMC 2722505. PMID 19666485.
28. Lin JC, Jeong S, Liang G, Takai D, Fatemi M, Tsai YC, Egger G, Gal-Yam EN, Jones PA (November 2007). Role of nucleosomal occupancy in the epigenetic silencing of the MLH1 CpG island. Cancer Cell. 12 (5): 432–44. doi:10.1016/j.ccr.2007.10.014. PMC 4657456. PMID 17996647.

29. Tabish AM, Poels K, Hoet P, Godderis L (2012). Chiariotti L, ed. Epigenetic factors in cancer risk: effect of chemical carcinogens on global DNA methylation pattern in human TK6 cells. PLoS ONE. 7 (4): e34674. doi:10.1371/journal.pone.0034674. PMC 3324488. PMID 22509344.

30. Burdge GC, Hoile SP, Uller T, Thomas NA, Gluckman PD, Hanson MA, Lillycrop KA (2011). Imhof A, ed. Progressive, transgenerational changes in offspring phenotype and epigenotype following nutritional transition. PLoS ONE. 6 (11): e28282. doi:10.1371/journal.pone.0028282. PMC 3227644. PMID 22140567.

31. Fang M, Chen D, Yang CS (January 2007). Dietary polyphenols may affect DNA methylation. J. Nutr. 137 (1 Suppl): 223S–228S. PMID 17182830.

32. Olaharski AJ, Rine J, Marshall BL, Babiarz J, Zhang L, Verdin E, Smith MT (December 2005). The flavoring agent dihydrocoumarin reverses epigenetic silencing and inhibits sirtuin deacetylases. PLoS Genet. 1 (6): e77. doi:10.1371/journal.pgen.0010077. PMC 1315280. PMID 16362078.

33. Kikuno N, Shiina H, Urakami S, Kawamoto K, Hirata H, Tanaka Y, Majid S, Igawa M, Dahiya R (August 2008). Genistein mediated histone acetylation and demethylation activates tumor suppressor genes in prostate cancer cells. Int. J. Cancer. 123 (3): 552–60. doi:10.1002/ijc.23590. PMID 18431742.

34. Davis JN, Kucuk O, Djuric Z, Sarkar FH (June 2001). Soy isoflavone supplementation in healthy men prevents NF-kappa B activation by TNF-alpha in blood lymphocytes. Free Radic. Biol. Med. 30 (11): 1293–302. doi:10.1016/S0891-5849(01)00535-4. PMID 11368927.

35. Djuric Z, Chen G, Doerge DR, Heilbrun LK, Kucuk O (October 2001). Effect of soy isoflavone supplementation on markers of oxidative stress in men and women. Cancer Lett. 172 (1): 1–6. doi:10.1016/S0304-3835(01)00627-9. PMID 11595123.

36. Kropat C, Mueller D, Boettler U, Zimmermann K, Heiss EH, Dirsch VM, Rogoll D, Melcher R, Richling E, Marko D (March 2013). Modulation of Nrf2-dependent gene transcription by bilberry anthocyanins in vivo. Mol Nutr Food Res. 57 (3): 545–50. doi:10.1002/mnfr.201200504. PMID 23349102.

37. Baron R, Vellore NA (2012). LSD1/CoREST is an allosteric nanoscale clamp regulated by H3-histone-tail molecular recognition. Proc Natl Acad Sci U S A. 109 (31): 12509–14. doi:10.1073/pnas.1207892109. PMC 3411975. PMID 22802671.

38. Jablonka E, Lamb MJ, Lachmann M (September 1992). Evidence, mechanisms and models for the inheritance of acquired characteristics. J. Theor. Biol. 158 (2): 245–268. doi:10.1016/S0022-5193(05)80722-2.

39. Ptashne M (April 2007). On the use of the word ‹epigenetic’. Curr. Biol. 17 (7): R233–6. doi:10.1016/j.cub.2007.02.030. PMID 17407749.

40. Elhamamsy, Amr Rafat (2016). DNA methylation dynamics in plants and mammals: overview of regulation and dysregulation. Cell Biochemistry and Function. 34 (5): 289–98. doi:10.1002/cbf.3183. PMID 27003927.

41. Jenuwein T, Laible G, Dorn R, Reuter G (January 1998). SET domain proteins modulate chromatin domains in eu- and heterochromatin. Cell. Mol. Life Sci. 54 (1): 80–93. doi:10.1007/s000180050127. PMID 9487389.

42. Slotkin RK, Martienssen R (April 2007). Transposable elements and the epigenetic regulation of the genome. Nature Reviews Genetics. 8 (4): 272–85. doi:10.1038/nrg2072. PMID 17363976.

43. Li E, Bestor TH, Jaenisch R (June 1992). Targeted mutation of the DNA methyltransferase gene results in embryonic lethality. Cell. 69 (6): 915–26. doi:10.1016/0092-8674(92)90611-F. PMID 1606615.

44. Robertson KD, Uzvolgyi E, Liang G, Talmadge C, Sumegi J, Gonzales FA, Jones PA (June 1999). The human DNA methyltransferases (DNMTs) 1, 3a and 3b: coordinate mRNA expression in normal tissues and overexpression in tumors. Nucleic Acids Res. 27 (11): 2291–8. doi:10.1093/nar/27.11.2291. PMC 148793. PMID 10325416.

45. Leonhardt H, Page AW, Weier HU, Bestor TH (November 1992). A targeting sequence directs DNA methyltransferase to sites of DNA replication in mammalian nuclei. Cell. 71 (5): 865–73. doi:10.1016/0092-8674(92)90561-P. PMID 1423634.

46. Chuang LS, Ian HI, Koh TW, Ng HH, Xu G, Li BF (September 1997). Human DNA-(cytosine-5) methyltransferase-PCNA complex as a target for p21WAF1. Science. 277 (5334): 1996–2000. doi:10.1126/science.277.5334.1996. PMID 9302295.

47. Robertson KD, Wolffe AP (October 2000). DNA methylation in health and disease. Nature Reviews Genetics. 1 (1): 11–9. doi:10.1038/35049533. PMID 11262868.

48. Li E, Beard C, Jaenisch R (November 1993). Role for DNA methylation in genomic imprinting. Nature. 366 (6453): 362–5. doi:10.1038/366362a0. PMID 8247133.

49. Viens A, Mechold U, Brouillard F, Gilbert C, Leclerc P, Ogryzko V (July 2006). Analysis of human histone H2AZ deposition in vivo argues against its direct role in epigenetic templating mechanisms. Mol. Cell. Biol. 26 (14): 5325–35. doi:10.1128/MCB.00584-06. PMC 1592707. PMID 16809769.

50. Ogryzko VV (2008). Erwin Schroedinger, Francis Crick and epigenetic stability. Biol. Direct. 3: 15. doi:10.1186/1745-6150-3-15. PMC 2413215. PMID 18419815.

51. Nottke A, Colaiácovo MP, Shi Y (March 2009). Developmental roles of the histone lysine demethylases. Development. 136 (6): 879–89. doi:10.1242/dev.020966. PMC 2692332. PMID 19234061.

52. Rosenfeld JA, Wang Z, Schones DE, Zhao K, DeSalle R, Zhang MQ (2009). Determination of enriched histone modifications in non-genic portions of the human genome. BMC Genomics. 10: 143. doi:10.1186/1471-2164-10-143. PMC 2667539. PMID 19335899.

53. Sneppen K, Micheelsen MA, Dodd IB (April 15, 2008). Ultrasensitive gene regulation by positive feedback loops in nucleosome modification. Molecular

systems biology. 4 (1): 182. doi:10.1038/msb.2008.21. PMC 2387233. PMID 18414483.

54. Dodd IB, Micheelsen MA, Sneppen K, Thon G (May 2007). Theoretical analysis of epigenetic cell memory by nucleosome modification. Cell. 129 (4): 813–22. doi:10.1016/j.cell.2007.02.053. PMID 17512413.

55. Morris KL (2008). Epigenetic Regulation of Gene Expression. RNA and the Regulation of Gene Expression: A Hidden Layer of Complexity. Norfolk, England: Caister Academic Press. ISBN 1-904455-25-5.

56. Mattick JS, Amaral PP, Dinger ME, Mercer TR, Mehler MF (January 2009). RNA regulation of epigenetic processes. BioEssays. 31 (1): 51–9. doi:10.1002/bies.080099. PMID 19154003.

57. Bernal JE, Duran C, Papiha SS (2012). Transcriptional and epigenetic regulation of human microRNAs. Cancer Lett. 331 (1): 1–10. doi:10.1016/j.canlet.2012.12.006. PMID 23246373.

58. Lim LP, Lau NC, Garrett-Engele P, Grimson A, Schelter JM, Castle J, Bartel DP, Linsley PS, Johnson JM (2005). Microarray analysis shows that some microRNAs downregulate large numbers of target mRNAs. Nature. 433 (7027): 769–773. doi:10.1038/nature03315. PMID 15685193.

59. Lee D, Shin C (2012). MicroRNA-target interactions: new insights from genome-wide approaches. Ann N Y Acad Sci. 1271: 118–28. doi:10.1111/j.1749-6632.2012.06745.x. PMID 23050973.

60. Friedman RC, Farh KK, Burge CB, Bartel DP (2009). Most mammalian mRNAs are conserved targets of microRNAs. Genome Res. 19 (1): 92–105. doi:10.1101/gr.082701.108. PMC 2612969. PMID 18955434.

61. Goll MG, Bestor TH (2005). Eukaryotic cytosine methyltransferases. Annu Rev Biochem. 74: 481–514. doi:10.1146/annurev.biochem.74.010904.153721. PMID 15952895.

62. Guifang Jia; Ye Fu; Xu Zhao; Qing Dai; Guanqun Zheng; Ying Yang; Chengqi Yi; Lindahl, Tomas; Tao Pan; Yun-Gui Yang; Chuan He (16 October 2011). N6-Methyladenosine in nuclear RNA is a major substrate of the obesity-associated FTO. Nature Chemical Biology. 7 (12): 885–887. doi:10.1038/nchembio.687. PMC 3218240. PMID 22002720.

63. Howden BP, Beaume M, Harrison PF, Hernandez D, Schrenzel J, Seemann T, Francois P, Stinear TP (August 2013). Analysis of the Small RNA Transcriptional Response in Multidrug-Resistant Staphylococcus aureus after Antimicrobial Exposure. Antimicrob. Agents Chemother. 57 (8): 3864–74. doi:10.1128/AAC.00263-13. PMC 3719707. PMID 23733475.

64. Yool A, Edmunds WJ (1998). Epigenetic inheritance and prions. Journal of Evolutionary Biology. 11 (2): 241–242. doi:10.1007/s000360050085.

65. Cox BS (1965). [PSI], a cytoplasmic suppressor of super-suppression in yeast. Heredity. 20 (4): 505–521. doi:10.1038/hdy.1965.65.

66. Lacroute F (May 1971). Non-Mendelian mutation allowing ureidosuccinic acid uptake in yeast. J. Bacteriol. 106 (2): 519–22. PMC 285125. PMID 5573734.

67. Liebman SW, Sherman F (September 1979). Extrachromosomal psi+ determinant suppresses nonsense mutations in yeast. J. Bacteriol. 139 (3): 1068–71. PMC 218059. PMID 225301.

68. True HL, Lindquist SL (September 2000). A yeast prion provides a mechanism for genetic variation and phenotypic diversity. Nature. 407 (6803): 477–83. doi:10.1038/35035005. PMID 11028992.

69. Shorter J, Lindquist S (June 2005). Prions as adaptive conduits of memory and inheritance. Nature Reviews Genetics. 6 (6): 435–50. doi:10.1038/nrg1616. PMID 15931169.

70. Giacomelli MG, Hancock AS, Masel J (2007). The conversion of 3′ UTRs into coding regions. Molecular Biology & Evolution. 24 (2): 457–464. doi:10.1093/molbev/msl172. PMC 1808353. PMID 17099057.

71. Lancaster AK, Bardill JP, True HL, Masel J (2010). The Spontaneous Appearance Rate of the Yeast Prion PSI+ and Its Implications for the Evolution of the Evolvability Properties of the PSI+ System. Genetics. 184 (2): 393–400. doi:10.1534/genetics.109.110213. PMC 2828720. PMID 19917766.

72. Sapp J (1991). Concepts of organization. The leverage of ciliate protozoa. Dev. Biol. (NY). 7: 229–58. doi:10.1007/978-1-4615-6823-0_11. PMID 1804215.

73. Sapp J (2003). Genesis: the evolution of biology. Oxford [Oxfordshire]: Oxford University Press. ISBN 0-19-515619-6.

74. Gray RD, Oyama S, Griffiths PE (2003). Cycles of Contingency: Developmental Systems and Evolution (Life and Mind: Philosophical Issues in Biology and Psychology). Cambridge, Mass: The MIT Press. ISBN 0-262-65063-0.

75. Teif VB, Beshnova DA, Vainshtein Y, Marth C, Mallm JP, Höfer T, Rippe K (8 May 2014). Nucleosome repositioning links DNA (de)methylation and differential CTCF binding during stem cell development. Genome Research. 24: 1285–1295. doi:10.1101/gr.164418.113. PMID 24812327.

76. Griesemer J, Haber MH, Yamashita G, Gannett L (March 2005). Critical Notice: Cycles of Contingency – Developmental Systems and Evolution. Biology & Philosophy. 20 (2–3): 517–544. doi:10.1007/s10539-004-0836-4.

77. Costa S, Shaw P (March 2007). 'Open minded' cells: how cells can change fate. Trends Cell Biol. 17 (3): 101–6. doi:10.1016/j.tcb.2006.12.005. PMID 17194589.

78. Cooney CA, Dave AA, Wolff GL (August 2002). Maternal methyl supplements in mice affect epigenetic variation and DNA methylation of offspring. J. Nutr. 132 (8 Suppl): 2393S–2400S. PMID 12163699.

79. Waterland RA, Jirtle RL (August 2003). Transposable elements: targets for early nutritional effects on epigenetic gene regulation. Mol. Cell. Biol. 23 (15): 5293–300. doi:10.1128/MCB.23.15.5293-5300.2003. PMC 165709. PMID 12861015.

80. Dias BG, Ressler KJ (2014). Parental olfactory experience influences behavior

and neural structure in subsequent generations. Nat. Neurosci. 17: 89–96. doi:10.1038/nn.3594. PMC 3923835. PMID 24292232.

81. Lamb MJ, Jablonka E (2005). Evolution in four dimensions: genetic, epigenetic, behavioral, and symbolic variation in the history of life. Cambridge, Mass: MIT Press. ISBN 0-262-10107-6.

82. Danchin É, Charmantier A, Champagne FA, Mesoudi A, Pujol B, Blanchet S (2011). Beyond DNA: integrating inclusive inheritance into an extended theory of evolution. Nature Reviews Genetics. 12: 475–486. doi:10.1038/nrg3028. PMID 21681209.

83. Maynard Smith J (1990). Models of a Dual Inheritance System. Journal of Theoretical Biology. 143 (1): 41–53. doi:10.1016/S0022-5193(05)80287-5. PMID 2359317.

84. Lynch M (2007). The frailty of adaptive hypotheses for the origins of organismal complexity. PNAS. 104 (suppl. 1): 8597–8604. doi:10.1073/pnas.0702207104. PMC 1876435. PMID 17494740.

85. Dickins, Thomas E.; Rahman, Qazi (2012). The extended evolutionary synthesis and the role of soft inheritance in evolution. Proceedings of the Royal Society B: Biological Sciences. 279 (1740): 2913–21. doi:10.1098/rspb.2012.0273. PMC 3385474. PMID 22593110.

86. Rando OJ, Verstrepen KJ (February 2007). Timescales of genetic and epigenetic inheritance. Cell. 128 (4): 655–68. doi:10.1016/j.cell.2007.01.023. PMID 17320504.

87. Lancaster AK, Masel J (1 September 2009). The evolution of reversible switches in the presence of irreversible mimics. Evolution. 63 (9): 2350–2362. doi:10.1111/j.1558-5646.2009.00729.x. PMC 2770902. PMID 19486147.

88. van der Graaf, Adriaan; Wardenaar, René; Neumann, Drexel A.; Taudt, Aaron; Shaw, Ruth G.; Jansen, Ritsert C.; Schmitz, Robert J.; Colomé-Tatché, Maria; Johannes, Frank (2015). Rate, spectrum, and evolutionary dynamics of spontaneous epimutations. Proceedings of the National Academy of Sciences. 112 (21): 6676–81. doi:10.1073/pnas.1424254112. PMC 4450394. PMID 25964364.

89. Úbeda, Francisco; Griswold, Cortland K.; Masel, Joanna (2009). Complex Adaptations Can Drive the Evolution of the Capacitor [PSI+], Even with Realistic Rates of Yeast Sex. PLoS Genetics. 5 (6): e1000517. doi:10.1371/journal.pgen.1000517. PMC 2686163. PMID 19521499.

90. Jablonka E, Raz G (June 2009). Transgenerational epigenetic inheritance: prevalence, mechanisms, and implications for the study of heredity and evolution. Q Rev Biol. 84 (2): 131–76. doi:10.1086/598822. PMID 19606595.

91. Davies, Hazel (2008). Do Butterflies Bite?: Fascinating Answers to Questions about Butterflies and Moths (Animals Q&A). Rutgers University Press.

92. Lewis ZA, Honda S, Khlafallah TK, Jeffress JK, Freitag M, Mohn F, Schübeler D, Selker EU (March 2009). Relics of repeat-induced point mutation direct heterochromatin formation in Neurospora crassa. Genome Res. 19 (3): 427–37. doi:10.1101/gr.086231.108. PMC 2661801. PMID 19092133.

93. Jorg Tost (2008). Epigenetics. Norfolk, England: Caister Academic Press. ISBN 1-904455-23-9.

94. Schadt EE, Banerjee O, Fang G, Feng Z, Wong WH, Zhang X, Kislyuk A, Clark TA, Luong K, Keren-Paz A, Chess A, Kumar V, Chen-Plotkin A, Sondheimer N, Korlach J, Kasarskis A (2012). Modeling kinetic rate variation in third generation DNA sequencing data to detect putative modifications to DNA bases. Genome Research. 23 (1): 129–41. doi:10.1101/gr.136739.111. PMC 3530673. PMID 23093720.

95. Davis BM, Chao MC, Waldor MK (2013). Entering the era of bacterial epigenomics with single molecule real time DNA sequencing. Current Opinion in Microbiology. 16 (2): 192–8. doi:10.1016/j.mib.2013.01.011. PMC 3646917. PMID 23434113.

96. Lluch-Senar M, Luong K, Lloréns-Rico V, Delgado J, Fang G, Spittle K, Clark TA, Schadt E, Turner SW, Korlach J, Serrano L (2013). Richardson PM, ed. Comprehensive Methylome Characterization of Mycoplasma genitalium and Mycoplasma pneumoniae at Single-Base Resolution. PLoS Genetics. 9 (1): e1003191. doi:10.1371/journal.pgen.1003191. PMC 3536716. PMID 23300489.

97. Murray IA, Clark TA, Morgan RD, Boitano M, Anton BP, Luong K, Fomenkov A, Turner SW, Korlach J, Roberts RJ (2012). The methylomes of six bacteria. Nucleic Acids Research. 40 (22): 11450–62. doi:10.1093/nar/gks891. PMC 3526280. PMID 23034806.

98. Fang G, Munera D, Friedman DI, Mandlik A, Chao MC, Banerjee O, Feng Z, Losic B, Mahajan MC, Jabado OJ, Deikus G, Clark TA, Luong K, Murray IA, Davis BM, Keren-Paz A, Chess A, Roberts RJ, Korlach J, Turner SW, Kumar V, Waldor MK, Schadt EE (2012). Genome-wide mapping of methylated adenine residues in pathogenic Escherichia coli using single-molecule real-time sequencing. Nature Biotechnology. 30 (12): 1232–9. doi:10.1038/nbt.2432. PMID 23138224.

99. Casadesús J, Low D (September 2006). Epigenetic gene regulation in the bacterial world. Microbiol. Mol. Biol. Rev. 70 (3): 830–56. doi:10.1128/MMBR.00016-06. PMC 1594586. PMID 16959970.

100. Manso AS, Chai MH, Atack JM, Furi L, De Ste Croix M, Haigh R, Trappetti C, Ogunniyi AD, Shewell LK, Boitano M, Clark TA, Korlach J, Blades M, Mirkes E, Gorban AN, Paton JC, Jennings MP, Oggioni MR (September 2014). A random six-phase switch regulates pneumococcal virulence via global epigenetic changes. Nature Communications. 5: 5055. doi:10.1038/ncomms6055. PMC 4190663. PMID 25268848.

101. Chahwan R, Wontakal SN, Roa S (March 2011). The multidimensional nature of epigenetic information and its role in disease. Discov Med. 11 (58): 233–43. PMID 21447282.

102. Fraga MF, Ballestar E, Paz MF, Ropero S, Setien F, Ballestar ML, Heine-Suñer D, Cigudosa JC, Urioste M, Benitez J, Boix-Chornet M, Sanchez-Aguilera A, Ling C, Carlsson E, Poulsen P, Vaag A, Stephan Z, Spector TD, Wu YZ, Plass

C, Esteller M (July 2005). Epigenetic differences arise during the lifetime of monozygotic twins. Proc. Natl. Acad. Sci. U.S.A. 102 (30): 10604–9. doi:10.1073/pnas.0500398102. PMC 1174919. PMID 16009939.

103. Kaminsky ZA, Tang T, Wang SC, Ptak C, Oh GH, Wong AH, Feldcamp LA, Virtanen C, Halfvarson J, Tysk C, McRae AF, Visscher PM, Montgomery GW, Gottesman II, Martin NG, Petronis A (February 2009). DNA methylation profiles in monozygotic and dizygotic twins. Nat. Genet. 41 (2): 240–5. doi:10.1038/ng.286. PMID 19151718.

104. O'Connor, Anahad (11 March 2008). The Claim: Identical Twins Have Identical DNA. New York Times.

105. Ballestar E (2010). Epigenetics lessons from twins: prospects for autoimmune disease. Clin Rev Allergy Immunol. 39 (1): 30–41. doi:10.1007/s12016-009-8168-4. PMID 19653134.

106. Wallace, Robert G.; Twomey, Laura C.; Custaud, Marc-Antoine; Moyna, Niall; Cummins, Philip M.; Mangone, Marco; Murphy, Ronan P. (2016). Potential Diagnostic and Prognostic Biomarkers of Epigenetic Drift within the Cardiovascular Compartment. BioMed Research International. 2016: 2465763. doi:10.1155/2016/2465763. PMC 4749768. PMID 26942189.

107. Kaminsky ZA, Tang T, Wang SC, Ptak C, Oh GH, Wong AH, Feldcamp LA, Virtanen C, Halfvarson J, Tysk C, McRae AF, Visscher PM, Montgomery GW, Gottesman II, Martin NG, Petronis A (2009). DNA methylation profiles in monozygotic and dizygotic twins. Nat Genet. 41 (2): 240–245. doi:10.1038/ng.286. PMID 19151718.

108. Wood AJ, Oakey RJ (November 2006). Genomic imprinting in mammals: emerging themes and established theories. PLoS Genet. 2 (11): e147. doi:10.1371/journal.pgen.0020147. PMC 1657038. PMID 17121465.

109. Knoll JH, Nicholls RD, Magenis RE, Graham JM, Lalande M, Latt SA (February 1989). Angelman and Prader-Willi syndromes share a common chromosome 15 deletion but differ in parental origin of the deletion. Am. J. Med. Genet. 32 (2): 285–90. doi:10.1002/ajmg.1320320235. PMID 2564739.

110. Pembrey ME, Bygren LO, Kaati G, Edvinsson S, Northstone K, Sjöström M, Golding J (February 2006). Sex-specific, male-line transgenerational responses in humans. Eur. J. Hum. Genet. 14 (2): 159–66. doi:10.1038/sj.ejhg.5201538. PMID 16391557.

111. NOVA | Transcripts | Ghost in Your Genes. PBS. 16 October 2007.

112. Wood LD, Parsons DW, Jones S, Lin J, Sjöblom T, Leary RJ, Shen D, Boca SM, Barber T, Ptak J, Silliman N, Szabo S, Dezso Z, Ustyanksky V, Nikolskaya T, Nikolsky Y, Karchin R, Wilson PA, Kaminker JS, Zhang Z, Croshaw R, Willis J, Dawson D, Shipitsin M, Willson JK, Sukumar S, Polyak K, Park BH, Pethiyagoda CL, Pant PV, Ballinger DG, Sparks AB, Hartigan J, Smith DR, Suh E, Papadopoulos N, Buckhaults P, Markowitz SD, Parmigiani G, Kinzler KW, Velculescu VE, Vogelstein B (2007). The genomic landscapes of human breast and colorectal cancers. Science. 318 (5853): 1108–1113. doi:10.1126/

science.1145720. PMID 17932254.

113. Jasperson KW, Tuohy TM, Neklason DW, Burt RW (2010). Hereditary and familial colon cancer. Gastroenterology. 138 (6): 2044–2058. doi:10.1053/j.gastro.2010.01.054. PMID 20420945.

114. Novak, Kris (20 December 2004). Epigenetics Changes in Cancer Cells. Medscape General Medicine. 6 (4): 17. PMC 1480584. PMID 15775844.

115. Banno K, Kisu I, Yanokura M, Tsuji K, Masuda K, Ueki A, Kobayashi Y, Yamagami W, Nomura H, Tominaga E, Susumu N, Aoki D (2012). Epimutation and cancer: a new carcinogenic mechanism of Lynch syndrome (Review). Int. J. Oncol. 41 (3): 793–7. doi:10.3892/ijo.2012.1528. PMC 3582986. PMID 22735547.

116. Robison AJ, Nestler EJ (November 2011). Transcriptional and epigenetic mechanisms of addiction. Nat. Rev. Neurosci. 12 (11): 623–637. doi:10.1038/nrn3111. PMC 3272277. PMID 21989194.

117. Nestler EJ (December 2013). Cellular basis of memory for addiction. Dialogues Clin. Neurosci. 15 (4): 431–443. PMC 3898681. PMID 24459410.

118. Ruffle JK (November 2014). Molecular neurobiology of addiction: what's all the (Δ)FosB about?. Am J Drug Alcohol Abuse. 40 (6): 428–437. doi:10.3109/00952990.2014.933840. PMID 25083822.

119. Biliński P, Wojtyła A, Kapka-Skrzypczak L, Chwedorowicz R, Cyranka M, Studziński T (2012). Epigenetic regulation in drug addiction. Ann. Agric. Environ. Med. 19 (3): 491–496. PMID 23020045.

120. Vassoler FM, Sadri-Vakili G (2014). Mechanisms of transgenerational inheritance of addictive-like behaviors. Neuroscience. 264: 198–206. doi:10.1016/j.neuroscience.2013.07.064. PMC 3872494. PMID 23920159.

121. Yuan TF, Li A, Sun X, Ouyang H, Campos C, Rocha NB, Arias-Carrión O, Machado S, Hou G, So KF (2015). Transgenerational Inheritance of Paternal Neurobehavioral Phenotypes: Stress, Addiction, Ageing and Metabolism. Mol. Neurobiol. doi:10.1007/s12035-015-9526-2. PMID 26572641.

122. Short, A K; Fennell, K A; Perreau, V M; Fox, A; O'Bryan, M K; Kim, J H; Bredy, T W; Pang, T Y; Hannan, A J (2016). Elevated paternal glucocorticoid exposure alters the small noncoding RNA profile in sperm and modifies anxiety and depressive phenotypes in the offspring. Translational Psychiatry. 6 (6): e837. doi:10.1038/tp.2016.109. PMC 4931607. PMID 27300263.

123. Szyl, Moshe Lamarck Revisited: Epigenetic Inheritance of Ancestral Odor Fear Conditioning Nature Neuroscience 17.1 (2013): 2-4 | PMID 24369368

124. Chahwan R, Wontakal SN, Roa S (October 2010). Crosstalk between genetic and epigenetic information through cytosine deamination. Trends Genet. 26 (10): 443–8. doi:10.1016/j.tig.2010.07.005. PMID 20800313.

125. Nonantibiotic Effects of Fluoroquinolones in Mammalian Cells.. J Biol Chem. 290: 22287–97. Sep 2015. doi:10.1074/jbc.M115.671222. PMID 26205818.

126. mezentseva, nadejda; yang, jinpu; kaur, keerat; eisenberg, carol; eisenberg, leonard (2012). The histone methyltransferase inhibitor BIX01294 enhances the

cardiac potential of bone marrow cells.. Stem Cells Dev. 22: 654–67. doi:10.1089/scd.2012.0181. PMID 22994322.

127. yang, jinpu; kaur, keerat; ong, lilin; eisenberg, carol; eisenberg, leonard (2015). Inhibition of G9a Histone Methyltransferase Converts Bone Marrow Mesenchymal Stem Cells to Cardiac Competent Progenitors.. stem cell international. 2015: 270428. doi:10.1155/2015/270428. PMID 26089912.

CHAPTER 7

Mutation

Chapter Contents

A gene mutation is a permanent alteration in the DNA sequence that makes up a gene, such that the sequence differs from what is found in most people. Mutations range in size; they can affect anywhere from a single DNA building block (base pair) to a large segment of a chromosome that includes multiple genes.

INTRODUCTION

In biology, a mutation is the permanent alteration of the nucleotide sequence of the genome of an organism, virus, or extrachromosomal DNA or other genetic elements. Mutations result from errors during DNA replication or other types of damage to DNA, which then may undergo error-prone repair (especially microhomology-mediated end joining), or cause an error during other forms of repair, or else may cause an error during replication (translesion synthesis). Mutations may also result from insertion or deletion of segments of DNA due to mobile genetic elements.Mutations may or may not produce discernible changes in the observable characteristics (phenotype) of an organism. Mutations play a part in both normal and abnormal biological processes including: evolution, cancer, and the development of the immune system, including junctional diversity.

The genomes of RNA viruses are based on RNA rather than DNA. The RNA viral genome can be double stranded (as in DNA) or single stranded. In some of these viruses (such as the single stranded human immunodeficiency virus) replication occurs quickly and there are no mechanisms to check the genome for accuracy. This error-prone process often results in mutations.

Mutation can result in many different types of change in sequences. Mutations in genes can either have no effect, alter the product of a gene, or prevent the gene from functioning properly or completely. Mutations can also occur in nongenic regions. One study on genetic variations between different species of *Drosophila* suggests that, if a mutation changes a protein produced by a gene, the result is likely to be harmful, with an estimated 70 percent of amino acidpolymorphisms that have damaging effects, and the remainder being either neutral or marginally beneficial. Due to the damaging effects that mutations can have on genes, organisms have mechanisms such as DNA repair to prevent or correct mutations by reverting the mutated sequence back to its original state.

DESCRIPTION

Mutations can involve the duplication of large sections of DNA, usually through genetic recombination. These duplications are a major source of raw material for evolving new genes, with tens to hundreds of genes duplicated in animal genomes every million years. Most genes belong to larger gene families of shared ancestry, known as homology. Novel genes are produced by several methods, commonly through the duplication and mutation of an ancestral gene, or by recombining parts of different genes to form new combinations with new functions.

Here, protein domains act as modules, each with a particular and independent function, that can be mixed together to produce genes encoding new proteins with novel properties. For example, the human eye uses four genes to make structures that sense light: three for cone cell or color vision and one for rod cell or night vision; all four arose from a single ancestral gene. Another advantage of duplicating a gene (or even an entire genome) is that this increases engineering redundancy; this allows one gene in the pair to acquire a new function while the other copy performs the original function. Other types of mutation occasionally create new genes from previously noncoding DNA.

Changes in chromosome number may involve even larger mutations, where segments of the DNA within chromosomes break and then rearrange. For example, in the Homininae, two chromosomes fused to produce human chromosome 2; this fusion did not occur in the lineage of the other apes, and they retain these separate chromosomes. In evolution, the most important role of such chromosomal rearrangements may be to accelerate the divergence of a population into new species by making populations less likely to interbreed, thereby preserving genetic differences between these populations.

Sequences of DNA that can move about the genome, such as transposons, make up a major fraction of the genetic material of plants and animals, and may have been important in the evolution of genomes. For example, more than a million copies of the Alu sequence are present in the human genome, and these sequences have now been recruited to perform functions such as regulating gene expression. Another effect of these mobile DNA sequences is that when they move within a genome, they can mutate or delete existing genes and thereby produce genetic diversity.

Nonlethal mutations accumulate within the gene pool and increase the amount of genetic variation. The abundance of some genetic changes within the gene pool can be reduced by natural selection, while other "more favorable" mutations may accumulate and result in adaptive changes.

Prodryas persephone, a Late Eocene butterfly

For example, a butterfly may produce offspring with new mutations. The majority of these mutations will have no effect; but one might change the color of one of the butterfly's offspring, making it harder (or easier) for predators to see. If this color change is advantageous, the chance of this butterfly's surviving and producing its own offspring are a little better, and over time the number of butterflies with this mutation may form a larger percentage of the population.

Neutral mutations are defined as mutations whose effects do not influence the fitness of an individual. These can accumulate over time due to genetic drift. It is believed that the overwhelming majority of mutations have no significant effect on an organism›s fitness. Also, DNA repair mechanisms are able to mend most changes before they become permanent mutations, and many organisms have mechanisms for eliminating otherwise-permanently mutated somatic cells.

Beneficial mutations can improve reproductive success.

CAUSES

Four classes of mutations are (1) spontaneous mutations (molecular decay), (2) mutations due to error-prone replication bypass of naturally occurring DNA damage (also called error-prone translesion synthesis), (3) errors introduced during DNA repair, and (4) induced mutations caused by mutagens. Scientists may also deliberately introduce mutant sequences through DNA manipulation for the sake of scientific experimentation.

Spontaneous Mutation

- Tautomerism — A base is changed by the repositioning of a hydrogen atom, altering the hydrogen bonding pattern of that base, resulting in incorrect base pairing during replication.
- Depurination — Loss of a purine base (A or G) to form an apurinic site (AP site).
- Deamination — Hydrolysis changes a normal base to an atypical base containing a keto group in place of the original amine group. Examples include C → U and A → HX (hypoxanthine), which can be corrected by DNA repair mechanisms; and 5MeC (5-methylcytosine) → T, which is less likely to be detected as a mutation because thymine is a normal DNA base.
- Slipped strand mispairing — Denaturation of the new strand from the template during replication, followed by renaturation in a different spot ("slipping"). This can lead to insertions or deletions.

Error-prone Replication Bypass

There is increasing evidence that the majority of spontaneously arising mutations are due to error-prone replication (translesion synthesis) past DNA damage in the template strand. Naturally occurring oxidative DNA damages arise at least 10,000 times per cell per day in humans and 50,000 times or more per cell per day in rats. In mice, the majority of mutations are caused by translesion synthesis. Likewise, in yeast, Kunz et al. found that more than 60% of the spontaneous single base pair substitutions and deletions were caused by translesion synthesis.

Errors Introduced During DNA Repair

Although naturally occurring double-strand breaks occur at a relatively low frequency in DNA, their repair often causes mutation. Non-homologous end joining (NHEJ) is a major pathway for repairing double-strand breaks. NHEJ involves removal of a few nucleotides to allow somewhat inaccurate alignment of the two ends for rejoining followed by addition of nucleotides to fill in gaps. As a consequence, NHEJ often introduces mutations.

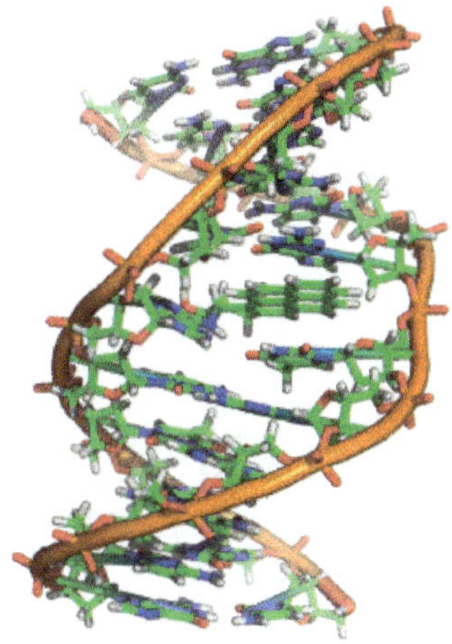

A covalent adduct between the metabolite of benzo[*a*]pyrene, the major mutagen in tobacco smoke, and DNA

Induced Mutation

Induced mutations on the molecular level can be caused by:-

- Chemicals
    - Hydroxylamine
    - Base analogs (e.g., Bromodeoxyuridine (BrdU))
    - Alkylating agents (e.g., *N*-ethyl-*N*-nitrosourea (ENU)). These agents can mutate both replicating and non-replicating DNA. In contrast, a base analog can mutate the DNA only when the analog is incorporated in replicating the DNA. Each of these classes of chemical mutagens has certain effects that then lead to transitions, transversions, or deletions.
    - Agents that form DNA adducts (e.g., ochratoxin A)
    - DNA intercalating agents (e.g., ethidium bromide)
    - DNA crosslinkers
    - Oxidative damage
    - Nitrous acid converts amine groups on A and C to diazo groups, altering their hydrogen bonding patterns, which leads to incorrect base pairing during replication.
- Radiation
    - Ultraviolet light (UV) (non-ionizing radiation). Two nucleotide bases in DNA—cytosine and thymine—are most vulnerable to radiation that can change their properties. UV light can induce adjacent pyrimidine bases in a DNA strand to become covalently joined as a pyrimidine dimer. UV radiation, in particular longer-wave UVA, can also cause oxidative damage to DNA.

CLASSIFICATION OF MUTATION TYPES

By Effect on Structure

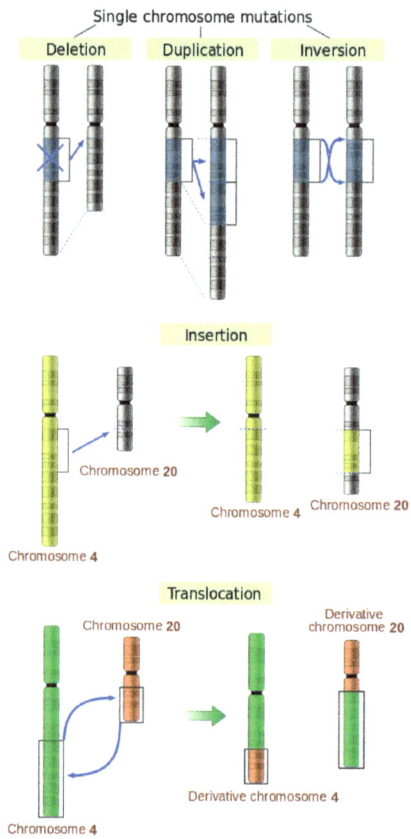

Illustrations of five types of chromosomal mutations.

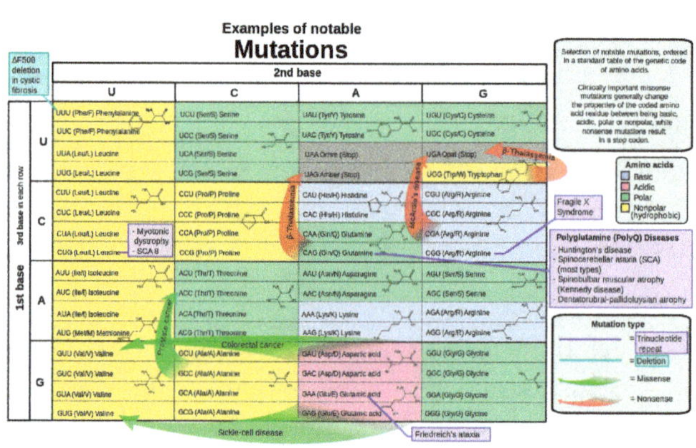

Selection of disease-causing mutations, in a standard table of the genetic code of amino acids.

The sequence of a gene can be altered in a number of ways. Gene mutations have varying effects on health depending on where they occur and whether they alter the function of essential proteins. Mutations in the structure of genes can be classified as:

- Small-scale mutations, such as those affecting a small gene in one or a few nucleotides, including:
    - **Substitution mutations**, often caused by chemicals or malfunction of DNA replication, exchange a single nucleotide for another. These changes are classified as transitions or transversions. Most common is the transition that exchanges a purine for a purine (A ↔ G) or a pyrimidine for a pyrimidine, (C ↔ T). A transition can be caused by nitrous acid, base mis-pairing, or mutagenic base analogs such as BrdU. Less common is a transversion, which exchanges a purine for a pyrimidine or a pyrimidine for a purine (C/T ↔ A/G). An example of a transversion is the conversion of adenine (A) into a cytosine (C). A point mutation can be reversed by another point mutation, in which the nucleotide is changed back to its original state (true reversion) or by second-site reversion (a complementary mutation elsewhere that results in regained gene functionality). Point mutations that occur within the protein coding region of a gene may be classified into three kinds, depending upon what the erroneous codon codes for:
        - Silent mutations, which code for the same (or a sufficiently similar) amino acid.
        - Missense mutations, which code for a different amino acid.
        - Nonsense mutations, which code for a stop codon and can truncate the protein.
    - **Insertions** add one or more extra nucleotides into the DNA. They are usually caused by transposable elements, or errors during replication of repeating elements. Insertions in the coding region of a gene may alter splicing of the mRNA (splice site mutation), or cause a shift in the reading frame (frameshift), both of which can significantly alter the gene product. Insertions can be reversed by excision of the transposable element.
    - **Deletions** remove one or more nucleotides from the DNA. Like insertions, these mutations can alter the reading frame of the gene. In general, they are irreversible: Though exactly the same sequence might in theory be restored by an insertion, transposable elements able to revert a very short deletion (say 1–2 bases) in *any* location either are highly unlikely to exist or do not exist at all.
- Large-scale mutations in chromosomal structure, including:
    - **Amplifications** (or gene duplications) leading to multiple copies of all chromosomal regions, increasing the dosage of the genes located within them.
    - **Deletions** of large chromosomal regions, leading to loss of the genes within those regions.

- Mutations whose effect is to juxtapose previously separate pieces of DNA, potentially bringing together separate genes to form functionally distinct fusion genes (e.g., bcr-abl). These include:
    - **Chromosomal translocations**: interchange of genetic parts from nonhomologous chromosomes.
    - **Interstitial deletions**: an intra-chromosomal deletion that removes a segment of DNA from a single chromosome, thereby apposing previously distant genes. For example, cells isolated from a human astrocytoma, a type of brain tumor, were found to have a chromosomal deletion removing sequences between the Fused in Glioblastoma (FIG) gene and the receptor tyrosine kinase (ROS), producing a fusion protein (FIG-ROS). The abnormal FIG-ROS fusion protein has constitutively active kinase activity that causes oncogenic transformation (a transformation from normal cells to cancer cells).
    - **Chromosomal inversions**: reversing the orientation of a chromosomal segment.
- **Loss of heterozygosity**: loss of one allele, either by a deletion or a genetic recombination event, in an organism that previously had two different alleles.

By effect on Function

- Loss-of-function mutations, also called inactivating mutations, result in the gene product having less or no function (being partially or wholly inactivated). When the allele has a complete loss of function (null allele), it is often called an amorphic mutation in the Muller's morphs schema. Phenotypes associated with such mutations are most often recessive. Exceptions are when the organism is haploid, or when the reduced dosage of a normal gene product is not enough for a normal phenotype (this is called haploinsufficiency).
- Gain-of-function mutations, also called activating mutations, change the gene product such that its effect gets stronger (enhanced activation) or even is superseded by a different and abnormal function. When the new allele is created, a heterozygote containing the newly created allele as well as the original will express the new allele; genetically this defines the mutations as dominant phenotypes. Often called a neomorphic mutation.
- Dominant negative mutations (also called antimorphic mutations) have an altered gene product that acts antagonistically to the wild-type allele. These mutations usually result in an altered molecular function (often inactive) and are characterized by a dominant or semi-dominant phenotype. In humans, dominant negative mutations have been implicated in cancer (e.g., mutations in genes p53, ATM, CEBPA and PPARgamma). Marfan syndrome is caused by mutations in the *FBN1* gene, located on chromosome 15, which encodes fibrillin-1, a glycoprotein component of the extracellular matrix. Marfan syndrome is also

an example of dominant negative mutation and haploinsufficiency.

- Lethal mutations are mutations that lead to the death of the organisms that carry the mutations.
- A back mutation or reversion is a point mutation that restores the original sequence and hence the original phenotype.

By effect on fitness

In applied genetics, it is usual to speak of mutations as either harmful or beneficial.

- A harmful, or deleterious, mutation decreases the fitness of the organism.
- A beneficial, or advantageous mutation increases the fitness of the organism. Mutations that promotes traits that are desirable, are also called beneficial. In theoretical population genetics, it is more usual to speak of mutations as deleterious or advantageous than harmful or beneficial.
- A neutral mutation has no harmful or beneficial effect on the organism. Such mutations occur at a steady rate, forming the basis for the molecular clock. In the neutral theory of molecular evolution, neutral mutations provide genetic drift as the basis for most variation at the molecular level.
- A nearly neutral mutation is a mutation that may be slightly deleterious or advantageous, although most nearly neutral mutations are slightly deleterious.

Distribution of Fitness Effects

Attempts have been made to infer the distribution of fitness effects (DFE) using mutagenesis experiments and theoretical models applied to molecular sequence data. DFE, as used to determine the relative abundance of different types of mutations (i.e., strongly deleterious, nearly neutral or advantageous), is relevant to many evolutionary questions, such as the maintenance of genetic variation, the rate of genomic decay, the maintenance of outcrossing sexual reproduction as opposed to inbreeding and the evolution of sex and genetic recombination. In summary, the DFE plays an important role in predicting evolutionary dynamics. A variety of approaches have been used to study the DFE, including theoretical, experimental and analytical methods.

- **Mutagenesis experiment**: The direct method to investigate the DFE is to induce mutations and then measure the mutational fitness effects, which has already been done in viruses, bacteria, yeast, and *Drosophila*. For example, most studies of the DFE in viruses used site-directed mutagenesis to create point mutations and measure relative fitness of each mutant. In *Escherichia coli*, one study used transposon mutagenesis to directly measure the fitness of a random insertion of a derivative of Tn10. In yeast, a combined mutagenesis and deep sequencing approach has been developed to generate high-quality systematic mutant libraries and measure fitness in high throughput. However, given that many mutations have effects too small to be detected and that mutagenesis experiments can detect only mutations of moderately large effect; DNA sequence data analysis can provide valuable information about these mutations.

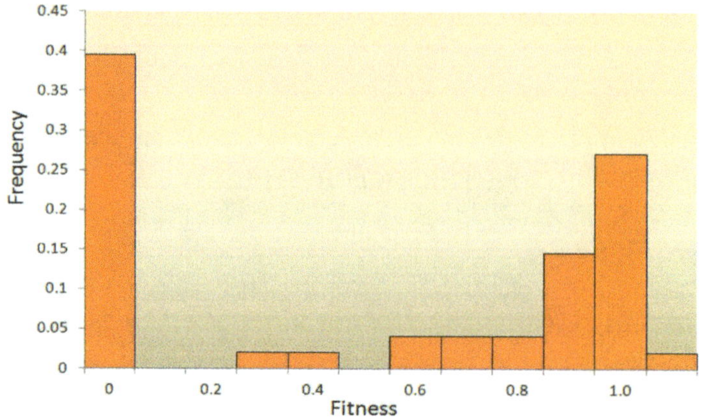

The distribution of fitness effects (DFE) of mutations in vesicular stomatitis virus. In this experiment, random mutations were introduced into the virus by site-directed mutagenesis, and the fitness of each mutant was compared with the ancestral type. A fitness of zero, less than one, one, more than one, respectively, indicates that mutations are lethal, deleterious, neutral, and advantageous.

- **Molecular sequence analysis**: With rapid development of DNA sequencing technology, an enormous amount of DNA sequence data is available and even more is forthcoming in the future. Various methods have been developed to infer the DFE from DNA sequence data. By examining DNA sequence differences within and between species, we are able to infer various characteristics of the DFE for neutral, deleterious and advantageous mutations. To be specific, the DNA sequence analysis approach allows us to estimate the effects of mutations with very small effects, which are hardly detectable through mutagenesis experiments.

One of the earliest theoretical studies of the distribution of fitness effects was done by Motoo Kimura, an influential theoretical population geneticist. His neutral theory of molecular evolution proposes that most novel mutations will be highly deleterious, with a small fraction being neutral. Hiroshi Akashi more recently proposed a bimodal model for the DFE, with modes centered around highly deleterious and neutral mutations. Both theories agree that the vast majority of novel mutations are neutral or deleterious and that advantageous mutations are rare, which has been supported by experimental results. One example is a study done on the DFE of random mutations in vesicular stomatitis virus.Out of all mutations, 39.6% were lethal, 31.2% were non-lethal deleterious, and 27.1% were neutral. Another example comes from a high throughput mutagenesis experiment with yeast. In this experiment it was shown that the overall DFE is bimodal, with a cluster of neutral mutations, and a broad distribution of deleterious mutations.

Though relatively few mutations are advantageous, those that are play an important role in evolutionary changes. Like neutral mutations, weakly selected advantageous mutations can be lost due to random genetic drift, but strongly selected advantageous mutations are more likely to be fixed. Knowing the DFE of advantageous mutations may lead to increased ability to predict the evolutionary dynamics. Theoretical work on the DFE for advantageous mutations has been done by John H. Gillespie and H. Allen Orr.

They proposed that the distribution for advantageous mutations should be exponential under a wide range of conditions, which, in general, has been supported by experimental studies, at least for strongly selected advantageous mutations.

In general, it is accepted that the majority of mutations are neutral or deleterious, with rare mutations being advantageous; however, the proportion of types of mutations varies between species. This indicates two important points: first, the proportion of effectively neutral mutations is likely to vary between species, resulting from dependence on effective population size; second, the average effect of deleterious mutations varies dramatically between species. In addition, the DFE also differs between coding regions and noncoding regions, with the DFE of noncoding DNA containing more weakly selected mutations.

By Impact on Protein Sequence

- A **frameshift mutation** is a mutation caused by insertion or deletion of a number of nucleotides that is not evenly divisible by three from a DNA sequence. Due to the triplet nature of gene expression by codons, the insertion or deletion can disrupt the reading frame, or the grouping of the codons, resulting in a completely different translation from the original. The earlier in the sequence the deletion or insertion occurs, the more altered the protein produced is.

In contrast, any insertion or deletion that is evenly divisible by three is termed an *in-frame mutation*

- A **nonsense mutation** is a point mutation in a sequence of DNA that results in a premature stop codon, or a *nonsense codon* in the transcribed mRNA, and possibly a truncated, and often nonfunctional protein product. (See Stop codon.)
- **Missense mutations** or *nonsynonymous mutations* are types of point mutations where a single nucleotide is changed to cause substitution of a different amino acid. This in turn can render the resulting protein nonfunctional. Such mutations are responsible for diseases such as Epidermolysis bullosa, sickle-cell disease, and SOD1-mediated ALS.
- A **neutral mutation** is a mutation that occurs in an amino acid codon that results in the use of a different, but chemically similar, amino acid. The similarity between the two is enough that little or no change is often rendered in the protein. For example, a change from AAA to AGA will encode arginine, a chemically similar molecule to the intended lysine.
- **Silent mutations** are mutations that do not result in a change to the amino acid sequence of a protein, unless the changed amino acid is sufficiently similar to the original. They may occur in a region that does not code for a protein, or they may occur within a codon in a manner that does not alter the final amino acid sequence. The phrase *silent mutation* is often used interchangeably with the phrase *synonymous mutation*; however, synonymous mutations are a subcategory of the former, occurring only within exons (and necessarily exactly preserving the amino acid sequence of the protein). Synonymous mutations occur due to the degenerate nature of the genetic code.

By Inheritance

A mutation has caused this garden moss rose to produce flowers of different colors. This is a somatic mutation that may also be passed on in the germline.

In multicellular organisms with dedicated reproductive cells, mutations can be subdivided into germline mutations, which can be passed on to descendants through their reproductive cells, and somatic mutations (also called acquired mutations),which involve cells outside the dedicated reproductive group and which are not usually transmitted to descendants.

A germline mutation gives rise to a *constitutional mutation* in the offspring, that is, a mutation that is present in every cell. A constitutional mutation can also occur very soon after fertilisation, or continue from a previous constitutional mutation in a parent.

The distinction between germline and somatic mutations is important in animals that have a dedicated germline to produce reproductive cells. However, it is of little value in understanding the effects of mutations in plants, which lack dedicated germline. The distinction is also blurred in those animals that reproduce asexually through mechanisms such as budding, because the cells that give rise to the daughter organisms also give rise to that organism›s germline. A new germline mutation that was not inherited from either parent is called a *de novo* mutation.

Diploid organisms (e.g., humans) contain two copies of each gene—a paternal and a maternal allele. Based on the occurrence of mutation on each chromosome, we may classify mutations into three types.

- A **heterozygous mutation** is a mutation of only one allele.
- A **homozygous mutation** is an identical mutation of both the paternal and maternal alleles.
- **Compound heterozygous** mutations or a **genetic compound** consists of two different mutations in the paternal and maternal alleles.

- A **wild type** or **homozygous non-mutated** organism is one in which neither allele is mutated.

Special classes

Conditional mutation is a mutation that has wild-type (or less severe) phenotype under certain "permissive" environmental conditions and a mutant phenotype under certain "restrictive" conditions. For example, a temperature-sensitive mutation can cause cell death at high temperature (restrictive condition), but might have no deleterious consequences at a lower temperature (permissive condition).

- Replication timing quantitative trait loci affects DNA replication.

Nomenclature

In order to categorize a mutation as such, the "normal" sequence must be obtained from the DNA of a "normal" or "healthy" organism (as opposed to a "mutant" or "sick" one), it should be identified and reported; ideally, it should be made publicly available for a straightforward nucleotide-by-nucleotide comparison, and agreed upon by the scientific community or by a group of expert geneticists and biologists, who have the responsibility of establishing the *standard* or so-called "consensus" sequence. This step requires a tremendous scientific effort. Once the consensus sequence is known, the mutations in a genome can be pinpointed, described, and classified. The committee of the Human Genome Variation Society (HGVS) has developed the standard human sequence variant nomenclature, which should be used by researchers and DNA diagnostic centers to generate unambiguous mutation descriptions. In principle, this nomenclature can also be used to describe mutations in other organisms. The nomenclature specifies the type of mutation and base or amino acid changes.

- Nucleotide substitution (e.g., 76A>T) — The number is the position of the nucleotide from the 5› end; the first letter represents the wild-type nucleotide, and the second letter represents the nucleotide that replaced the wild type. In the given example, the adenine at the 76th position was replaced by a thymine.
  - If it becomes necessary to differentiate between mutations in genomic DNA, mitochondrial DNA, and RNA, a simple convention is used. For example, if the 100th base of a nucleotide sequence mutated from G to C, then it would be written as g.100G>C if the mutation occurred in genomic DNA, m.100G>C if the mutation occurred in mitochondrial DNA, or r.100g>c if the mutation occurred in RNA. Note that, for mutations in RNA, the nucleotide code is written in lower case.
- Amino acid substitution (e.g., D111E) — The first letter is the one letter code of the wild-type amino acid, the number is the position of the amino acid from the N-terminus, and the second letter is the one letter code of the amino acid present in the mutation. Nonsense mutations are represented with an X for the second amino acid (e.g. D111X).

- Amino acid deletion (e.g., ΔF508) — The Greek letter Δ (delta) indicates a deletion. The letter refers to the amino acid present in the wild type and the number is the position from the N terminus of the amino acid were it to be present as in the wild type.

MUTATION RATES

Mutation rates vary substantially across species, and the evolutionary forces that generally determine mutation are the subject of ongoing investigation.

HARMFUL MUTATIONS

Changes in DNA caused by mutation can cause errors in protein sequence, creating partially or completely non-functional proteins. Each cell, in order to function correctly, depends on thousands of proteins to function in the right places at the right times. When a mutation alters a protein that plays a critical role in the body, a medical condition can result. A condition caused by mutations in one or more genes is called a Sheik disorder. Some mutations alter a gene's DNA base sequence but do not change the function of the protein made by the gene. One study on the comparison of genes between different species of *Drosophila* suggests that if a mutation does change a protein, this will probably be harmful, with an estimated 70 percent of amino acid polymorphisms having damaging effects, and the remainder being either neutral or weakly beneficial.]] Studies have shown that only 7% of point mutations in noncoding DNA of yeast are deleterious and 12% in coding DNA are deleterious. The rest of the mutations are either neutral or slightly beneficial.

If a mutation is present in a germ cell, it can give rise to offspring that carries the mutation in all of its cells. This is the case in hereditary diseases. In particular, if there is a mutation in a DNA repair gene within a germ cell, humans carrying such germline mutations may have an increased risk of cancer. A list of 34 such germline mutations is given in the article DNA repair-deficiency disorder. An example of one is albinism, a mutation that occurs in the OCA1 or OCA2 gene. Individuals with this disorder are more prone to many types of cancers, other disorders and have impaired vision. On the other hand, a mutation may occur in a somatic cell of an organism. Such mutations will be present in all descendants of this cell within the same organism, and certain mutations can cause the cell to become malignant, and, thus, cause cancer.

A DNA damage can cause an error when the DNA is replicated, and this error of replication can cause a gene mutation that, in turn, could cause a genetic disorder. DNA damages are repaired by the DNA repair system of the cell. Each cell has a number of pathways through which enzymes recognize and repair damages in DNA. Because DNA can be damaged in many ways, the process of DNA repair is an important way in which the body protects itself from disease. Once DNA damage has given rise to a mutation, the mutation cannot be repaired. DNA repair pathways can only recognize and act on "abnormal" structures in the DNA. Once a mutation occurs in a gene sequence it then has normal DNA structure and cannot be repaired.

BENEFICIAL MUTATIONS

Although mutations that cause changes in protein sequences can be harmful to an organism, on occasions the effect may be positive in a given environment. In this case, the mutation may enable the mutant organism to withstand particular environmental stresses better than wild-type organisms, or reproduce more quickly. In these cases a mutation will tend to become more common in a population through natural selection.

For example, a specific 32 base pair deletion in human CCR5 (CCR5-Δ32) confers HIV resistance to homozygotes and delays AIDS onset in heterozygotes. One possible explanation of the etiology of the relatively high frequency of CCR5-Δ32 in the European population is that it conferred resistance to the bubonic plague in mid-14th century Europe. People with this mutation were more likely to survive infection; thus its frequency in the population increased. This theory could explain why this mutation is not found in Southern Africa, which remained untouched by bubonic plague. A newer theory suggests that the selective pressure on the CCR5 Delta 32 mutation was caused by smallpox instead of the bubonic plague.

An example of a harmful mutation is sickle-cell disease, a blood disorder in which the body produces an abnormal type of the oxygen-carrying substance hemoglobin in the red blood cells. One-third of all indigenous inhabitants of Sub-Saharan Africa carry the gene, because, in areas where malaria is common, there is a survival value in carrying only a single sickle-cell gene (sickle cell trait). Those with only one of the two alleles of the sickle-cell disease are more resistant to malaria, since the infestation of the malaria *Plasmodium* is halted by the sickling of the cells that it infests.

PRION MUTATIONS

Prions are proteins and do not contain genetic material. However, prion replication has been shown to be subject to mutation and natural selection just like other forms of replication. The human gene PRNP codes for the major prion protein, PrP, and is subject to mutations that can give rise to disease-causing prions.

SOMATIC MUTATIONS

A change in the genetic structure that is not inherited from a parent, and also not passed to offspring, is called a *somatic cell genetic mutation* or *acquired mutation*.

Cells with heterozygous mutations (one good copy of gene and one mutated copy) may function normally with the unmutated copy until the good copy has been spontaneously somatically mutated. This kind of mutation happens all the time in living organisms, but it is difficult to measure the rate. Measuring this rate is important in predicting the rate at which people may develop cancer.

Point mutations may arise from spontaneous mutations that occur during DNA replication. The rate of mutation may be increased by mutagens. Mutagens can be physical, such as radiation from UV rays, X-rays or extreme heat, or chemical (molecules that misplace base pairs or disrupt the helical shape of DNA). Mutagens associated with cancers are often studied to learn about cancer and its prevention.

MUTATION RATE

In genetics, the **mutation rate** is a measure of the rate at which various types of mutations occur over time. Mutation rates are typically given for a specific class of mutation, for instance point mutations, small or large scale insertions or deletions. The rate of substitutions can be further subdivided into a mutation spectrum which describes the influence of genetic context on the mutation rate.

There are several natural units of time for each of these rates, with rates being characterized either as mutations per base pair per cell division, per gene per generation, or per genome per generation. The mutation rate of an organism is an evolved characteristic and is strongly influenced by the genetics of each organism, in addition to strong influence from the environment. The upper and lower limits to which mutation rates can evolve is the subject of ongoing investigation.

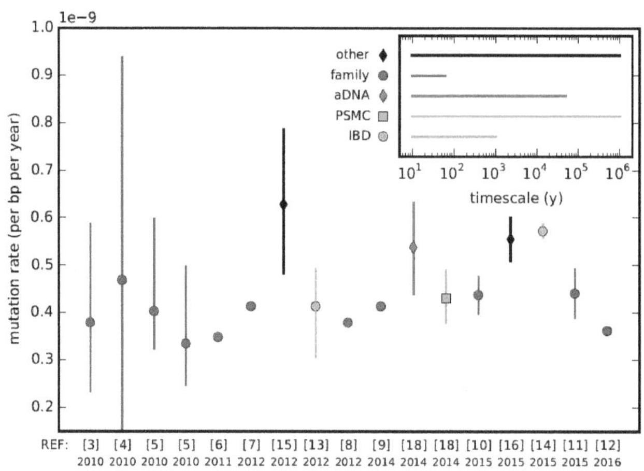

Recently reported estimates of the human genome-wide mutation rate. The human germline mutation rate is approximately 0.5×10^{-9} per basepair per year.

Different genetic variants within a species are referred to as alleles, and so a new mutation is said to create a new allele. In population genetics, each allele is characterized by a selection coefficient, which measures the expected change in an allele›s frequency over time. The selection coefficient can either be negative, corresponding to an expected decrease, positive, corresponding to an expected increase, or zero, corresponding to no expected change. The distribution of fitness effects of new mutations is an important parameter in population genetics and has been the subject of extensive investigation Although measurements of this distribution have been inconsistent in the past, it is now generally thought that the majority of mutations are mildly deleterious, that many have little effect on an organism›s fitness, and that a few can be favorable. As a result of natural selection, unfavorable mutations will typically be eliminated from a population while favorable changes are quickly fixed, and neutral changes accumulate at the rate they are created by mutations.

Measurement

An organism's mutation rates can be measured by a number of techniques.

Substitution Rates

Many sites in an organism's genome may not admit mutations with large fitness effects. These sites are typically called neutral sites. Theoretically mutations under no selection become fixed between organisms at precisely the mutation rate. Fixed synonymous mutations, i.e. synonymous substitutions, are changes to the sequence of a gene that do not change the protein produced by that gene. They are often used as estimates of that mutation rate, despite the fact that some synonymous mutations have fitness effects. As an example, mutation rates have been directly inferred from the whole genome sequences of experimentally evolved replicate lines of *Escherichia coli* B.

Mutation Accumulation Lines

A particularly labor-intensive way of characterizing the mutation rate is the mutation accumulation line.

Mutation accumulation lines have been used to characterize mutation rates with the Bateman-Mukai Method and direct sequencing of intestinal bacteria, round-worms, yeast, fruit flies, small annual plants,

Variation in Mutation Rates

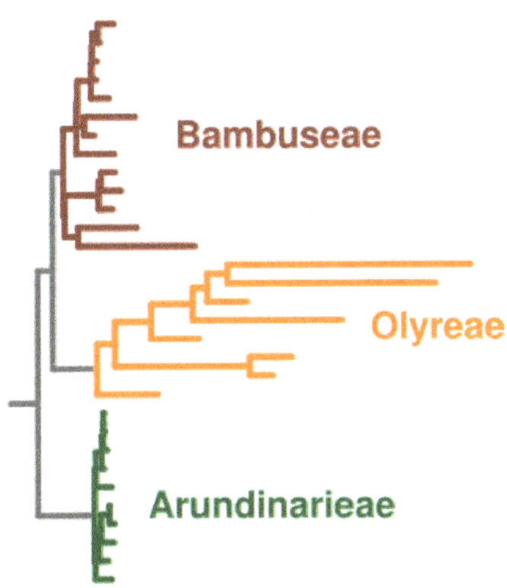

Generation time affects mutation rates: The long-lived woody bamboos (tribes Arundinarieae and Bambuseae) have lower mutation rates (short branches in the phylogenetic tree) than the fast-evolving herbaceous bamboos (Olyreae).

Mutation rates differ between species and even between different regions of the genome of a single species. These different rates of nucleotide substitution are measured in substitutions (fixed mutations) per base pair per generation. For example, mutations in intergenic, or non-coding, DNA tend to accumulate at a faster rate than mutations in DNA that is actively in use in the organism (gene expression). That is not necessarily due to a higher mutation rate, but to lower levels of purifying selection. A region which mutates at predictable rate is a candidate for use as a molecular clock.

If the rate of neutral mutations in a sequence is assumed to be constant (clock-like), and if most differences between species are neutral rather than adaptive, then the number of differences between two different species can be used to estimate how long ago two species diverged (see molecular clock). In fact, the mutation rate of an organism may change in response to environmental stress. For example, UV light damages DNA, which may result in error prone attempts by the cell to perform DNA repair.

The human mutation rate is higher in the male germ line (sperm) than the female (egg cells), but estimates of the exact rate have varied by an order of magnitude or more.

In general, the mutation rate in unicellular eukaryotes and bacteria is roughly 0.003 mutations per genome per *cell* generation. This means that a human genome accumulates around 64 new mutations per generation because each full generation involves a number of cell divisions to generate gametes. The highest per base pair per generation mutation rates are found in viruses, which can have either RNA or DNA genomes. DNA viruses have mutation rates between 10^{-6} to 10^{-8} mutations per base per generation, and RNA viruses have mutation rates between 10^{-3} to 10^{-5} per base per generation. Human mitochondrial DNA has been estimated to have mutation rates of ~3× or ~2.7×10^{-5} per base per 20 year generation (depending on the method of estimation);these rates are considered to be significantly higher than rates of human genomic mutation at ~2.5×10^{-8} per base per generation. Using data available from whole genome sequencing, the human genome mutation rate is similarly estimated to be ~1.1×10^{-8} per site per generation.

The rate for other forms of mutation also differs greatly from point mutations. An individual microsatellite locus often has a mutation rate on the order of 10^{-4}, though this can differ greatly with length.

Some sequences of DNA may be more susceptible to mutation. For example, stretches of DNA in human sperm which lack methylation are more prone to mutation.

Mutational Spectrum

The mutation spectrum of an organism is the rate at which different mutations occur at different sites. Typically two sites are considered, each of which may have three mutations, resulting in six total rates for most mutation spectra. The two sites are the two correct pairs possible in DNA: A:T pairs and C:G pairs;

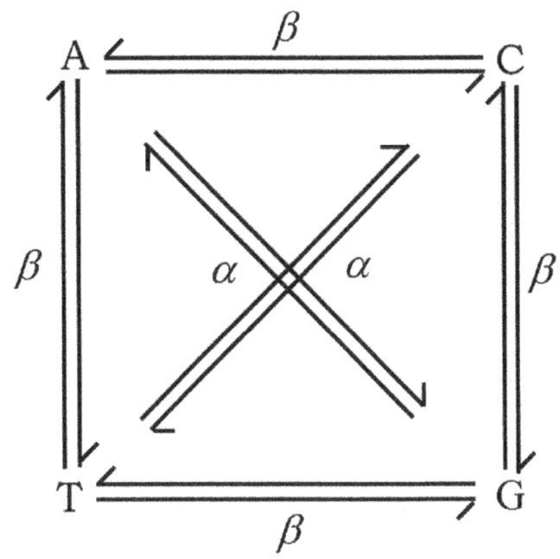

Transitions (Alpha) and transversions (Beta).

There is a systematic difference in rates for transitions (Alpha) and transversions (Beta).

Evolution

Theory on the evolution of mutation rates identifies three principal forces involved: the generation of more deleterious mutations with higher mutation, the generation of more advantageous mutations with higher mutation, and the metabolic costs and reduced replication rates that are required to prevent mutations. Different conclusions are reached based on the relative importance attributed to each force. The optimal mutation rate of organisms may be determined by a trade-off between costs of a high mutation rate, such as deleterious mutations, and the metabolic costs of maintaining systems to reduce the mutation rate (such as increasing the expression of DNA repair enzymes. or, as reviewed by Bernstein et al. having increased energy use for repair, coding for additional gene products and/or having slower replication). Second, higher mutation rates increase the rate of beneficial mutations, and evolution may prevent a lowering of the mutation rate in order to maintain optimal rates of adaptation. Finally, natural selection may fail to optimize the mutation rate because of the relatively minor benefits of lowering the mutation rate, and thus the observed mutation rate is the product of neutral processes.

Studies have shown that treating RNA viruses such as poliovirus with ribavirin produce results consistent with the idea that the viruses mutated too frequently to maintain the integrity of the information in their genomes. This is termed error catastrophe.

EVOLVABILITY

Evolvability is defined as the capacity of a system for adaptive evolution. Evolvability is the ability of a population of organisms to not merely generate genetic diversity, but to generate *adaptive* genetic diversity, and thereby evolve through natural selection.

In order for a biological organism to evolve by natural selection, there must be a certain minimum probability that new, heritable variants are beneficial. Random mutations, unless they occur in DNA sequences with no function, are expected to be mostly detrimental. Beneficial mutations are always rare, but if they are too rare, then adaptation cannot occur. Early failed efforts to evolve computer programs by random mutation and selection showed that evolvability is not a given, but depends on the representation of the program. Analogously, the evolvability of organisms depends on their genotype-phenotype map. This means that biological genomes are structured in ways that make beneficial changes less unlikely than they would otherwise be. This has been taken as evidence that evolution has created not just fitter organisms, but populations of organisms that are better able to evolve.

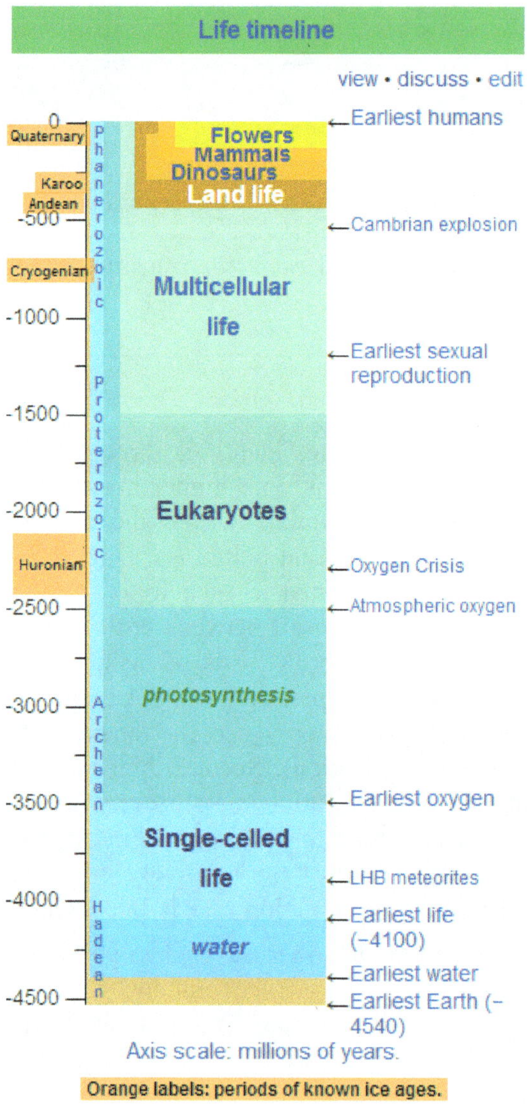

Andreas Wagner describes two definitions of evolvability. According to the first definition, a biological system is evolvable:

- if its properties show heritable genetic variation, and
- if natural selection can thus change these properties.

According to the second definition, a biological system is evolvable:

- if it can acquire novel functions through genetic change, functions that help the organism survive and reproduce.

For example, consider an enzyme with multiple alleles in the population. Each allele catalyzes the same reaction, but with a different level of activity. However, even after millions of years of evolution, exploring many sequences with similar function, no mutation might exist that gives this enzyme the ability to catalyze a different reaction. Thus, although the enzyme's activity is evolvable in the first sense, that does not mean that the enzyme›s function is evolvable in the second sense. However, every system evolvable in the second sense must also be evolvable in the first.

Pigliucci recognizes three classes of definition, depending on timescale. The first corresponds to Wagner's first, and represents the very short timescales that are described by quantitative genetics. He divides Wagner›s second definition into two categories, one representing the intermediate timescales that can be studied using population genetics, and one representing exceedingly rare long-term innovations of form.

Pigliucci›s second evolvability definition includes Altenberg›s quantitative concept of evolvability, being not a single number, but the entire upper tail of the fitness distribution of the offspring produced by the population. This quantity was considered a "local" property of the instantaneous state of a population, and its integration over the population›s evolutionary trajectory, and over many possible populations, would be necessary to give a more global measure of evolvability.

Generating more Variation

More heritable phenotypic variation means more evolvability. While mutation is the ultimate source of heritable variation, its permutations and combinations also make a big difference. Sexual reproduction generates more variation (and thereby evolvability) relative to asexual reproduction (see evolution of sexual reproduction). Evolvability is further increased by generating more variation when an organism is stressed, and thus likely to be less well adapted, but less variation when an organism is doing well. The amount of variation generated can be adjusted in many different ways, for example via the mutation rate, via the probability of sexual vs. asexual reproduction, via the probability of outcrossing vs. inbreeding, via dispersal, and via access to previously cryptic variants through the switching of an evolutionary capacitor. A large population size increases the influx of novel mutations each generation.

Enhancement of Selection

Rather than creating more phenotypic variation, some mechanisms increase the intensity and effectiveness with which selection acts on existing phenotypic variation. For example:

- Mating rituals that allow sexual selection on "good genes", and so intensify natural selection

- Large effective population size increasing the threshold value of the selection coefficient above which selection becomes an important player. This could happen through an increase in the census population size, decreasing genetic drift, through an increase in the recombination rate, decreasing genetic draft, or through changes in the probability distribution of the numbers of offspring.
- Recombination decreasing the importance of the Hill-Robertson effect, where different genotypes contain different adaptive mutations. Recombination brings the two alleles together, creating a super-genotype in place of two competing lineages.
- Shorter generation time

Robustness and evolvability

The relationship between robustness and evolvability depends on whether recombination can be ignored. Recombination can generally be ignored in asexual populations and for traits affected by single genes.

Without Recombination

Robustness will not increase evolvability in the first sense. In organisms with a high level of robustness, mutations will have smaller phenotypic effects than in organisms with a low level of robustness. Thus, robustness reduces the amount of heritable genetic variation on which selection can act. However, robustness may allow exploration of large regions of genotype space, increasing evolvability according to the second sense. Even without genetic diversity, some genotypes have higher evolvability than others, and selection for robustness can increase the "neighborhood richness" of phenotypes that can be accessed from the same starting genotype by mutation. For example, one reason many proteins are less robust to mutation is that they have marginal thermodynamic stability, and most mutations reduce this stability further. Proteins that are more thermostable can tolerate a wider range of mutations and are more evolvable. For polygenic traits, neighborhood richness contributes more to evolvability than does genetic diversity or "spread" across genotype space.

With recombination

Temporary robustness, or canalisation, may lead to the accumulation of significant quantities of cryptic genetic variation. In a new environment or genetic background, this variation may be revealed and sometimes be adaptive.

Exploration Ahead of Time

When mutational robustness exists, many mutants will persist in a cryptic state. Mutations tend to fall into two categories, having either a very bad effect or very little effect: few mutations fall somewhere in between. Sometimes, these mutations will not be completely invisible, but still have rare effects, with very low penetrance. When this happens, natural selection weeds out the very bad mutations, while leaving the others relatively unaffected. While evolution has no "foresight" to know which environment will be encountered in the future, some mutations cause major disruption to a basic biological process, and will never be adaptive in any environment. Screening these out in advance leads to preadapted stocks of cryptic genetic variation.

Another way that phenotypes can be explored, prior to strong genetic commitment, is through learning. An organism that learns gets to "sample" several different phenotypes during its early development, and later sticks to whatever worked best. Later in evolution, the optimal phenotype can be genetically assimilated so it becomes the default behavior rather than a rare behavior. This is known as the Baldwin effect, and it can increase evolvability.

Learning biases phenotypes in a beneficial direction. But an exploratory flattening of the fitness landscape can also increase evolvability even when it has no direction, for example when the flattening is a result of random errors in molecular and/or developmental processes. This increase in evolvability can happen when evolution is faced with crossing a "valley" in an adaptive landscape. This means that two mutations exist that are deleterious by themselves, but beneficial in combination. These combinations can evolve more easily when the landscape is first flattened, and the discovered phenotype is then fixed by genetic assimilation.

Modularity

If every mutation affected every trait, then a mutation that was an improvement for one trait would be a disadvantage for other traits. This means that almost no mutations would be beneficial overall. But if pleiotropy is restricted to within functional modules, then mutations affect only one trait at a time, and adaptation is much less constrained. In a modular gene network, for example, a gene that induces a limited set of other genes that control a specific trait under selection may evolve more readily than one that also induces other gene pathways controlling traits not under selection. Individual genes also exhibit modularity. A mutation in one cis-regulatory element of a gene›s promoter region may allow the expression of the gene to be altered only in specific tissues, developmental stages, or environmental conditions rather than changing gene activity in the entire organism simultaneously.

Evolution of Evolvability

While variation yielding high evolvability could be useful in the long term, in the short term most of that variation is likely to be a disadvantage. For example, naively it would seem that increasing the mutation rate via a mutator allele would increase evolvability. But as an extreme example, if the mutation rate is too high then all individuals will be dead or at least carry a heavy mutation load. Short-term selection for low variation most of the time is usually thought likely to be more powerful than long-term selection for evolvability, making it difficult for natural selection to cause the evolution of evolvability. Other forces of selection also affect the generation of variation; for example, mutation and recombination may in part be byproducts of mechanisms to cope with DNA damage.

When recombination is low, mutator alleles may still sometimes hitchhike on the success of adaptive mutations that they cause. In this case, selection can take place at the level of the lineage. This may explain why mutators are often seen during experimental evolution of microbes. Mutator alleles can also evolve more easily when they only increase mutation rates in nearby DNA sequences, not across the whole genome: this is known as a contingency locus.

The evolution of evolvability is less controversial if it occurs via the evolution of sexual reproduction, or via the tendency of variation-generating mechanisms to become more active when an organism is stressed. The yeast prion [PSI+] may also be an example of the evolution of evolvability through evolutionary capacitance. An evolutionary capacitor is a switch that turns genetic variation on and off. This is very much like bet-hedging the risk that a future environment will be similar or different. Theoretical models also predict the evolution of evolvability via modularity. When the costs of evolvability are sufficiently short-lived, more evolvable lineages may be the most successful in the long-term. However, the hypothesis that evolvability is an adaptation is often rejected in favor of alternative hypotheses, e.g. minimization of costs.

Applications

The study of evolvability has fundamental importance for understanding very long term evolution of protein superfamilies and organism phyla and kingdoms. A thorough understanding of the details of long term evolution will likely form part of the Extended Evolutionary Synthesis (the update to the Modern Synthesis). In addition, these phenomena have two main practical applications. For protein engineering we wish to increase evolvability, and in medicine and agriculture we wish to decrease it.

Firstly, for protein engineering it is important to understand the factors that determine how much a protein function can be altered. In particular, both rational design and directed evolution approaches aim to create changes rapidly through mutations with large effects. Such mutations, however, commonly destroy enzyme function or at least reduce tolerance to further mutations. Identifying evolvable proteins and manipulating their evolvability is becoming increasingly necessary in order to achieve ever larger functional modification of enzymes.

Many human diseases are not static phenomena, but capable of evolution. Viruses, bacteria, fungi and cancers evolve to be resistant to host immune defences, as well as pharmaceutical drugs. These same problems occur in agriculture with pesticide and herbicide resistance. It is possible that we are facing the end of the effective life of most of available antibiotics and predicting the evolution and evolvability of our pathogens and devising strategies to slow or circumvent it is requiring deeper knowledge of the complex forces driving evolution at the molecular level.

DNA DAMAGE (NATURALLY OCCURRING)

DNA damage is an alteration in the chemical structure of DNA, such as a break in a strand of DNA, a base missing from the backbone of DNA, or a chemically changed base such as 8-OHdG. Damage to DNA that occurs naturally can result from metabolic or hydrolytic processes. Metabolism releases compounds that damage DNA including reactive oxygen species, reactive nitrogen species, reactive carbonyl species, lipid peroxidation products and alkylating agents, among others, while hydrolysis cleaves chemical bonds in DNA. Naturally occurring oxidative DNA damages arise at least 10,000 times per cell per day in humans and 50,000 times or more per cell per day in rats, as documented below.

DNA damage is distinctly different from mutation, although both are types of error in DNA. DNA damage is an abnormal chemical structure in DNA, while a mutation is a change in the sequence of standard base pairs.

DNA damage and mutation have different biological consequences. While most DNA damages can undergo DNA repair, such repair is not 100% efficient. Un-repaired DNA damages accumulate in non-replicating cells, such as cells in the brains or muscles of adult mammals and can cause aging. (Also see DNA damage theory of aging.) In replicating cells, such as cells lining the colon, errors occur upon replication of past damages in the template strand of DNA or during repair of DNA damages. These errors can give rise to mutations or epigenetic alterations. Both of these types of alteration can be replicated and passed on to subsequent cell generations. These alterations can change gene function or regulation of gene expression and possibly contribute to progression to cancer.

A Major Problem for Life

One indication that DNA damages are a major problem for life is that DNA repair processes, to cope with ubiquitously occurring DNA damages, have been found in all cellular organisms in which DNA repair has been investigated. For example, in bacteria, a regulatory network aimed at repairing DNA damages (called the SOS response in *Escherichia coli*) has been found in many bacterial species. *E. coli* RecA, a key enzyme in the SOS response pathway, is the defining member of a ubiquitous class of DNA strand-exchange proteins that are essential for homologous recombination, a pathway that maintains genomic integrity by repairing broken DNA. Genes homologous to *RecA* and to other central genes in the SOS response pathway are found in almost all the bacterial genomes sequenced to date, covering a large number of phyla, suggesting both an ancient origin and a widespread occurrence of recombinational repair of DNA damage. Eukaryotic recombinases that are homologues of RecA are also widespread in eukaryotic organisms. For example, in fission yeast and humans, RecA homologues promote duplex-duplex DNA-strand exchange needed for repair of many types of DNA lesions.

Another indication that DNA damages are a major problem for life is that cells make large investments in DNA repair processes. As pointed out by Hoeijmakers,repairing just one double-strand break could require more than 10,000 ATP molecules, as used in signaling the presence of the damage, the generation of repair foci, and the formation (in humans) of the RAD51 nucleofilament (an intermediate in homologous recombinational repair). (RAD51 is a homologue of bacterial RecA.)

Frequencies

The list below shows some frequencies with which new naturally occurring DNA damages arise per day, due to endogenous cellular processes.

- Oxidative damages

- Humans, per cell per day
    - 10,000
      11,500
      2,800 specific damages 8-oxoGua, 8-oxodG plus 5-HMUra
      2,800 specific damages 8-oxoGua, 8-oxodG plus 5-HMUra
- Rats, per cell per day
    - 74,000
      86,000
      100,000
- Mice, per cell per day
    - 34,000 specific damages 8-oxoGua, 8-oxodG plus 5-HMUra
      47,000 specific damages oxo8dG in mouse liver
      28,000 specific damages 8-oxoGua, 8-oxodG, 5-HMUra

■ Depurinations
- Mammalian cells, per cell per day
    - 2,000 to 10,000
      9,000
      12,000
      13,920

■ Depyrimidinations
- Mammalian cells, per cell per day
    - 600
      696

■ Single-strand breaks
- Mammalian cells, per cell per day
    - 55,200

■ Double-strand breaks
- Human cells, per cell cycle
    - 10
      50

■ O6-methylguanines
- Mammalian cells, per cell per day
    - 3,120

■ Cytosine deamination
    - Mammalian cells, per cell per day

- 192

Another important endogenous DNA damage is M1dG, short for (3-(2'-deoxy-beta-D-erythro-pentofuranosyl)-pyrimido[1,2-a]-purin-10(3H)-one). The excretion in urine (likely reflecting rate of occurrence) of M1dG may be as much as 1,000-fold lower than that of 8-oxodG. However, a more important measure may be the steady-state level in DNA, reflecting both rate of occurrence and rate of DNA repair. The steady-state level of M1dG is higher than that of 8-oxodG. This points out that some DNA damages produced at a low rate may be difficult to repair and remain in DNA at a high steady-state level. Both M1dG and 8-oxodG are mutagenic.

Steady-state Levels

Steady-state levels of DNA damages represent the balance between formation and repair. More than 100 types of oxidative DNA damage have been characterized, and 8-oxodG constitutes about 5% of the steady state oxidative damages in DNA. Helbock et al. estimated that there were 24,000 steady state oxidative DNA adducts per cell in young rats and 66,000 adducts per cell in old rats. This reflects the accumulation of DNA damage with age. DNA damage accumulation with age is further described in DNA damage theory of aging.

Swenberg et al. measured average amounts of selected steady state endogenous DNA damages in mammalian cells. The seven most common damages they evaluated are shown in Table 1.

Table 1. Steady-state amounts of endogenous DNA damages

Endogenous lesions	Number per cell
Abasic sites	30,000
N7-(2-hydroxethyl)guanine (7HEG)	3,000
8-hydroxyguanine	2,400
7-(2-oxoethyl)guanine	1,500
Formaldehyde adducts	960
Acrolein-deoxyguanine	120
Malondialdehyde-deoxyguanine	60

Evaluating steady-state damages in specific tissues of the rat, Nakamura and Swenberg indicated that the number of abasic sites varied from about 50,000 per cell in liver, kidney and lung to about 200,000 per cell in the brain.

Consequences

Differentiated somatic cells of adult mammals generally replicate infrequently or not at all. Such cells, including, for example, brain neurons and muscle myocytes, have little or no cell turnover. Non-replicating cells do not generally generate mutations due to DNA damage-induced errors of replication. These non-replicating cells do not commonly give rise to cancer, but they do accumulate DNA damages with time that likely contribute to aging (*see DNA damage theory of aging*). In a non-replicating cell, a single-strand break or other type of damage in the transcribed strand of DNA can block RNA polymerase II catalysed transcription. This would interfere with the synthesis of the protein coded for by the gene in which the blockage occurred.

Brasnjevic et al. summarized the evidence showing that single-strand breaks accumulate with age in the brain (though accumulation differed in different regions of the brain) and that single-strand breaks are the most frequent steady-state DNA damages in the brain. As discussed above, these accumulated single-strand breaks would be expected to block transcription of genes. Consistent with this, as reviewed by Hetman et al., 182 genes were identified and shown to have reduced transcription in the brains of individuals older than 72 years, compared to transcription in the brains of those less than 43 years old. When 40 particular proteins were evaluated in a muscle of rats, the majority of the proteins showed significant decreases during aging from 18 months (mature rat) to 30 months (aged rat) of age.

Another type of DNA damage, the double strand break, was shown to cause cell death (loss of cells) through apoptosis. This type of DNA damage would not accumulate with age, since once a cell was lost through apoptosis, its double strand damage would be lost with it.

CHROMOSOMAL TRANSLOCATION

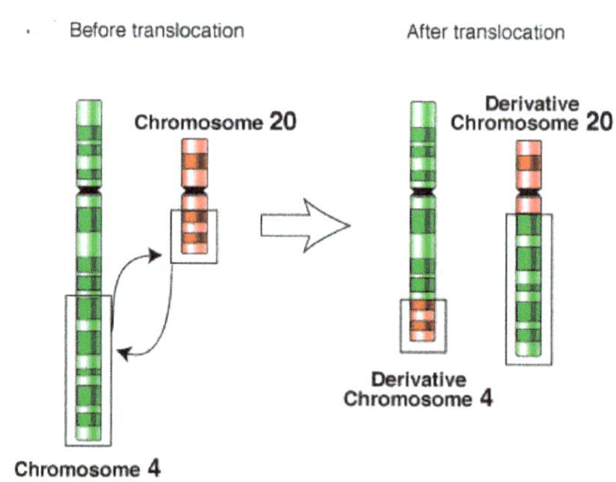

Chromosomal reciprocal translocation of the 4th and 20th chromosome.

In genetics, a **chromosome translocation** is a chromosome abnormality caused by rearrangement of parts between nonhomologous chromosomes. A gene fusion may be created when the translocation joins two otherwise-separated genes, it is detected on cytogenetics or a karyotype of affected cells. Translocations can be **balanced** (in an even exchange of material with no genetic information extra or missing, and ideally full functionality) or **unbalanced** (where the exchange of chromosome material is unequal resulting in extra or missing genes).

Translocations can be reciprocal (balanced or unbalanced) or nonreciprocal (balanced or unbalanced).

Reciprocal Translocations

Reciprocal translocations are usually an exchange of material between nonhomologous chromosomes. Estimates of incidence range from about 1 in 500 to 1 in 625 human newborns. Such translocations are usually harmless and may be found through prenatal diagnosis. However, carriers of balanced reciprocal translocations have increased risks of creating gametes with unbalanced chromosome translocations, leading to miscarriages or children with abnormalities. Genetic counseling and genetic testing are often offered to families that may carry a translocation. Most balanced translocation carriers are healthy and do not have any symptoms. But about 6% of them have a range of symptoms that may include autism, intellectual disability, or congenital anomalies. A gene disrupted or disregulated at the breakpoint of the translocation carrier is likely the cause of these symptoms.

It is important to distinguish between chromosomal translocations occurring in gametogenesis, due to errors in meiosis, and translocations that occur in cellular division of somatic cells, due to errors in mitosis. The former results in a chromosomal abnormality featured in all cells of the offspring, as in translocation carriers. Somatic translocations, on the other hand, result in abnormalities featured only in the affected cell line, as in chronic myelogenous leukemia with the Philadelphia chromosome translocation.

Robertsonian Translocations

Robertsonian translocation is a type of translocation caused by breaks at or near the centromeres of two acrocentric chromosomes. The reciprocal exchange of parts gives rise to one large metacentric chromosome and one extremely small chromosome that may be lost from the organism with little effect because it contains so few genes. The resulting karyotype in humans leaves only 45 chromosomes, since two chromosomes have fused together. This has no direct effect on the phenotype, since the only genes on the short arms of acrocentrics are common to all of them and are present in variable copy number (nucleolar organiser genes).

Robertsonian translocations have been seen involving all combinations of acrocentric chromosomes. The most common translocation in humans involves chromosomes 13 and 14 and is seen in about 0.97 / 1000 newborns. Carriers of Robertsonian translocations are not associated with any phenotypic abnormalities, but there is a risk of unbalanced gametes that lead to miscarriages or abnormal offspring. For example, carriers of Robertsonian translocations involving chromosome 21 have a higher chance of having a child with Down syndrome. This is known as a ‹translocation Downs›. This is due to a mis-segregation (nondisjunction) during gametogenesis. The mother has a higher (10%) risk of transmission than the father (1%). Robertsonian translocations involving chromosome 14 also carry a slight risk of uniparental disomy 14 due to trisomy rescue.

Role in Disease

Some human diseases caused by translocations are:

- Cancer: Several forms of cancer are caused by acquired translocations (as opposed to those present from conception); this has been described mainly in leukemia (acute myelogenous leukemia and chronic myelogenous leukemia). Translocations have also been described in solid malignancies such as Ewing›s sarcoma.
- Infertility: One of the would-be parents carries a *balanced translocation*, where the parent is asymptomatic but conceived fetuses are not viable.
- Down syndrome is caused in a minority (5% or less) of cases by a Robertsonian translocation of the chromosome 21 long arm onto the long arm of chromosome 14.

Chromosomal translocations between the sex chromosomes can also result in a number of genetic conditions, such as

- XX male syndrome: caused by a translocation of the SRY gene from the Y to the X chromosome

By Chromosome

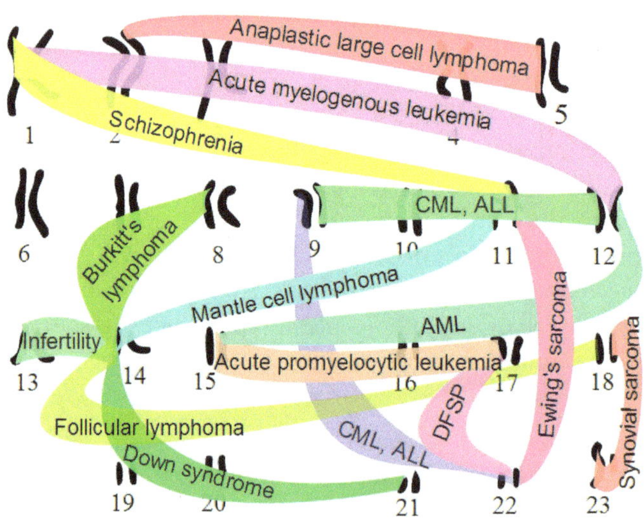

Overview of some chromosomal translocations involved in different cancers, as well as implicated in some other conditions, e.g. schizophrenia, with chromosomes arranged in standard karyogram order. Abbreviations:

LL - Acute lymphoblastic leukemia

AML - Acute myeloid leukemia

CML - Chronic myelogenous leukemia

DFSP - Dermatofibrosarcoma protuberans

Denotation

The International System for Human Cytogenetic Nomenclature (ISCN) is used to denote a translocation between chromosomes. The designation **t(A;B)(p1;q2)** is used to denote a translocation between chromosome A and chromosome B. The information in the second set of parentheses, when given, gives the precise location within the chromosome for chromosomes A and B respectively—with *p* indicating the short arm of the chromosome, *q* indicating the long arm, and the numbers after p or q refers to regions, bands and subbands seen when staining the chromosome with a staining dye. See also the definition of a genetic locus. The translocation is the mechanism that can cause a gene to move from one linkage group to another.

Examples

For an explanation of the symbols and abbreviations used in these examples, see Cytogenetic notation.

Translocation	Associated diseases	Fused genes/proteins	
		First	Second
t(8;14)(q24;q32)	Burkitt's lymphoma	c-myc on chromosome 8, gives the fusion protein lymphocyte-proliferative ability	IGH@ (immunoglobulin heavy locus) on chromosome 14, induces massive transcription of fusion protein
t(11;14)(q13;q32)	Mantle cell lymphoma	cyclin D1 on chromosome 11, gives fusion protein cell-proliferative ability	IGH@ (immunoglobulin heavy locus) on chromosome 14, induces massive transcription of fusion protein
t(14;18)(q32;q21)	Follicular lymphoma (~90% of cases)	IGH@(immunoglobulin heavy locus) on chromosome 14, induces massive transcription of fusion protein	Bcl-2 on chromosome 18, gives fusion protein anti-apoptotic abilities
t(10;(various)) (q11;(various))	Papillary thyroid cancer	RET proto-oncogene on chromosome 10	PTC (*Papillary Thyroid Cancer*) - Placeholder for any of several other genes/proteins
t(2;3)(q13;p25)	Follicular thyroid cancer	PAX8 - paired box gene 8on chromosome 2	PPARγ1 (peroxisome proliferator-activated receptor γ 1) on chromosome 3
t(8;21)(q22;q22)	Acute myeloblastic leukemia with maturation	ETO on chromosome 8	AML1 on chromosome 21 found in ~7% of new cases of AML, carries a favorable prognosis and predicts good response to cytosine arabinoside therapy
t(9;22)(q34;q11) Philadelphia chromosome	Chronic myelogenous leukemia (CML), acute lymphoblastic leukemia (ALL)	*Abl1* gene on chromosome 9	*BCR* ("breakpoint cluster region" on chromosome 22

t(15;17)(q22;q21)	Acute promyelocytic leukemia	PML protein on chromosome 15	RAR-α on chromosome 17 persistent laboratory detection of the PML-RARA transcript is strong predictor of relapse
t(12;15)(p13;q25)	Acute myeloid leukemia, congenital fibrosarcoma, secretory breast carcinoma, mammary analogue secretory carcinoma of salivary glands	TEL on chromosome 12	TrkC receptor on chromosome 15
t(9;12)(p24;p13)	CML, ALL	JAK on chromosome 9	TEL on chromosome 12
t(12;21)(p12;q22)	ALL	TEL on chromosome 12	AML1 on chromosome 21
t(11;18)(q21;q21)	MALT lymphoma	API-2	MLT
t(1;11)(q42.1;q14.3)	Schizophrenia		
t(2;5)(p23;q35)	Anaplastic large cell lymphoma	ALK	NPM1
t(11;22) (q24;q11.2-12)	Ewing's sarcoma	FLI1	EWS
t(17;22)	DFSP	Collagen I on chromosome 17	Platelet derived growth factor B on chromosome 22
t(1;12)(q21;p13)	Acute myelogenous leukemia		
t(X;18)(p11.2;q11.2)	Synovial sarcoma		
t(1;19)(q10;p10)	Oligodendroglioma and oligoastrocytoma		
t(17;19)(q22;p13)	ALL		
t(7,16) (q32-34;p11) or t(11,16) (p11;p11)	Low-grade fibromyxoid sarcoma	FUS	CREB3L2 or CREB3L1

Chromosome Abnormality

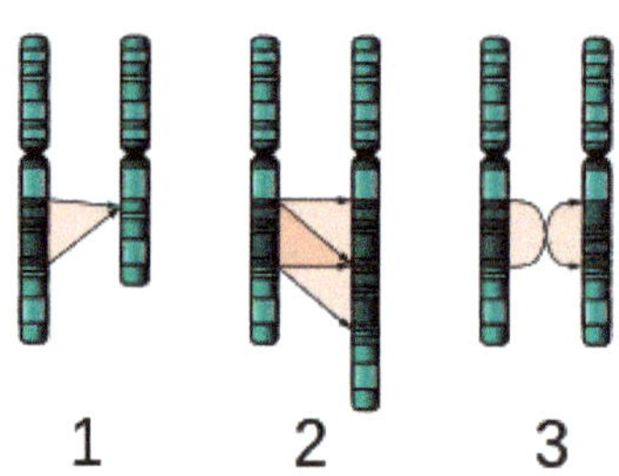

The three major single chromosome mutations; deletion (1), duplication (2) and inversion (3).

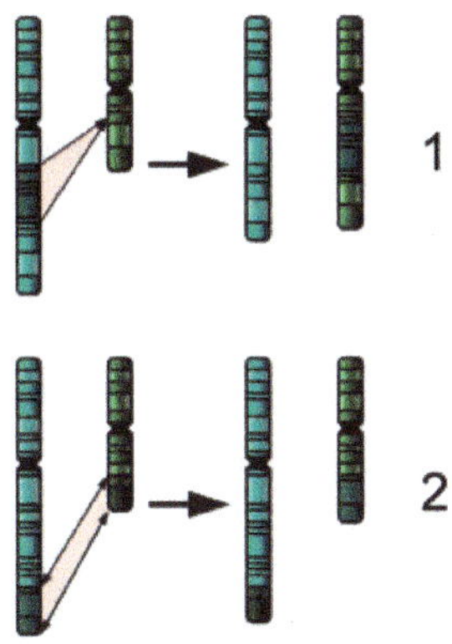

The two major two chromosome mutations; insertion (1) and Translocation (2).

A **chromosome anomaly**, **abnormality**, **aberration**, or **mutation** is a missing, extra, or irregular portion of chromosomal DNA. It can be from an atypical number of chromosomes or a structural abnormality in one or more chromosomes. **Chromosome mutation** was formerly used in a strict sense to mean a change in a chromosomal segment, involving more than one gene. A karyotype refers to a full set of chromosomes from an individual that can be compared to a "normal" karyotype for the species via genetic testing. A chromosome anomaly may be detected or confirmed in this manner. Chromosome anomalies usually occur when there is an error in cell division following meiosis or mitosis. There are many types of chromosome anomalies. They can be organized into two basic groups, numerical and structural anomalies.

Numerical Disorders

This is called aneuploidy (an abnormal number of chromosomes), and occurs when an individual either is missing a chromosome from a pair (monosomy) or has more than two chromosomes of a pair (trisomy, tetrasomy, etc.).

In humans, an example of a condition caused by a numerical anomaly is Down Syndrome, also known as Trisomy 21 (an individual with Down Syndrome has three copies of chromosome 21, rather than two). The frequency of Trisomy 21 has been determined to be a function of Advanced maternal age.

An example of monosomy is Turner syndrome, where the individual is born with only one sex chromosome, an X.

Sperm Aneuploidy

Exposure of males to certain lifestyle, environmental and/or occupational hazards may increase the risk of aneuploid spermatozoa. In particular, risk of aneuploidy is increased by tobacco smoking, and occupational exposure to benzene, insecticides, and perfluorinated compounds. Increased aneuploidy is often associated with increased DNA damage in spermatozoa.

Structural Abnormalities

When the chromosome's structure is altered, this can take several forms:

- Deletions: A portion of the chromosome is missing or deleted. Known disorders in humans include Wolf-Hirschhorn syndrome, which is caused by partial deletion of the short arm of chromosome 4; and Jacobsen syndrome, also called the terminal 11q deletion disorder.
- Duplications: A portion of the chromosome is duplicated, resulting in extra genetic material. Known human disorders include Charcot-Marie-Tooth disease type 1A, which may be caused by duplication of the gene encoding peripheral myelin protein 22 (PMP22) on chromosome 17.
- Translocations: A portion of one chromosome is transferred to another chromosome. There are two main types of translocations:
    - Reciprocal translocation: Segments from two different chromosomes have been exchanged.
    - Robertsonian translocation: An entire chromosome has attached to another at the centromere - in humans these only occur with chromosomes 13, 14, 15, 21, and 22.
- Inversions: A portion of the chromosome has broken off, turned upside down, and reattached, therefore the genetic material is inverted.
- Insertions: A portion of one chromosome has been deleted from its normal place and inserted into another chromosome.
- Rings: A portion of a chromosome has broken off and formed a circle or ring. This can happen with or without loss of genetic material.
- Isochromosome: Formed by the mirror image copy of a chromosome segment including the centromere.

Chromosome instability syndromes are a group of disorders characterized by chromosomal instability and breakage. They often lead to an increased tendency to develop certain types of malignancies.

Inheritance

Most chromosome abnormalities occur as an accident in the egg cell or sperm, and therefore the anomaly is present in every cell of the body. Some anomalies, however, can happen after conception, resulting in Mosaicism (where some cells have the anomaly and some do not). Chromosome anomalies can be inherited from a parent or be "de novo". This is why chromosome studies are often performed on parents when a child is found to have an anomaly. If the parents do not possess the abnormality it was not initially inherited; however it may be transmitted to subsequent generations.

Acquired Chromosome Abnormalities

Most cancers, if not all, could cause chromosome abnormalities, with either the formation of hybrid genes and fusion proteins, deregulation of genes and overexpresson of proteins, or loss of tumor suppressor genes (see the "Mitelman Database" and the Atlas of Genetics and Cytogenetics in Oncology and Haematology,).

Chromosomal Inversion

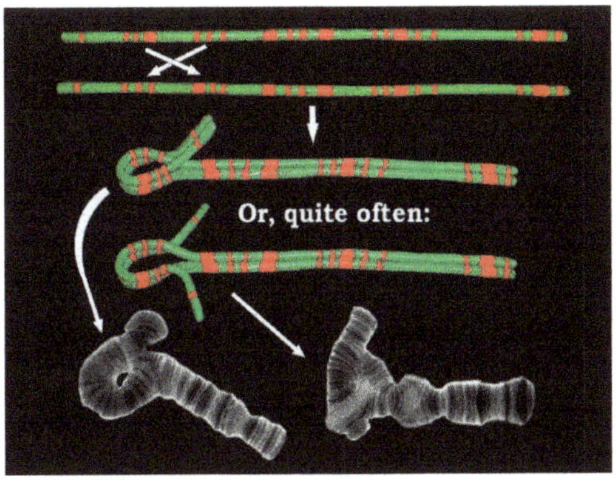

A clay model showing why heterozygous inversion loops are visible in polytene chromosome preparations

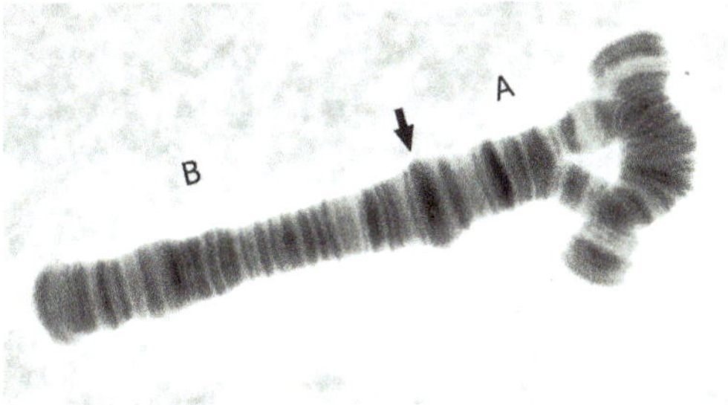

An inversion loop in the A arm of a chromosome from an *Axarus* species midge

An inversion is a chromosome rearrangement in which a segment of a chromosome is reversed end to end. An inversion occurs when a single chromosome undergoes breakage and rearrangement within itself. Inversions are of two types: paracentric and pericentric.

Paracentric inversions do not include the centromere and both breaks occur in one arm of the chromosome. Pericentric inversions include the centromere and there is a break point in each arm.

Cytogenetic techniques may be able to detect inversions, or inversions may be inferred from genetic analysis. Nevertheless, in most species small inversions go undetected.

In insects with polytene chromosomes, for example Drosophila, preparations of larval salivary gland chromosomes allow inversions to be seen when they are heterozygous. This useful characteristic of polytene chromosomes was first advertised by Theophilus Shickel Painter in 1933.

Inversions usually do not cause any abnormalities in carriers as long as the rearrangement is balanced with no extra or missing DNA. However, in individuals which are heterozygous for an inversion, there is an increased production of abnormal chromatids (this occurs when crossing-over occurs within the span of the inversion). This leads to lowered fertility due to production of unbalanced gametes.

The most common inversion seen in humans is on chromosome 9, at inv(9)(p12q13). This inversion is generally considered to have no harmful effects, but there is some suspicion it could lead to an increased risk for miscarriage or infertility for some affected individuals. An inversion does not involve a loss of genetic information, but simply rearranges the linear gene sequence.

Families that may be carriers of inversions may be offered genetic counseling and genetic testing.

REFERENCES

1. *Sharma S, Javadekar SM, Pandey M, Srivastava M, Kumari R, Raghavan SC (2015).* "Homology and enzymatic requirements of microhomology-dependent alternative end joining". Cell Death Dis. 6: e1697. doi:10.1038/cddis.2015.58. PMC 4385936. PMID 25789972.
2. Chen J, Miller BF, Furano AV (2014). "Repair of naturally occurring mismatches can induce mutations in flanking DNA". Elife. 3: e02001. doi:10.7554/elife.02001. PMC 3999860. PMID 24843013.
3. Rodgers K, McVey M (2016). "Error-Prone Repair of DNA Double-Strand Breaks". J. Cell. Physiol. 231 *(1): 15–24. doi:10.1002/jcp.25053. PMID 26033759.*
4. Bertram, John S. (December 2000). "The molecular biology of cancer". Molecular Aspects of Medicine. Amsterdam, the Netherlands: Elsevier. 21 *(6): 167–223.* doi:10.1016/S0098-2997(00)00007-8. ISSN 0098-2997. PMID 11173079.
5. Aminetzach, Yael T.; Macpherson, J. Michael; Petrov, Dmitri A. (July 29, 2005). "Pesticide Resistance via Transposition-Mediated Adaptive Gene Truncation in Drosophila". Science. Washington, D.C.: American Association for the Advancement of Science. 309 *(5735): 764–767. Bibcode*:2005Sci...309..764A. doi:10.1126/science.1112699. ISSN 0036-8075. PMID 16051794.
6. Burrus, Vincent; Waldor, Matthew K. (June 2004). "Shaping bacterial genomes with integrative and conjugative elements". Research in Microbiology. Amsterdam, the Netherlands: Elsevier. 155 *(5): 376–386. doi:10.1016/j.resmic.2004.01.012. ISSN 0923-2508. PMID 15207870.*
7. Sawyer, Stanley A.; Parsch, John; Zhi Zhang; et al. (April 17, 2007). "Prevalence of positive selection among nearly neutral amino acid replacements

inDrosophila". Proc. Natl. Acad. Sci. U.S.A. Washington, D.C.: National Academy of Sciences. 104 *(16): 6504–6510. Bibcode*:2007PNAS..104.6504S. doi:10.1073/pnas.0701572104. ISSN 0027-8424. PMC 1871816. PMID 17409186.

8. Hastings, P. J.; Lupski, James R.; Rosenberg, Susan M.; et al. (August 2009). "Mechanisms of change in gene copy number". Nature Reviews Genetics. London: Nature Publishing Group. 10 *(8): 551–564.* doi:10.1038/nrg2593. ISSN 1471-0056. PMC 2864001. PMID 19597530.
9. Harrison, Paul M.; Gerstein, Mark (May 17, 2002). "Studying Genomes Through the Aeons: Protein Families, Pseudogenes and Proteome Evolution". Journal of Molecular Biology. Amsterdam, the Netherlands: Elsevier. 318 *(5): 1155–1174.* doi:10.1016/S0022-2836(02)00109-2. ISSN 0022-2836. PMID 12083509.
10. Orengo, Christine A.; Thornton, Janet M. (July 2005). "Protein families and their evolution—a structural perspective". Annual Review of Biochemistry. Palo Alto, CA: Annual Reviews. 74: 867–900. doi:10.1146/annurev.biochem.74.082803.133029. ISSN 0066-4154. PMID 15954844.
11. Manyuan Long; Betrán, Esther; Thornton, Kevin; et al. (November 2003). "The origin of new genes: glimpses from the young and old". Nature Reviews Genetics. London: Nature Publishing Group. 4 *(11): 865–875.* doi:10.1038/nrg1204. ISSN 1471-0056. PMID 14634634.
12. Minglei Wang; Caetano-Anollés, Gustavo (January 14, 2009). "The Evolutionary Mechanics of Domain Organization in Proteomes and the Rise of Modularity in the Protein World". Structure. Cambridge, MA: Cell Press. 17 *(1): 66–78. doi:10.1016/j.str.2008.11.008. ISSN 0969-2126. PMID 19141283.*
13. Bowmaker, James K. (May 1998). "Evolution of colour vision in vertebrates". Eye. London: Nature Publishing Group. 12 *(Pt 3b): 541–547. doi:10.1038/eye.1998.143. ISSN 0950-222X. PMID 9775215.*
14. Gregory, T. Ryan; Hebert, Paul D. N. (April 1999). "The Modulation of DNA Content: Proximate Causes and Ultimate Consequences". Genome Research. Cold Spring Harbor, NY: Cold Spring Harbor Laboratory Press. 9 *(4): 317–324. doi:10.1101/gr.9.4.317 (inactive 2015-01-09). ISSN 1088-9051. PMID 10207154.*
15. Hurles, Matthew (July 13, 2004). "Gene Duplication: The Genomic Trade in Spare Parts". PLOS Biology. San Francisco, CA: Public Library of Science. 2 *(7): E206. doi:10.1371/journal.pbio.0020206. ISSN 1545-7885. PMC 449868. PMID 15252449.*
16. Na Liu; Okamura, Katsutomo; Tyler, David M.; et al. (October 2008). "The evolution and functional diversification of animal microRNA genes". Cell Research. London: Nature Publishing Group on behalf of the Shanghai Institutes for Biological Sciences. 18 *(10): 985–996. doi:10.1038/cr.2008.278. ISSN 1001-0602. PMC 2712117. PMID 18711447.*
17. Siepel, Adam (October 2009). "Darwinian alchemy: Human genes from noncoding DNA". Genome Research. Cold Spring Harbor, NY: Cold Spring Harbor Laboratory Press. 19 *(10): 1693–1695. doi:10.1101/gr.098376.109. ISSN 1088-9051. PMC 2765273. PMID 19797681.*

18. Jianzhi Zhang; Xiaoxia Wang; Podlaha, Ondrej (May 2004). "Testing the Chromosomal Speciation Hypothesis for Humans and Chimpanzees". Genome Research. Cold Spring Harbor, NY: Cold Spring Harbor Laboratory Press. 14 *(5): 845–851. doi:10.1101/gr.1891104. ISSN 1088-9051. PMC 479111. PMID 15123584.*

19. Ayala, Francisco J.; Coluzzi, Mario (May 3, 2005). "Chromosome speciation: Humans, Drosophila, and mosquitoes". Proc. Natl. Acad. Sci. U.S.A. Washington, D.C.: National Academy of Sciences. 102 *(Suppl 1): 6535–6542. Bibcode*:2005PNAS..102.6535A. doi:10.1073/pnas.0501847102. ISSN 0027-8424. PMC 1131864. PMID 15851677.

20. Hurst, Gregory D. D.; Werren, John H. (August 2001). "The role of selfish genetic elements in eukaryotic evolution". Nature Reviews Genetics. London: Nature Publishing Group. 2 *(8): 597–606.* doi:10.1038/35084545. ISSN 1471-0056. PMID 11483984.

21. Häsler, Julien; Strub, Katharina (November 2006). "Alu elements as regulators of gene expression". Nucleic Acids Research. Oxford, UK: Oxford University Press. 34 *(19): 5491–5497. doi:10.1093/nar/gkl706. ISSN 0305-1048. PMC 1636486. PMID 17020921.*

22. Eyre-Walker, Adam; Keightley, Peter D. (August 2007). "The distribution of fitness effects of new mutations" (PDF). Nature Reviews Genetics. London: Nature Publishing Group. 8 *(8): 610–618.* doi:10.1038/nrg2146. ISSN 1471-0056. PMID 17637733.

23. Montelone, Beth A. (1998). "Mutation, Mutagens, and DNA Repair". www-personal.ksu.edu.

24. Bernstein C, Prasad AR, Nfonsam V, Bernstein H. (2013). DNA Damage, DNA Repair and Cancer, New Research Directions in DNA Repair, Prof. Clark Chen (Ed.), ISBN 978-953-51-1114-6, InTech, http://www.intechopen.com/books/new-research-directions-in-dna-repair/dna-damage-dna-repair-and-cancer

25. Stuart, Gregory R.; Oda, Yoshimitsu; de Boer, Johan G.; Glickman, Barry W. (March 2000). "Mutation Frequency and Specificity With Age in Liver, Bladder and Brain of lacI Transgenic Mice". Genetics. Bethesda, MD: Genetics Society of America. 154 *(3): 1291–1300. ISSN 0016-6731. PMC 1460990. PMID 10757770.*

26. Kunz, Bernard A.; Ramachandran, Karthikeyan; Vonarx, Edward J. (April 1998). "DNA Sequence Analysis of Spontaneous Mutagenesis in Saccharomyces cerevisiae". Genetics. Bethesda, MD: Genetics Society of America. 148 *(4): 1491–1505. ISSN 0016-6731. PMC 1460101. PMID 9560369.*

27. Lieber, Michael R. (July 2010). "The Mechanism of Double-Strand DNA Break Repair by the Nonhomologous DNA End-Joining Pathway". Annual Review of Biochemistry. Palo Alto, CA: Annual Reviews. 79: 181–211. doi:10.1146/annurev.biochem.052308.093131. ISSN 0066-4154. PMC 3079308. PMID 20192759.

28. Pfohl-Leszkowicz, Annie; Manderville, Richard A. (January 2007). "Ochratoxin A: An overview on toxicity and carcinogenicity in animals and humans". Molecular Nutrition & Food Research. Hoboken, NJ: Wiley-Blackwell. 51 *(1):*

61–99. doi:10.1002/mnfr.200600137. ISSN 1613-4125. PMID 17195275.

29. Kozmin, Stanislav; Slezak, Guenaelle; Reynaud-Angelin, Anne; et al. (September 20, 2005). "UVA radiation is highly mutagenic in cells that are unable to repair 7,8-dihydro-8-oxoguanine in Saccharomyces cerevisiae". Proc. Natl. Acad. Sci. U.S.A. Washington, D.C.: National Academy of Sciences. 102 *(38): 13538–13543.* doi:10.1073/pnas.0504497102. ISSN 0027-8424. PMC 1224634. PMID 16157879.
30. Freese, Ernst (April 15, 1959). "The difference between Spontaneous and Base-Analogue Induced Mutations of Phage T4". Proc. Natl. Acad. Sci. U.S.A. Washington, D.C.: National Academy of Sciences. 45 *(4): 622–633. doi:10.1073/pnas.45.4.622. ISSN 0027-8424. PMC 222607. PMID 16590424.*
31. Freese, Ernst (June 1959). "The specific mutagenic effect of base analogues on Phage T4". Journal of Molecular Biology. Amsterdam, the Netherlands: Elsevier. 1 *(2): 87–105.* doi:10.1016/S0022-2836(59)80038-3. ISSN 0022-2836.
32. McClean, Phillip (1999). "Types of Mutations". Genes and Mutations. Course material from PLSC 431/631 - Intermediate Genetics.
33. Goh, Amanda M.; Coffill, Cynthia R; Lane, David P. (January 2011). "The role of mutant p53 in human cancer". The Journal of Pathology. Hoboken, NJ: John Wiley & Sons. 223 *(2): 116–126. doi:10.1002/path.2784. ISSN 0022-3417. PMID 21125670.*
34. Chenevix-Trench G, Spurdle AB, Gatei M, et al. (February 6, 2002). "Dominant Negative ATM Mutations in Breast Cancer Families". Journal of the National Cancer Institute. Oxford, UK: Oxford University Press. 94 *(3): 205–215. doi:10.1093/jnci/94.3.205. ISSN 0027-8874. PMID 11830610.*
35. Paz-Priel, Ido; Friedman, Alan (2011). "C/EBPα dysregulation in AML and ALL". Critical Reviews in Oncogenesis. Redding, CT: Begell House. 16 *(1–2): 93–102. doi:10.1615/critrevoncog.v16.i1-2.90. ISSN 0893-9675. PMC 3243939. PMID 22150310.*
36. Capaccio D, Ciccodicola A, Sabatino L, et al. (June 2010). "A novel germline mutation in Peroxisome Proliferator-Activated Receptor γ gene associated with large intestine polyp formation and dyslipidemia". Biochimica et Biophysica Acta (BBA) - Molecular Basis of Disease. Amsterdam, the Netherlands: Elsevier. 1802 *(6): 572–581. doi:10.1016/j.bbadis.2010.01.012. ISSN 0925-4439. PMID 20123124.*
37. McKusick, Victor A. (July 25, 1991). "The defect in Marfan syndrome". Nature. London: Nature Publishing Group. 352 *(6333): 279–281.* doi:10.1038/352279a0. ISSN 0028-0836. PMID 1852198.
38. Judge DP, Biery NJ, Keene DR, et al. (July 15, 2004). "Evidence for a critical contribution of haploinsufficiency in the complex pathogenesis of Marfan syndrome". Journal of Clinical Investigation. Ann Arbor, MI: American Society for Clinical Investigation. 114 *(2): 172–181.* doi:10.1172/JCI20641. ISSN 0021-9738. PMC 449744 PMID 15254584.

39. Judge, Daniel P.; Dietz, Harry C. (December 3, 2005). "Marfan›s syndrome". The Lancet. Amsterdam, the Netherlands: Elsevier. 366 *(9501): 1965–1976.* doi:10.1016/S0140-6736(05)67789-6. ISSN 0140-6736. PMC 1513064. PMID 16325700.

40. Ellis, Nathan A.; Ciocci, Susan; German, James (February 2001). "Back mutation can produce phenotype reversion in Bloom syndrome somatic cells". Human Genetics. Berlin; New York: Springer-Verlag. 108 *(2): 167–173.* doi:10.1007/s004390000447. ISSN 0340-6717. PMID 11281456.

41. Charlesworth, Deborah; Charlesworth, Brian; Morgan, Martin T. (December 1995). "The Pattern of Neutral Molecular Variation Under the Background Selection Model". Genetics. Bethesda, MD: Genetics Society of America. 141 *(4): 1619–1632. ISSN 0016-6731. PMC 1206892. PMID 8601499.*

42. Loewe, Laurence (April 2006). "Quantifying the genomic decay paradox due to Muller›s ratchet in human mitochondrial DNA". Genetical Research. London; New York: Cambridge University Press. 87 *(2): 133–159.* doi:10.1017/S0016672306008123. ISSN 0016-6723. PMID 16709275.

43. Bernstein, H; Hopf, FA; Michod, RE (1987). "The molecular basis of the evolution of sex.". *Adv Genet.* 24: 323–70. doi:10.1016/s0065-2660(08)60012-7. PMID 3324702.

44. Peck, Joel R.; Barreau, Guillaume; Heath, Simon C. (April 1997). "Imperfect Genes, Fisherian Mutation and the Evolution of Sex". Genetics. Bethesda, MD: Genetics Society of America. 145 *(4): 1171–1199. ISSN 0016-6731. PMC 1207886. PMID 9093868.*

45. Keightley, Peter D.; Lynch, Michael (March 2003). "Toward a Realistic Model of Mutations Affecting Fitness". Evolution. Hoboken, NJ: John Wiley & Sons for the Society for the Study of Evolution. 57 *(3): 683–689.* doi:10.1554/0014-3820(2003)057[0683:tarmom]2.0.co;2. ISSN 0014-3820. JSTOR 3094781. PMID 12703958.

46. Barton, Nicholas H.; Keightley, Peter D. (January 2002). "Understanding quantitative genetic variation". Nature Reviews Genetics. London: Nature Publishing Group. 3 *(1): 11–21.* doi:10.1038/nrg700. ISSN 1471-0056. PMID 11823787.

47. Sanjuán, Rafael; Moya, Andrés; Elena, Santiago F. (June 1, 2004). "The distribution of fitness effects caused by single-nucleotide substitutions in an RNA virus". Proc. Natl. Acad. Sci. U.S.A. Washington, D.C.: National Academy of Sciences. 101 *(22): 8396–8401. doi:10.1073/pnas.0400146101. ISSN 0027-8424. PMC 420405. PMID 15159545.*

48. Carrasco, Purificación; de la Iglesia, Francisca; Elena, Santiago F. (December 2007). "Distribution of Fitness and Virulence Effects Caused by Single-Nucleotide Substitutions in Tobacco Etch Virus". Journal of Virology. Washington, D.C.: American Society for Microbiology. 81 *(23): 12979–12984. doi:10.1128/JVI.00524-07. ISSN 0022-538X. PMC 2169111. PMID 17898073.*

49. Sanjuán, Rafael (June 27, 2010). "Mutational fitness effects in RNA and single-stranded DNA viruses: common patterns revealed by site-directed mutagenesis studies". Philosophical Transactions of the Royal Society B. London: Royal Society. 365 *(1548): 1975–1982. doi:10.1098/rstb.2010.0063. ISSN 0962-8436. PMC 2880115. PMID 20478892.*

50. Peris, Joan B.; Davis, Paulina; Cuevas, José M.; et al. (June 2010). "Distribution of Fitness Effects Caused by Single-Nucleotide Substitutions in Bacteriophage f1". Genetics. Bethesda, MD: Genetics Society of America. 185 *(2): 603–609. doi:10.1534/genetics.110.115162. ISSN 0016-6731. PMC 2881140.* PMID 20382832.

51. Elena, Santiago F.; Ekunwe, Lynette; Hajela, Neerja; et al. (March 1998). "Distribution of fitness effects caused by random insertion mutations in Escherichia coli". Genetica. Kluwer Academic Publishers. 102–103 *(1–6): 349–358.* doi:10.1023/A:1017031008316. ISSN 0016-6707. PMID 9720287.

52. Hietpas, Ryan T.; Jensen, Jeffrey D.; Bolon, Daniel N. A. (May 10, 2011). "Experimental illumination of a fitness landscape". Proc. Natl. Acad. Sci. U.S.A. Washington, D.C.: National Academy of Sciences. 108 *(19): 7896–7901. doi:10.1073/pnas.1016024108. ISSN 0027-8424. PMC 3093508. PMID 21464309.*

53. Davies, Esther K.; Peters, Andrew D.; Keightley, Peter D. (September 10, 1999). "High Frequency of Cryptic Deleterious Mutations in Caenorhabditis elegans". Science. Washington, D.C.: American Association for the Advancement of Science. 285 *(5434): 1748–1751. doi:10.1126/science.285.5434.1748. ISSN 0036-8075. PMID 10481013.*

54. Loewe, Laurence; Charlesworth, Brian (September 22, 2006). "Inferring the distribution of mutational effects on fitness in Drosophila". Biology Letters. London: Royal Society. 2 *(3): 426–430. doi:10.1098/rsbl.2006.0481. ISSN 1744-9561. PMC 1686194. PMID 17148422.*

55. Eyre-Walker, Adam; Woolfit, Megan; Phelps, Ted (June 2006). "The Distribution of Fitness Effects of New Deleterious Amino Acid Mutations in Humans". Genetics. Bethesda, MD: Genetics Society of America. 173 *(2): 891–900. doi:10.1534/genetics.106.057570. ISSN 0016-6731. PMC 1526495. PMID 16547091.*

56. Sawyer, Stanley A.; Kulathinal, Rob J.; Bustamante, Carlos D.; et al. (August 2003). "Bayesian Analysis Suggests that Most Amino Acid Replacements in Drosophila Are Driven by Positive Selection". Journal of Molecular Evolution. New York: Springer-Verlag. 57 *(1): S154–S164.* doi:10.1007/s00239-003-0022-3. ISSN 0022-2844. PMID 15008412.

57. Piganeau, Gwenaël; Eyre-Walker, Adam (September 2, 2003). "Estimating the distribution of fitness effects from DNA sequence data: implications for the molecular clock". Proc. Natl. Acad. Sci. U.S.A. Washington, D.C.: National Academy of Sciences. 100 *(18): 10335–10340. doi:10.1073/pnas.1833064100. ISSN 0027-8424. PMC 193562. PMID 12925735.*

58. Kimura, Motoo (February 17, 1968). "Evolutionary Rate at the Molecular Level". Nature. London: Nature Publishing Group. 217 *(5129): 624–626.* doi:10.1038/217624a0. ISSN 0028-0836. PMID 5637732.

59. Akashi, Hiroshi (September 30, 1999). "Within- and between-species DNA sequence variation and the ‹footprint› of natural selection". Gene. Amsterdam, the Netherlands: Elsevier. 238 *(1): 39–51.* doi:10.1016/S0378-1119(99)00294-2. ISSN 0378-1119. PMID 10570982.

60. Eyre-Walker, Adam (October 2006). "The genomic rate of adaptive evolution". Trends in Ecology & Evolution. Cambridge, MA: Cell Press. 21 *(10): 569–575. doi:10.1016/j.tree.2006.06.015. ISSN 0169-5347. PMID 16820244.*

61. Gillespie, John H. (September 1984). "Molecular Evolution Over the Mutational Landscape". Evolution. Hoboken, NJ: John Wiley & Sons for the Society for the Study of Evolution. 38 *(5): 1116–1129.* doi:10.2307/2408444. ISSN 0014-3820. JSTOR 2408444.

62. Orr, H. Allen (April 2003). "The Distribution of Fitness Effects Among Beneficial Mutations". Genetics. Bethesda, MD: Genetics Society of America. 163 *(4): 1519–1526. ISSN 0016-6731. PMC 1462510. PMID 12702694. Retrieved 2015-10-07.*

63. Kassen, Rees; Bataillon, Thomas (April 2006). "Distribution of fitness effects among beneficial mutations before selection in experimental populations of bacteria". Nature Genetics. London: Nature Publishing Group. 38 *(4): 484–488.* doi:10.1038/ng1751. ISSN 1061-4036. PMID 16550173.

64. Rokyta, Darin R.; Joyce, Paul; Caudle, S. Brian; et al. (April 2005). "An empirical test of the mutational landscape model of adaptation using a single-stranded DNA virus". Nature Genetics. London: Nature Publishing Group. 37 *(4): 441–444.* doi:10.1038/ng1535. ISSN 1061-4036. PMID 15778707.

65. Imhof, Marianne; Schlotterer, Christian (January 30, 2001). "Fitness effects of advantageous mutations in evolving Escherichia coli populations". Proc. Natl. Acad. Sci. U.S.A. Washington, D.C.: National Academy of Sciences. 98 *(3): 1113–1137. doi:10.1073/pnas.98.3.1113. ISSN 0027-8424. PMC 14717. PMID 11158603.*

66. Hogan, C. Michael (October 12, 2010). "Mutation". In Monosson, Emily. Encyclopedia of Earth. Washington, D.C.: Environmental Information Coalition, National Council for Science and the Environment. OCLC 72808636. Retrieved 2015-10-08.

67. Boillée, Séverine; Vande Velde, Christine; Cleveland, Don W. (October 5, 2006). "ALS: A Disease of Motor Neurons and Their Nonneuronal Neighbors". Neuron. Cambridge, MA: Cell Press. 52 *(1): 39–59. doi:10.1016/j.neuron.2006.09.018. ISSN 0896-6273. PMID 17015226.*

68. den Dunnen, Johan T.; Antonarakis, Stylianos E. (January 2000). "Mutation Nomenclature Extensions and Suggestions to Describe Complex Mutations: A Discussion". Human Mutation. Hoboken, NJ: Wiley-Liss, Inc. 15 *(1): 7–12.* doi:10.1002/(SICI)1098-1004(200001)15:1<7::AID-HUMU4>3.0.CO;2-N. ISSN

1059-7794. PMID 10612815.

69. Doniger, Scott W.; Hyun Seok Kim; Swain, Devjanee; et al. (August 29, 2008). Pritchard, Jonathan K., ed. "A Catalog of Neutral and Deleterious Polymorphism in Yeast". PLOS Genetics. San Francisco, CA: Public Library of Science. 4 *(8): e1000183. doi:10.1371/journal.pgen.1000183. ISSN 1553-7404. PMC 2515631. PMID 18769710.*

70. Ionov, Yurij; Peinado, Miguel A.; Malkhosyan, Sergei; et al. (June 10, 1993). "Ubiquitous somatic mutations in simple repeated sequences reveal a new mechanism for colonic carcinogenesis". Nature. London: Nature Publishing Group. 363 *(6429): 558–561.* doi:10.1038/363558a0. ISSN 0028-0836. PMID 8505985.

71. Sullivan, Amy D.; Wigginton, Janis; Kirschner, Denise (August 28, 2001). "The coreceptor mutation CCR5Δ32 influences the dynamics of HIV epidemics and is selected for by HIV". Proc. Natl. Acad. Sci. U.S.A. Washington, D.C.: National Academy of Sciences. 95 *(18): 10214–10219.* doi:10.1073/pnas.181325198. ISSN 0027-8424. PMC 56941. PMID 11517319.

72. «Mystery of the Black Death". Secrets of the Dead. Season 3. Episode 2. October 30, 2002. PBS. Retrieved 2015-10-10. Episode background.

73. Galvani, Alison P.; Slatkin, Montgomery (December 9, 2003). "Evaluating plague and smallpox as historical selective pressures for the CCR5-Δ32 HIV-resistance allele". Proc. Natl. Acad. Sci. U.S.A. Washington, D.C.: National Academy of Sciences. 100 *(25): 15276*–15279. doi:10.1073/pnas.2435085100. ISSN 0027-8424. PMC 299980. PMID 14645720.

74. Araten, David J.; Golde, David W.; Rong H. Zhang; et al. (September 15, 2005). "A Quantitative Measurement of the Human Somatic Mutation Rate". Cancer Research. Philadelphia, PA: American Association for Cancer Research. 65 *(18): 8111–8117. doi:10.1158/0008-5472.CAN-04-1198. ISSN 0008-5472. PMID 16166284.*

CHAPTER 8

Molecular Basis of Diseases and Ageing

Chapter Contents

Theories suggesting that deterioration and death due to ageing are a purposeful result of an organism's evolved design (such as Weismann's "programmed death" theory) are referred to as theories of programmed ageing or adaptive ageing.

INTRODUCTION

Ageing, also spelled aging, is the process of becoming older. The term refers especially to human beings, many animals, and fungi, whereas for example bacteria, perennial plants and some simple animals are potentially immortal. In the broader sense, ageing can refer to single cells within an organism which have ceased dividing (cellular senescence) or to the population of a species (population ageing).

In humans, ageing represents the accumulation of changes in a human being over time, encompassing physical, psychological, and social change. Reaction time, for example, may slow with age, while knowledge of world events and wisdom may expand. Ageing is among the greatest known risk factors for most human diseases: of the roughly 150,000 people who die each day across the globe, about two thirds die from age-related causes.

The causes of ageing are unknown; current theories are assigned to the damage concept, whereby the accumulation of damage (such as DNA breaks, oxidised DNA and/or mitochondrial malfunctions) may cause biological systems to fail, or to the programmed ageing concept, whereby internal processes (such as DNA telomere shortening) may cause ageing. Programmed ageing should not be confused with programmed cell death (apoptosis).

The discovery, in 1934, that calorie restriction can extend lifespan by 50% in rats has motivated research into delaying and preventing ageing.

AGEING VERSUS IMMORTALITY

Human beings and members of other species, especially animals, necessarily experience ageing and mortality. Fungi, too, can age. In contrast, many species can be considered immortal: for example, bacteria fission to produce daughter cells, strawberry plants grow runners to produce clones of themselves, and animals in the genus *Hydra* have a regenerative ability with which they avoid dying of old age.

Early life forms on Earth, starting at least 3.7 billion years ago, were single-celled organisms. Such single-celled organisms (prokaryotes, protozoans, algae) multiply by fissioning into daughter cells, thus do not age and are innately immortal.

Immortal *Hydra*, a relative of the jellyfish

Ageing and mortality of the individual organism became possible with the evolution of sexual reproduction, which occurred with the emergence of the fungal/animal kingdoms approximately a billion years ago, and with the evolution of flowering plants 160 million years ago. The sexual organism could henceforth pass on some of its genetic material to produce new individuals and itself could become disposable with regards to the survival of its species. This classic biological idea has however been perturbed recently by the discovery that the bacterium *E. coli* may split into distinguishable daughter cells, which opens the theoretical possibility of "age classes" among bacteria.

Even within humans and other mortal species, there are cells with the potential for immortality: cancer cells which have lost the ability to die when maintained in cell culture such as the HeLa cell line, and specific stem cells such as germ cells (producing ova and spermatozoa). In artificial cloning, adult cells can be rejuvenated back to embryonic status and then used to grow a new tissue or animal without ageing. Normal human cells however die after about 50 cell divisions in laboratory culture (the Hayflick Limit, discovered by Leonard Hayflick in 1961).

EFFECTS OF AGEING

Enlarged ears and noses of old humans are sometimes blamed on continual cartilage growth, but the cause is more probably gravity.

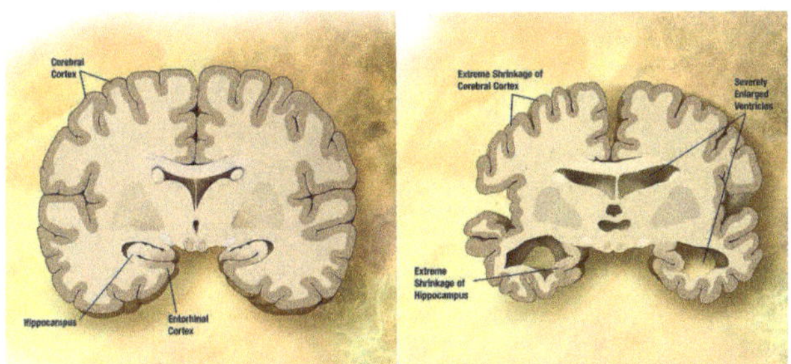

Comparison of a normal aged brain (left) and a brain affected by Alzheimer's disease (right).

A number of characteristic ageing symptoms are experienced by a majority or by a significant proportion of humans during their lifetimes.

- Teenagers lose the young child's ability to hear high-frequency sounds above 20 kHz.
- Some cognitive decline begins in the mid-20s.
- Wrinkles develop mainly due to photoageing, particularly affecting sun-exposed areas (face).
- After peaking in the mid-20s, female fertility declines.
- People over 35 years old are at risk for developing presbyopia, and most people benefit from reading glasses by age 45–50. The cause is lens hardening by decreasing levels of α-crystallin, a process which may be sped up by higher temperatures.
- Hair turns grey with age. Pattern hair loss by the age of 50 affects about half of males and a quarter of females.

- Menopause typically occurs between 49 and 52 years of age.
- In the 60–64 age cohort, the incidence of osteoarthritis rises to 53%. Only 20% however report disabling osteoarthritis at this age.
- Around a third of people between 65 and 74 have hearing loss and almost half of people older than 75.
- By age 80, more than half of all Americans either have a cataract or have had cataract surgery.
- Frailty, defined as loss of muscle mass and mobility, affects 25% of those over 85.
- Atherosclerosis is classified as an ageing disease. It leads to cardiovascular disease (for example stroke and heart attack) which globally is the most common cause of death.
- The maximum human lifespan is suggested to be 115 years "for the foreseeable future". The oldest reliably recorded human was Jeanne Calment who attained 122 years and died in 1997.

Dementia becomes more common with age. About 3% of people between the ages of 65–74 have dementia, 19% between 75 and 84 and nearly half of those over 85 years of age. The spectrum includes mild cognitive impairment and the neurodegenerative diseases of Alzheimer's disease, cerebrovascular disease, Parkinson's disease and Lou Gehrig's disease. Furthermore, many types of memory may decline with ageing, but not semantic memory or general knowledge such as vocabulary definitions, which typically increases or remains steady until late adulthood . Intelligence may decline with age, though the rate may vary depending on the type and may in fact remain steady throughout most of the lifespan, dropping suddenly only as people near the end of their lives. Individual variations in rate of cognitive decline may therefore be explained in terms of people having different lengths of life. There might be changes to the brain: after 20 years of age there may be a 10% reduction each decade in the total length of the brain's myelinated axons.

Age can result in visual impairment, whereby non-verbal communication is reduced, which can lead to isolation and possible depression. Macular degeneration causes vision loss and increases with age, affecting nearly 12% of those above the age of 80. This degeneration is caused by systemic changes in the circulation of waste products and by growth of abnormal vessels around the retina.

A distinction can be made between "proximal ageing" (age-based effects that come about because of factors in the recent past) and "distal ageing" (age-based differences that can be traced back to a cause early in person's life, such as childhood poliomyelitis).

Ageing is among the greatest known risk factors for most human diseases. Of the roughly 150,000 people who die each day across the globe, about two thirds—100,000 per day—die from age-related causes. In industrialised nations, the proportion is higher, reaching 90%.

BIOLOGICAL BASIS OF AGEING

95-year-old woman holding a five-month-old boy

At present, researchers are only just beginning to understand the biological basis of ageing even in relatively simple and short-lived organisms such as yeast. Less still is known about mammalian ageing, in part due to the much longer lives in even small mammals such as the mouse (around 3 years). A primary model organism for studying ageing is the nematode*C. elegans*, thanks to its short lifespan of 2–3 weeks, the ability to easily perform genetic manipulations or suppress gene activity with RNA interference, and other factors. Most known mutations and RNA interference targets that extend lifespan were first discovered in *C. elegans*.

Factors that are proposed to influence biological ageing fall into two main categories, *programmed* and *damage-related*. Programmed factors follow a biological timetable, perhaps a continuation of the one that regulates childhood growth and development. This regulation would depend on changes in gene expression that affect the systems responsible for maintenance, repair and defence responses. Damage-related factors include internal and environmental assaults to living organisms that induce cumulative damage at various levels.

There are three main metabolic pathways which can influence the rate of ageing:

- the FOXO3/Sirtuin pathway, probably responsive to caloric restriction
- the Growth hormone/Insulin-like growth factor 1 signalling pathway

- the activity levels of the electron transport chain in mitochondria and (in plants) in chloroplasts.

It is likely that most of these pathways affect ageing separately, because targeting them simultaneously leads to additive increases in lifespan.

Programmed Factors

The rate of ageing varies substantially across different species, and this, to a large extent, is genetically based. For example, numerous perennial plants ranging from strawberries and potatoes to willow trees typically produce clones of themselves by vegetative reproduction and are thus potentially immortal, while annual plants such as wheat and watermelons die each year and reproduce by sexual reproduction. In 2008 it was discovered that inactivation of only two genes in the annual plant *Arabidopsis thaliana* leads to its conversion into a potentially immortal perennial plant.

Clonal immortality apart, there are certain species whose individual lifespans stand out among Earth's life-forms, including the bristlecone pine at 5062 years(however Hayflick states that the bristlecone pine has no cells older than 30 years), invertebrates like the hard clam (known as *quahog* in New England) at 508 years, the Greenland shark at 400 years, fish like the sturgeon and the rockfish, and the sea anemone and lobster. Such organisms are sometimes said to exhibit negligible senescence. The genetic aspect has also been demonstrated in studies of human centenarians.

In laboratory settings, researchers have demonstrated that selected alterations in specific genes can extend lifespan quite substantially in yeast and roundworms, less so in fruit flies and less again in mice. Some of the targeted genes have homologues across species and in some cases have been associated with human longevity.

- Telomeres: In humans and other animals, cellular senescence has been attributed to the shortening of telomeres at each cell division; when telomeres become too short, the cells senesce and die or cease multiplying. The length of telomeres is therefore the "molecular clock", predicted by Hayflick.However, telomere length in wild mouse strains is unrelated to lifespan, and mice lacking the telomerase enzyme do not have a dramatically reduced lifespan. Laboratory mice's telomeres are many times longer than human ones. Another caveat is that a study following nearly 1000 humans for ten years showed that while some humans do shorten their telomeres over time, a third of the participants did not.
- A variation in the gene FOXO3A has a positive effect on the life expectancy of humans, and is found much more often in people living to 100 and beyond – moreover, this appears to be true worldwide. FOXO3A acts on the sirtuin family of genes which also have a significant effect on lifespan in yeast and in nematodes. Sirtuin' in turn inhibits mTOR.
- Caloric restriction leads to longer lifespans in various species, an effect that is unclear, but probably mediated by the nutrient-sensing function of the mTOR pathway.
- mTOR, a protein that inhibits autophagy, has been linked to ageing through the insulin signalling pathway. mTOR functions through nutrient and growth cues

leading scientists to believe that dietary restriction and mTOR are related in terms of longevity. When organisms restrict their diet, mTOR activity is reduced, which allows an increased level of autophagy. This recycles old or damaged cell parts, which increases longevity and decreases the chances of being obese. This is thought to prevent spikes of glucose concentration in the blood, leading to reduced insulin signalling. This has been linked to less mTOR activation as well. Therefore, longevity has been connected to caloric restriction and insulin sensitivity inhibiting mTOR, which in turns allows autophagy to occur more frequently. It may be that mTOR inhibition and autophagy reduce the effects of reactive oxygen species on the body, which damage DNA and other organic material, so longevity would be increased.

- A decreased Growth hormone/Insulin-like Growth Factor 1 signalling pathway has been associated with increased life span in various organisms including fruit flies, nematodes and mice. The precise mechanism by which decreased GH/IGF-1 signalling increases longevity is unknown, but various mouse strains with decreased GH and/or IGF-1 induced signalling share a similar phenotype which includes increased insulin sensitivity, enhanced stress resistance and protection from carcinogenesis. The studied mouse strains with decreased GH signalling showed between 20% and 68% increased longevity, and mouse strains with decreased IGF-1 induced signalling revealed a 19 to 33% increase in life span when compared to control mice.
- DNA methylation: The strong effect of age on DNA methylation levels has been known since the late 1960s. Horvath hypothesised that DNA methylation age measures the cumulative effect of an epigenetic maintenance system but details are unknown. The fact that DNA methylation age of blood predicts all-cause mortality in later life suggests that it relates to a process that causes ageing.
- Over-expression of the Ras2 gene increases lifespan in yeast by 30%. A yeast mutant lacking the genes sch9 and ras2 has recently been shown to have a tenfold increase in lifespan under conditions of calorie restriction and is the largest increase achieved in any organism.
- Evolutionary theories of ageing: Many have argued that life span, like other phenotypes, is selected. Traits that benefit early survival and reproduction will be selected for even if they contribute to an earlier death. Such a genetic effect is called the antagonistic pleiotropy effect when referring to a gene (pleiotropy signifying the gene has a double function – enabling reproduction at a young age but costing the organism life expectancy in old age) and is called the disposable soma effect when referring to an entire genetic programme (the organism diverting limited resources from maintenance to reproduction). Some evidence is provided by oxygen-deprived bacterial cultures. This would explain why the autosomal dominant disease, Huntington's disease, can persist even though it is inexorably lethal. Also, it has been suggested that some of the genetic variants that increase fertility in the young increase cancer risk in the old. Such variants occur in genes p53 and BRCA1. The biological mechanisms which regulate lifespan evolved several hundred million years ago.

- Reproductive-cell cycle theory: The idea that ageing is regulated by reproductive hormones that act in an antagonistic pleiotropic manner via cell cycle signalling, promoting growth and development early in life to achieve reproduction, but becoming dysregulated later in life, driving senescence (dyosis) in a futile attempt to maintain reproductive ability. The endocrine dyscrasia that follows the loss of follicles with menopause, and the loss of Leydig and Sertoli cells during andropause, drive aberrant cell cycle signalling that leads to cell death and dysfunction, tissue dysfunction (disease) and ultimately death. Moreover, the hormones that regulate reproduction also regulate cellular metabolism, explaining the increases in fat deposition during pregnancy through to the deposition of centralised adiposity with the dysregulation of the HPG axis following menopause and during andropause (Atwood and Bowen, 2006). This theory, which introduced a new definition of ageing, has facilitated the conceptualisation of why and how ageing occurs at the evolutionary, physiological and molecular levels.
- Autoimmunity: The idea that ageing results from an increase in autoantibodies that attack the body's tissues. A number of diseases associated with ageing, such as atrophic gastritis and Hashimoto's thyroiditis, are probably autoimmune in this way. However, while inflammation is very much evident in old mammals, even completely immunodeficient mice raised in pathogen-free laboratory conditions still experience senescence.

An elderly Somali woman

- The cellular balance between energy generation and consumption (energy homeostasis) requires tight regulation during ageing. In 2011, it was demonstrated that acetylation levels of AMP-activated protein kinase change with age in yeast and that preventing this change slows yeast ageing.

Damage-Related Factors

- DNA damage theory of ageing: DNA damage is thought to be the common basis of both cancer and ageing, and it has been argued that intrinsic causes of DNA damage are the most important drivers of ageing. Genetic damage (aberrant structural alterations of the DNA), mutations (changes in the DNA sequence), and epimutations (methylation of gene promoter regions or alterations of the DNA scaffolding which regulate gene expression), can cause abnormal gene expression. DNA damage causes the cells to stop dividing or induces apoptosis, often affecting stem cell pools and hence hindering regeneration. However, lifelong studies of mice suggest that most mutations happen during embryonic and childhood development, when cells divide often, as each cell division is a chance for errors in DNA replication.

- Genetic instability: In heart muscle cells, dogs annually lose approximately 3.3% of the DNA in their heart muscle cells while humans lose approximately 0.6% of their heart muscle DNA each year. These numbers are close to the ratio of the maximum longevities of the two species (120 years vs. 20 years, a 6/1 ratio). The comparative percentage is also similar between the dog and human for yearly DNA loss in the brain and lymphocytes. As stated by lead author, Bernard L. Strehler, "... genetic damage (particularly gene loss) is almost certainly (or probably the) central cause of ageing."
- Accumulation of waste:
- A buildup of waste products in cells presumably interferes with metabolism. For example, a waste product called lipofuscin is formed by a complex reaction in cells that binds fat to proteins. This waste accumulates in the cells as small granules, which increase in size as a person ages.
- The hallmark of ageing yeast cells appears to be overproduction of certain proteins.
- Autophagy induction can enhance clearance of toxic intracellular waste associated with neurodegenerative diseases and has been comprehensively demonstrated to improve lifespan in yeast, worms, flies, rodents and primates. The situation, however, has been complicated by the identification that autophagy up-regulation can also occur during ageing. Autophagy is enhanced in obese mice by caloric restriction, exercise, and a low fat diet (but in these mice is evidently not related with the activation of AMP-activated protein kinase, see above).
- Wear-and-tear theory: The very general idea that changes associated with ageing are the result of chance damage that accumulates over time.
- Accumulation of errors: The idea that ageing results from chance events that escape proof reading mechanisms, which gradually damages the genetic code.
- Cross-linkage: The idea that ageing results from accumulation of cross-linked compounds that interfere with normal cell function.
- Free-radical theory: Damage by free radicals, or more generally reactive oxygen species or oxidative stress, create damage that may gives rise to the symptoms we recognise as ageing. Michael Ristow's group has provided evidence that the effect of calorie restriction may be due to increased formation of free radicals within the mitochondria, causing a secondary induction of increased antioxidant defence capacity.
- DNA oxidation and caloric restriction: Caloric restriction reduces 8-OH-dG

DNA damage in organs of ageing rats and mice. Thus, reduction of oxidative DNA damage is associated with a slower rate of ageing and increased lifespan.

MANAGEMENT

Lifestyle

Caloric restriction substantially affects lifespan in many animals, including the ability to delay or prevent many age-related diseases. Typically, this involves caloric intake of 60–70% of what an *ad libitum* animal would consume, while still maintaining proper nutrient intake. In rodents, this has been shown to increase lifespan by up to 50%; similar effects occur for yeast and *Drosophila*. No lifespan data exist for humans on a calorie-restricted diet, but several reports support protection from age-related diseases. Two major ongoing studies on rhesus monkeys initially revealed disparate results; while one study, by the University of Wisconsin, showed that caloric restriction does extend lifespan, the second study, by the National Institute on Ageing (NIA), found no effects of caloric restriction on longevity. Both studies nevertheless showed improvement in a number of health parameters. Notwithstanding the similarly low calorie intake, the diet composition differed between the two studies (notably a high sucrose content in the Wisconsin study), and the monkeys have different origins (India, China), initially suggesting that genetics and dietary composition, not merely a decrease in calories, are factors in longevity. However, in a comparative analysis in 2014, the Wisconsin researchers found that the allegedly non-starved NIA control monkeys in fact are moderately underweight when compared with other monkey populations, and argued this was due to the NIA's apportioned feeding protocol in contrast to Wisconsin's truly unrestricted *ad libitum* feeding protocol. They conclude that moderate calorie restriction rather than extreme calorie restriction is sufficient to produce the observed health and longevity benefits in the studied rhesus monkeys.

In his book *How and Why We Age*, Hayflick says that caloric restriction may not be effective in humans, citing data from the Baltimore Longitudinal Study of Aging which shows that being thin does not favour longevity. Similarly, it is sometimes claimed that moderate *obesity* in later life may improve survival, but newer research has identified confounding factors such as weight loss due to terminal disease. Once these factors are accounted for, the optimal body weight above age 65 corresponds to a leaner body mass index of 23 to 27.

Alternatively, the benefits of dietary restriction can also be found by changing the macro nutrient profile to reduce protein intake without any changes to calorie level, resulting in similar increases in longevity. Dietary protein restriction not only inhibits mTOR activity but also IGF-1, two mechanisms implicated in ageing.Specifically, reducing leucine intake is sufficient to inhibit mTOR activity, achievable through reducing animal food consumption.

The Mediterranean diet is credited with lowering the risk of heart disease and early death. The major contributors to mortality risk reduction appear to be a higher consumption of vegetables, fish, fruits, nuts and monounsaturated fatty acids, i.e., olive oil.

The amount of sleep has an impact on mortality. People who live the longest report sleeping for six to seven hours each night. Lack of sleep (<5 hours) more than doubles the risk of death from cardiovascular disease, but too much sleep (>9 hours) is associated with a doubling of the risk of death, though not primarily from cardiovascular disease. Sleeping more than 7 to 8 hours per day has been consistently associated with increased mortality, though the cause is probably other factors such as depression and socioeconomic status, which would correlate statistically. Sleep monitoring of hunter-gatherer tribes from Africa and from South America has shown similar sleep patterns across continents: their average sleeping duration is 6.4 hours (with a summer/winter difference of 1 hour), afternoon naps (siestas) are uncommon, and insomnia is very rare (tenfold less than in industrial societies).

Physical exercise may increase life expectancy. People who participate in moderate to high levels of physical exercise have a lower mortality rate compared to individuals who are not physically active. Moderate levels of exercise have been correlated with preventing aging and improving quality of life by reducing inflammatory potential. The majority of the benefits from exercise are achieved with around 3500 metabolic equivalent (MET) minutes per week. For example, climbing stairs 10 minutes, vacuuming 15 minutes, gardening 20 minutes, running 20 minutes, and walking or bicycling for 25 minutes on a daily basis would *together* achieve about 3000 MET minutes a week.

Avoidance of chronic stress (as opposed to acute stress) is associated with a slower loss of telomeres in most but not all studies, and with decreased cortisol levels. A chronically high cortisol level compromises the immune system, causes cardiac damage/ arterosclerosis and is associated with facial ageing, and the latter in turn is a marker for increased morbidity and mortality. Stress can be countered by social connection, spirituality, and (for men more clearly than for women) married life, all of which are associated with longevity.

Medical Intervention

The following drugs and interventions have been shown to retard or reverse the biological effects of ageing in animal models, but none has yet been proven to do so in humans.

Evidence in both animals and humans suggests that resveratrol may be a caloric restriction mimetic.

As of 2015 metformin was under study for its potential effect on slowing ageing in the worm *C.elegans* and the cricket. Its effect on otherwise healthy humans is unknown.

Rapamycin was first shown to extend lifespan in eukaryotes in 2006 by Powers *et al.* who showed a dose-responsive effect of rapamycin on lifespan extension in yeast cells. In a 2009 study, the lifespans of mice fed rapamycin were increased between 28 and 38% from the beginning of treatment, or 9 to 14% in total increased maximum lifespan. Of particular note, the treatment began in mice aged 20 months, the equivalent of 60 human years. Rapamycin has subsequently been shown to extend mouse lifespan in several separate experiments, and is now being tested for this purpose in nonhuman primates (the marmoset monkey).

Cancer geneticist Ronald A. DePinho and his colleagues published research in mice where telomerase activity was first genetically removed. Then, after the mice had prematurely aged, they restored telomerase activity by reactivating the telomerase gene. As a result, the mice were rejuvenated: Shrivelled testes grew back to normal and the animals regained their fertility. Other organs, such as the spleen, liver, intestines and brain, recuperated from their degenerated state. "offers the possibility that normal human ageing could be slowed by reawakening the enzyme in cells where it has stopped working" says Ronald DePinho. However, activating telomerase in humans could potentially encourage the growth of tumours.

Most known genetic interventions in *C. elegans* increase lifespan by 1.5 to 2.5-fold. As of 2009, the record for lifespan extension in C. *elegans* is a single-gene mutation which increases adult survival by tenfold. The strong conservation of some of the mechanisms of ageing discovered in model organisms imply that they may be useful in the enhancement of human survival. However, the benefits may not be proportional; longevity gains are typically greater in *C. elegans* than fruit flies, and greater in fruit flies than in mammals. One explanation for this is that mammals, being much longer-lived, already have many traits which promote lifespan.

Research Projects and Prizes

Some research effort is directed to slow ageing and extend healthy lifespan.

The US National Institute on Aging currently funds an intervention testing programme, whereby investigators nominate compounds (based on specific molecular ageing theories) to have evaluated with respect to their effects on lifespan and age-related biomarkers in outbred mice. Previous age-related testing in mammals has proved largely irreproducible, because of small numbers of animals and lax mouse husbandry conditions. The intervention testing programme aims to address this by conducting parallel experiments at three internationally recognised mouse ageing-centres, the Barshop Institute at UTHSCSA, the University of Michigan at Ann Arbor and the Jackson Laboratory.

Several companies and organisations, such as Google Calico, Human Longevity, Craig Venter, Gero, SENS Research Foundation, and Science for Life Extension in Russia, declared stopping or delaying ageing as their goal.

Prizes for extending lifespan and slowing ageing in mammals exist. The Methuselah Foundation offers the Mprize. Recently, the $1 Million Palo Alto Longevity Prize was launched. It is a research incentive prize to encourage teams from all over the world to compete in an all-out effort to "hack the code" that regulates our health and lifespan. It was founded by Joon Yun.

SOCIETY AND CULTURE

Different cultures express age in different ways. The age of an adult human is commonly measured in whole years since the day of birth. Arbitrary divisions set to mark periods of life may include: juvenile (via infancy, childhood, preadolescence, adolescence), early adulthood, middle adulthood, and late adulthood. More casual terms may include "teenagers," "tweens," "twentysomething", "thirtysomething", etc. as well as "vicenarian", "tricenarian", "quadragenarian", etc.

An elderly man

Most legal systems define a specific age for when an individual is allowed or obliged to do particular activities. These age specifications include voting age, drinking age, age of consent, age of majority, age of criminal responsibility, marriageable age, age of candidacy, and mandatory retirement age. Admission to a movie for instance, may depend on age according to a motion picture rating system. A bus fare might be discounted for the young or old. Each nation, government and non-governmental organisation has different ways of classifying age. In other words, chronological ageing may be distinguished from "social ageing" (cultural age-expectations of how people should act as they grow older) and "biological ageing" (an organism's physical state as it ages).

In a UNFPA report about ageing in the 21st century, it highlighted the need to "Develop a new rights-based culture of ageing and a change of mindset and societal attitudes towards ageing and older persons, from welfare recipients to active, contributing members of society." UNFPA said that this "requires, among others, working towards the development of international human rights instruments and their translation into national laws and regulations and affirmative measures that challenge age discrimination and recognise older people as autonomous subjects." Older persons make contributions to society including caregiving and volunteering. For example, "A study of Bolivian migrants who [had] moved to Spain found that 69% left their children at home, usually with grandparents. In rural China, grandparents care for 38% of children aged under five whose parents have gone to work in cities."

Economics

Population Ageing

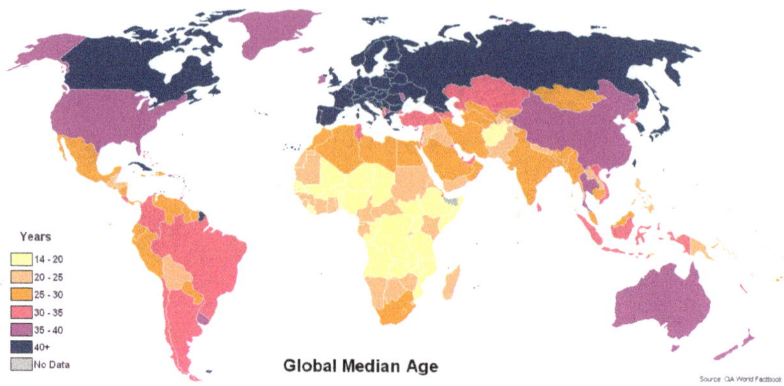

A map showing median age figures for 2001

Population ageing is the increase in the number and proportion of older people in society. Population ageing has three possible causes: migration, longer life expectancy (decreased death rate) and decreased birth rate. Ageing has a significant impact on society. Young people tend to have fewer legal privileges (if they are below the age of majority), they are more likely to push for political and social change, to develop and adopt new technologies, and to need education. Older people have different requirements from society and government, and frequently have differing values as well, such as for property and pension rights.

In the 21st century, one of the most significant population trends is ageing. Currently, over 11% of the world's current population are people aged 60 and older and the United Nations Population Fund (UNFPA) estimates that by 2050 that number will rise to approximately 22%. Ageing has occurred due to development which has enabled better nutrition, sanitation, health care, education and economic well-being. Consequently, fertility rates have continued to decline and life expectancy have risen. Life expectancy at birth is over 80 now in 33 countries. Ageing is a "global phenomenon," that is occurring fastest in developing countries, including those with large youth populations, and poses social and economic challenges to the work which can be overcome with "the right set of policies to equip individuals, families and societies to address these challenges and to reap its benefits."

As life expectancy rises and birth rates decline in developed countries, the median age rises accordingly. According to the United Nations, this process is taking place in nearly every country in the world. A rising median age can have significant social and economic implications, as the workforce gets progressively older and the number of old workers and retirees grows relative to the number of young workers. Older people generally incur more health-related costs than do younger people in the workplace and can also cost more in worker's compensation and pension liabilities. In most developed countries an older workforce is somewhat inevitable. In the United States for instance, the Bureau of Labor Statistics estimates that one in four American workers will be 55 or older by 2020.

Among the most urgent concerns of older persons worldwide is income security. This poses challenges for governments with ageing populations to ensure investments in pension systems continues in order to provide economic independence and reduce poverty in old age. These challenges vary for developing and developed countries. UNFPA stated that, "Sustainability of these systems is of particular concern, particularly in developed countries, while social protection and old-age pension coverage remain a challenge for developing countries, where a large proportion of the labour force is found in the informal sector."

The global economic crisis has increased financial pressure to ensure economic security and access to health care in old age. In order to elevate this pressure "social protection floors must be implemented in order to guarantee income security and access to essential health and social services for all older persons and provide a safety net that contributes to the postponement of disability and prevention of impoverishment in old age."

It has been argued that population ageing has undermined economic development. Evidence suggests that pensions, while making a difference to the well-being of older persons, also benefit entire families especially in times of crisis when there may be a shortage or loss of employment within households. A study by the Australian Government in 2003 estimated that "women between the ages of 65 and 74 years contribute A$16 billion per year in unpaid caregiving and voluntary work. Similarly, men in the same age group contributed A$10 billion per year."

Due to increasing share of the elderly in the population, health care expenditures will continue to grow relative to the economy in coming decades. This has been considered as a negative phenomenon and effective strategies like labour productivity enhancement should be considered to deal with negative consequences of ageing.

Sociology

In the field of sociology and mental health, ageing is seen in five different views: ageing as maturity, ageing as decline, ageing as a life-cycle event, ageing as generation, and ageing as survival. Positive correlates with ageing often include economics, employment, marriage, children, education, and sense of control, as well as many others. The social science of ageing includes disengagement theory, activity theory, selectivity theory, and continuity theory. Retirement, a common transition faced by the elderly, may have both positive and negative consequences. As cyborgs currently are on the rise some theorists argue there is a need to develop new definitions of ageing and for instance a bio-techno-social definition of ageing has been suggested.

Health Care Demand

With age inevitable biological changes occur that increase the risk of illness and disability. UNFPA states that,

"A life-cycle approach to health care – one that starts early, continues through the reproductive years and lasts into old age – is essential for the physical and emotional well-being of older persons, and, indeed, all people. Public policies and programmes should additionally address the needs of older impoverished people who cannot afford health care."

Many societies in Western Europe and Japan have ageing populations. While the effects on society are complex, there is a concern about the impact on health care demand. The large number of suggestions in the literature for specific interventions to cope with the expected increase in demand for long-term care in ageing societies can be organised under four headings: improve system performance; redesign service delivery; support informal caregivers; and shift demographic parameters.

However, the annual growth in national health spending is not mainly due to increasing demand from ageing populations, but rather has been driven by rising incomes, costly new medical technology, a shortage of health care workers and informational asymmetries between providers and patients. A number of health problems become more prevalent as people get older. These include mental health problems as well as physical health problems, especially dementia.

It has been estimated that population ageing only explains 0.2 percentage points of the annual growth rate in medical spending of 4.3% since 1970. In addition, certain reforms to the Medicare system in the United States decreased elderly spending on home health care by 12.5% per year between 1996 and 2000.

Self-Perception of Ageing

Positive self-perception of health has been correlated with higher well-being and reduced mortality in the elderly. Various reasons have been proposed for this association; people who are objectively healthy may naturally rate their health better than that of their ill counterparts, though this link has been observed even in studies which have controlled for socioeconomic status, psychological functioning and health status. This finding is generally stronger for men than women, though this relationship is not universal across all studies and may only be true in some circumstances.

As people age, subjective health remains relatively stable, even though objective health worsens. In fact, perceived health improves with age when objective health is controlled in the equation. This phenomenon is known as the "paradox of ageing." This may be a result of social comparison; for instance, the older people get, the more they may consider themselves in better health than their same-aged peers. Elderly people often associate their functional and physical decline with the normal ageing process.

Successful Ageing

The concept of *successful ageing* can be traced back to the 1950s and was popularised in the 1980s. Traditional definitions of successful ageing have emphasised absence of physical and cognitive disabilities. In their 1987 article, Rowe and Kahn characterised successful ageing as involving three components: a) freedom from disease and disability, b) high cognitive and physical functioning, and c) social and productive engagement.

Tree stump after one year's regrowth

Cultural References

The ancient Greek dramatist Euripides (5th century BC) describes the multiply-headed mythological monster Hydra as having a regenerative capacity which makes it immortal, which is the historical background to the name of the biological genus Hydra. The Book of Job (c. 6th century BC) describes human lifespan as inherently limited and makes a comparison with the innate immortality that a felled tree may have when undergoing vegetative regeneration.

DIABETES

Other recent studies suggest that promoters of genes may be the primary cause of diabetes. Promoters of genes associated with diabetes by Genome-wide association studies (GWAS) show specific DNA patterns for each phenotype. This observation indicates that the promoters of these genes use specific transcription factors for each diabetes phenotype.

DISEASES ASSOCIATED WITH ABERRANT FUNCTION

Although OMIM is a major resource for gathering information on the relationship between mutations and natural variation in gene sequence and susceptibility to hundreds of diseases, a sophisticated search strategy is required to extract diseases associated with defects in transcriptional control where the promoter is believed to have direct involvement.

This is a list of diseases where evidence suggests some promoter malfunction, through either direct mutation of a promoter sequence or mutation in a transcription factor or transcriptional co-activator.

Most diseases are heterogeneous in etiology, meaning that one "disease" is often many different diseases at the molecular level, though symptoms exhibited and response to treatment may be identical. How diseases of different molecular origin respond to treatments is partially addressed in the discipline of pharmacogenomics.

Not listed here are the many kinds of cancers involving aberrant transcriptional regulation owing to creation of chimeric genes through pathological chromosomal translocation. Importantly, intervention in the number or structure of promoter-bound proteins is one key to treating a disease without affecting expression of unrelated genes sharing elements with the target gene. Genes where change is not desirable are capable of influencing the potential of a cell to become cancerous and form a tumor.

CpG Islands In Promoters

Regulation of transcription in cancer

In humans, about 70% of promoters located near the transcription start site of a gene (proximal promoters) contain a CpG island. CpG islands are generally 200 to 2000 base pairs long, have a C:G base pair content >50%, and have regions of DNA where a cytosine nucleotide is followed by a guanine nucleotide and this occurs frequently in the linear sequence of bases along its 5' → 3' direction.

Distal promoters also frequently contain CpG islands, such as the promoter of the DNA repair gene *ERCC1*, where the CpG island-containing promoter is located about 5,400 nucleotides upstream of the coding region of the *ERCC1* gene. CpG islands also occur frequently in promoters for functional noncoding RNAs such as microRNAs.

ASTHMA

Asthma is a common long term inflammatory disease of the airways of the lungs. It is characterized by variable and recurring symptoms, reversible airflow obstruction, and bronchospasm. Symptoms include episodes of wheezing, coughing, chest tightness, and shortness of breath. These episodes may occur a few times a day or a few times per week. Depending on the person they may become worse at night or with exercise.

Asthma is thought to be caused by a combination of genetic and environmental factors. Environmental factors include exposure to air pollution and allergens. Other potential triggers include medications such as aspirin and beta blockers. Diagnosis is usually based on the pattern of symptoms, response to therapy over time, and spirometry. Asthma is classified according to the frequency of symptoms, forced expiratory volume in one second (FEV1), and peak expiratory flow rate. It may also be classified as atopic or non-atopic where atopy refers to a predisposition toward developing a type 1 hypersensitivity reaction.

There is no cure for asthma. Symptoms can be prevented by avoiding triggers, such as allergens and irritants, and by the use of inhaled corticosteroids. Long-acting beta agonists (LABA) or antileukotriene agents may be used in addition to inhaled corticosteroids if asthma symptoms remain uncontrolled. Treatment of rapidly worsening symptoms is usually with an inhaled short-acting beta-2 agonist such as salbutamol and corticosteroids taken by mouth. In very severe cases, intravenous corticosteroids, magnesium sulfate, and hospitalization may be required.

In 2013, 242 million people globally had asthma up from 183 million in 1990. It caused about 489,000 deaths in 2013, most of which occurred in the developing world. It often begins in childhood. The rates of asthma have increased significantly since the 1960s. Asthma was recognized as early as Ancient Egypt. The word asthma is from the Greek σθμα, *ásthma* which means «panting».

Signs and Symptoms

Asthma is characterized by recurrent episodes of wheezing, shortness of breath, chest tightness, and coughing. Sputum may be produced from the lung by coughing but is often hard to bring up. During recovery from an attack, it may appear pus-like due to high levels of white blood cells called eosinophils. Symptoms are usually worse at night and in the early morning or in response to exercise or cold air. Some people with asthma rarely experience symptoms, usually in response to triggers, whereas others may have marked and persistent symptoms.

Associated Conditions

A number of other health conditions occur more frequently in those with asthma, including gastro-esophageal reflux disease (GERD), rhinosinusitis, and obstructive sleep apnea.Psychological disorders are also more common, with anxiety disorders occurring in between 16–52% and mood disorders in 14–41%. However, it is not known if asthma causes psychological problems or if psychological problems lead to asthma. Those with asthma, especially if it is poorly controlled, are at high risk for radiocontrast reactions.

Causes

Asthma is caused by a combination of complex and incompletely understood environmental and genetic interactions. These factors influence both its severity and its responsiveness to treatment. It is believed that the recent increased rates of asthma are due to changing epigenetics (heritable factors other than those related to the DNA sequence) and a changing living environment. Onset before age 12 is more likely due to genetic influence, while onset after 12 is more likely due to environmental influence.

Environmental

Many environmental factors have been associated with asthma's development and exacerbation including allergens, air pollution, and other environmental chemicals. Smoking during pregnancy and after delivery is associated with a greater risk of asthma-like symptoms. Low air quality from factors such as traffic pollution or high ozone levels has been associated with both asthma development and increased asthma severity. Over half of cases in children in the United States occur in areas with air quality below EPA standards. Exposure to indoor volatile organic compounds may be a trigger for asthma; formaldehyde exposure, for example, has a positive association. Also, phthalates in certain types of PVC are associated with asthma in children and adults.

There is an association between acetaminophen (paracetamol) use and asthma. The majority of the evidence does not, however, support a causal role. A 2014 review found that the association disappeared when respiratory infections were taken into account. Use by a mother during pregnancy is also associated with an increased risk as is psychological stress during pregnancy.

Asthma is associated with exposure to indoor allergens. Common indoor allergens include dust mites, cockroaches, animal dander (fragments of fur or feathers), and mold. Efforts to decrease dust mites have been found to be ineffective on symptoms in sensitized subjects. Certain viral respiratory infections, such as respiratory syncytial virus and rhinovirus,may increase the risk of developing asthma when acquired as young children. Certain other infections, however, may decrease the risk.

Hygiene Hypothesis

The hygiene hypothesis attempts to explain the increased rates of asthma worldwide as a direct and unintended result of reduced exposure, during childhood, to non-pathogenic bacteria and viruses. It has been proposed that the reduced exposure to bacteria and viruses is due, in part, to increased cleanliness and decreased family size in modern societies. Exposure to bacterial endotoxin in early childhood may prevent the development of asthma, but exposure at an older age may provoke bronchoconstriction. Evidence supporting the hygiene hypothesis includes lower rates of asthma on farms and in households with pets.

Use of antibiotics in early life has been linked to the development of asthma. Also, delivery via caesarean section is associated with an increased risk (estimated at 20–80%) of asthma—this increased risk is attributed to the lack of healthy bacterial colonization that the newborn would have acquired from passage through the birth canal. There is a link between asthma and the degree of affluence.

Genetic

CD14-endotoxin interaction based on CD14 SNP C-159T		
Endotoxin levels	**CC genotype**	**TT genotype**

High exposure	Low risk	High risk
Low exposure	High risk	Low risk

Family history is a risk factor for asthma, with many different genes being implicated. If one identical twin is affected, the probability of the other having the disease is approximately 25%. By the end of 2005, 25 genes had been associated with asthma in six or more separate populations, including GSTM1, IL10, CTLA-4, SPINK5, LTC4S, IL4R and ADAM33, among others. Many of these genes are related to the immune system or modulating inflammation. Even among this list of genes supported by highly replicated studies, results have not been consistent among all populations tested. In 2006 over 100 genes were associated with asthma in one genetic association study alone; more continue to be found.

Some genetic variants may only cause asthma when they are combined with specific environmental exposures. An example is a specific single nucleotide polymorphism in the CD14 region and exposure to endotoxin (a bacterial product). Endotoxin exposure can come from several environmental sources including tobacco smoke, dogs, and farms. Risk for asthma, then, is determined by both a person's genetics and the level of endotoxin exposure.

Medical Conditions

A triad of atopic eczema, allergic rhinitis and asthma is called atopy. The strongest risk factor for developing asthma is a history of atopic disease; with asthma occurring at a much greater rate in those who have either eczema or hay fever. Asthma has been associated with eosinophilic granulomatosis with polyangiitis (formerly known as Churg–Strauss syndrome), an autoimmune disease and vasculitis. Individuals with certain types of urticaria may also experience symptoms of asthma.

There is a correlation between obesity and the risk of asthma with both having increased in recent years. Several factors may be at play including decreased respiratory function due to a buildup of fat and the fact that adipose tissue leads to a pro-inflammatory state.

Beta blocker medications such as propranolol can trigger asthma in those who are susceptible. Cardioselective beta-blockers, however, appear safe in those with mild or moderate disease. Other medications that can cause problems in asthmatics are angiotensin-converting enzyme inhibitors, aspirin, and NSAIDs.

Exacerbation

Some individuals will have stable asthma for weeks or months and then suddenly develop an episode of acute asthma. Different individuals react to various factors in different ways. Most individuals can develop severe exacerbation from a number of triggering agents.

Home factors that can lead to exacerbation of asthma include dust, animal dander (especially cat and dog hair), cockroach allergens and mold. Perfumes are a common cause of acute attacks in women and children. Both viral and bacterial infections of the upper respiratory tract can worsen the disease. Psychological stress may worsen symptoms—it is thought that stress alters the immune system and thus increases the airway inflammatory response to allergens and irritants.

Pathophysiology

Pathophysiology of Asthma

Asthma is the result of chronic inflammation of the conducting zone of the airways (most especially the bronchi and bronchioles), which subsequently results in increased contractability of the surrounding smooth muscles. This among other factors leads to bouts of narrowing of the airway and the classic symptoms of wheezing. The narrowing is typically reversible with or without treatment. Occasionally the airways themselves change. Typical changes in the airways include an increase in eosinophils and thickening of the lamina reticularis. Chronically the airways' smooth muscle may increase in size along with an increase in the numbers of mucous glands. Other cell types involved include: T lymphocytes, macrophages, and neutrophils. There may also be involvement of other components of the immune system including: cytokines, chemokines, histamine, and leukotrienes among others.

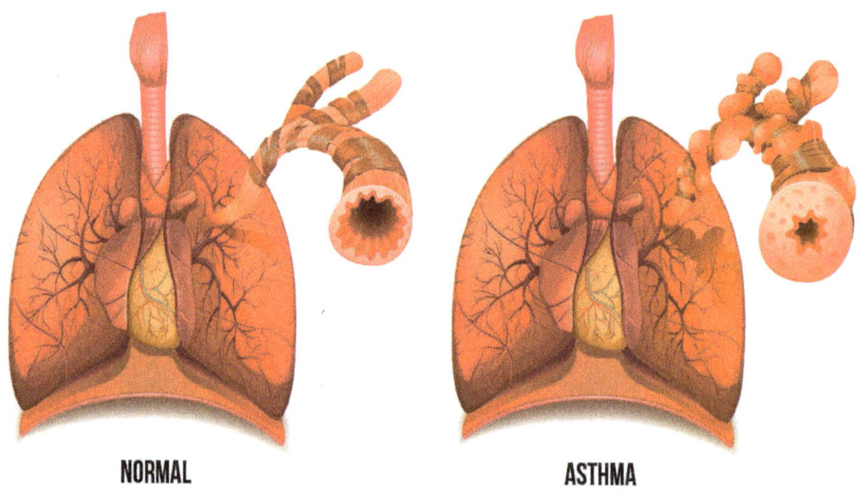

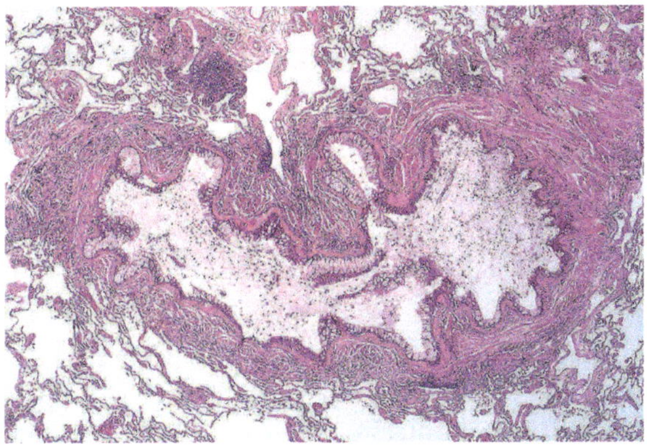

Obstruction of the lumen of a bronchiole by mucoid exudate, goblet cell metaplasia, and epithelial basement membrane thickening in a person with asthma.

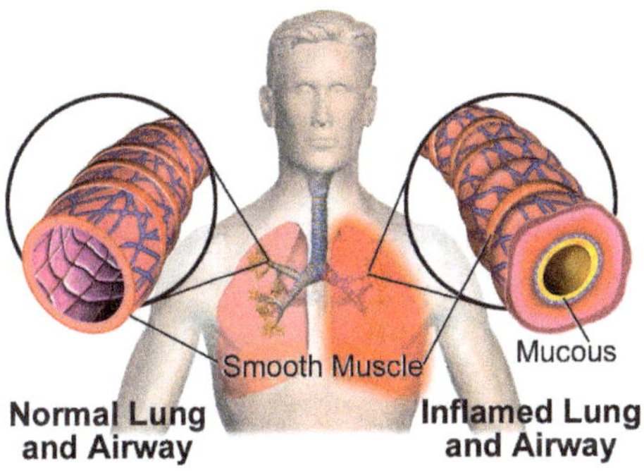

DIAGRAM OF ASTHMA

Diagnosis

While asthma is a well-recognized condition, there is not one universal agreed upon definition. It is defined by the Global Initiative for Asthma as "a chronic inflammatory disorder of the airways in which many cells and cellular elements play a role. The chronic inflammation is associated with airway hyper-responsiveness that leads to recurrent episodes of wheezing, breathlessness, chest tightness and coughing particularly at night or in the early morning. These episodes are usually associated with widespread but variable airflow obstruction within the lung that is often reversible either spontaneously or with treatment".

There is currently no precise test for the diagnosis, which is typically based on the pattern of symptoms and response to therapy over time. A diagnosis of asthma should be suspected if there is a history of recurrent wheezing, coughing or difficulty breathing and these symptoms occur or worsen due to exercise, viral infections, allergens or air pollution.Spirometry is then used to confirm the diagnosis. In children under the age of six the diagnosis is more difficult as they are too young for spirometry.

Spirometry

Spirometry is recommended to aid in diagnosis and management. It is the single best test for asthma. If the FEV1 measured by this technique improves more than 12% following administration of a bronchodilator such as salbutamol, this is supportive of the diagnosis. It however may be normal in those with a history of mild asthma, not currently acting up. As caffeine is a bronchodilator in people with asthma, the use of caffeine before a lung function test may interfere with the results. Single-breath diffusing capacity can help differentiate asthma from COPD. It is reasonable to perform spirometry every one or two years to follow how well a person's asthma is controlled.

Others

The methacholine challenge involves the inhalation of increasing concentrations of a substance that causes airway narrowing in those predisposed. If negative it means that a person does not have asthma; if positive, however, it is not specific for the disease.

Other supportive evidence includes: a ≥20% difference in peak expiratory flow rate on at least three days in a week for at least two weeks, a ≥20% improvement of peak flow following treatment with either salbutamol, inhaled corticosteroids or prednisone, or a ≥20% decrease in peak flow following exposure to a trigger. Testing peak expiratory flow is more variable than spirometry, however, and thus not recommended for routine diagnosis. It may be useful for daily self-monitoring in those with moderate to severe disease and for checking the effectiveness of new medications. It may also be helpful in guiding treatment in those with acute exacerbations.

Classification

Clinical classification (≥ 12 years old)					
Severity	**Symptom frequency**	**Night time symptoms**	**%FEV$_1$ of predicted**	**FEV$_1$ Variability**	**SABA use**
Intermittent	≤2/week	≤2/month	≥80%	<20%	≤2 days/ week
Mild persistent	>2/week	3–4/month	≥80%	20–30%	>2 days/ week
Moderate persistent	Daily	>1/week	60–80%	>30%	daily
Severe persistent	Continuously	Frequent (7×/week)	<60%	>30%	≥twice/day

Asthma is clinically classified according to the frequency of symptoms, forced expiratory volume in one second (FEV$_1$), and peak expiratory flow rate. Asthma may also be classified as atopic (extrinsic) or non-atopic (intrinsic), based on whether symptoms are precipitated by allergens (atopic) or not (non-atopic). While asthma is classified based on severity, at the moment there is no clear method for classifying different subgroups of asthma beyond this system. Finding ways to identify subgroups that respond well to different types of treatments is a current critical goal of asthma research.

Although asthma is a chronic obstructive condition, it is not considered as a part of chronic obstructive pulmonary disease as this term refers specifically to combinations of disease that are irreversible such as bronchiectasis, chronic bronchitis, and emphysema. Unlike these diseases, the airway obstruction in asthma is usually reversible; however, if left untreated, the chronic inflammation from asthma can lead the lungs to become irreversibly obstructed due to airway remodeling. In contrast to emphysema, asthma affects the bronchi, not the alveoli.

Asthma Exacerbation

<table>
<tr><th colspan="3">Severity of an acute exacerbation</th></tr>
<tr><td>Near-fatal</td><td colspan="2">High $PaCO_2$ and/or requiring mechanical ventilation</td></tr>
<tr><td rowspan="9">Life-threatening (any one of)</td><td></td><td></td></tr>
<tr><td>Clinical signs</td><td>Measurements</td></tr>
<tr><td>Altered level of consciousness</td><td>Peak flow < 33%</td></tr>
<tr><td>Exhaustion</td><td>Oxygen saturation < 92%</td></tr>
<tr><td>Arrhythmia</td><td>PaO_2 < 8 kPa</td></tr>
<tr><td>Low blood pressure</td><td>“Normal” $PaCO_2$</td></tr>
<tr><td>Cyanosis</td><td></td></tr>
<tr><td>Silent chest</td><td></td></tr>
<tr><td>Poor respiratory effort</td><td></td></tr>
<tr><td rowspan="5">Acute severe (any one of)</td><td></td><td></td></tr>
<tr><td colspan="2">Peak flow 33–50%</td></tr>
<tr><td colspan="2">Respiratory rate ≥ 25 breaths per minute</td></tr>
<tr><td colspan="2">Heart rate ≥ 110 beats per minute</td></tr>
<tr><td colspan="2">Unable to complete sentences in one breath</td></tr>
<tr><td rowspan="3">Moderate</td><td colspan="2">Worsening symptoms</td></tr>
<tr><td colspan="2">Peak flow 50–80% best or predicted</td></tr>
<tr><td colspan="2">No features of acute severe asthma</td></tr>
</table>

An acute asthma exacerbation is commonly referred to as an *asthma attack*. The classic symptoms are shortness of breath, wheezing, and chest tightness. The wheezing is most often when breathing out. While these are the primary symptoms of asthma, some people present primarily with coughing, and in severe cases, air motion may be significantly impaired such that no wheezing is heard. In children, chest pain is often present.

Signs which occur during an asthma attack include the use of accessory muscles of respiration (sternocleidomastoid and scalene muscles of the neck), there may be a paradoxical pulse (a pulse that is weaker during inhalation and stronger during exhalation), and over-inflation of the chest. A blue color of the skin and nails may occur from lack of oxygen.

In a mild exacerbation the peak expiratory flow rate (PEFR) is ≥200 L/min or ≥50% of the predicted best. Moderate is defined as between 80 and 200 L/min or 25% and 50% of the predicted best while severe is defined as ≤ 80 L/min or ≤25% of the predicted best.

Acute severe asthma, previously known as status asthmaticus, is an acute exacerbation of asthma that does not respond to standard treatments of bronchodilators and corticosteroids. Half of cases are due to infections with others caused by allergen, air pollution, or insufficient or inappropriate medication use.

Brittle asthma is a kind of asthma distinguishable by recurrent, severe attacks. Type 1 brittle asthma is a disease with wide peak flow variability, despite intense medication. Type 2 brittle asthma is background well-controlled asthma with sudden severe exacerbations.

Exercise-Induced

Exercise can trigger bronchoconstriction both in people with or without asthma. It occurs in most people with asthma and up to 20% of people without asthma. Exercise-induced bronchoconstriction is common in professional athletes. The highest rates are among cyclists (up to 45%), swimmers, and cross-country skiers. While it may occur with any weather conditions it is more common when it is dry and cold. Inhaled beta2-agonists do not appear to improve athletic performance among those without asthma however oral doses may improve endurance and strength.

Occupational

Asthma as a result of (or worsened by) workplace exposures, is a commonly reported occupational disease. Many cases however are not reported or recognized as such. It is estimated that 5–25% of asthma cases in adults are work–related. A few hundred different agents have been implicated with the most common being: isocyanates, grain and wood dust, colophony, soldering flux, latex, animals, and aldehydes. The employment associated with the highest risk of problems include: those who spray paint, bakers and those who process food, nurses, chemical workers, those who work with animals, welders, hairdressers and timber workers.

Aspirin-Induced Asthma

Aspirin-exacerbated respiratory disease, also known as aspirin-induced asthma, affects up to 9% of asthmatics. Reactions may also occur to other NSAIDs. People affected often also have trouble with nasal polyps. In people who are affected low doses paracetamol or COX-2 inhibitors are generally safe.

Alcohol-Induced Asthma

Alcohol may worsen asthmatic symptoms in up to a third of people. This may be even more common in some ethnic groups such as the Japanese and those with aspirin-induced asthma. Other studies have found improvement in asthmatic symptoms from alcohol.

Nonallergic Asthma

Nonallergic asthma, also known as intrinsic or nonatopic asthma makes up between 10 and 33% of cases. There is negative skin test to common inhalant allergens and normal serum concentrations of IgE. Often it starts later in life and women are more commonly affected than men. Usual treatments may not work as well.

Differential Diagnosis

Many other conditions can cause symptoms similar to those of asthma. In children, other upper airway diseases such as allergic rhinitis and sinusitis should be considered as well as other causes of airway obstruction including: foreign body aspiration, tracheal stenosis or laryngotracheomalacia, vascular rings, enlarged lymph nodes or neck masses. Bronchiolitis and other viral infections may also produce wheezing. In adults, COPD, congestive heart failure, airway masses, as well as drug-induced coughing due to ACE inhibitors should be considered. In both populations vocal cord dysfunction may present similarly.

Chronic obstructive pulmonary disease can coexist with asthma and can occur as a complication of chronic asthma. After the age of 65, most people with obstructive airway disease will have asthma and COPD. In this setting, COPD can be differentiated by increased airway neutrophils, abnormally increased wall thickness, and increased smooth muscle in the bronchi. However, this level of investigation is not performed due to COPD and asthma sharing similar principles of management: corticosteroids, long-acting beta-agonists, and smoking cessation. It closely resembles asthma in symptoms, is correlated with more exposure to cigarette smoke, an older age, less symptom reversibility after bronchodilator administration, and decreased likelihood of family history of atopy.

Prevention

The evidence for the effectiveness of measures to prevent the development of asthma is weak. Some show promise including: limiting smoke exposure both in utero and after delivery, breastfeeding, and increased exposure to daycare or large families but none are well supported enough to be recommended for this indication. Early pet exposure may be useful.Results from exposure to pets at other times are inconclusive and it is only recommended that pets be removed from the home if a person has allergic symptoms to said pet. Dietary restrictions during pregnancy or when breast feeding have not been found to be effective and thus are not recommended. Reducing or eliminating compounds known to sensitive people from the work place may be effective. It is not clear if annual influenza vaccinations effects the risk of exacerbations. Immunization; however, is recommended by the World Health Organization. Smoking bans are effective in decreasing exacerbations of asthma.

Management

While there is no cure for asthma, symptoms can typically be improved. A specific, customized plan for proactively monitoring and managing symptoms should be created. This plan should include the reduction of exposure to allergens, testing to assess the severity of symptoms, and the usage of medications. The treatment plan should be written down and advise adjustments to treatment according to changes in symptoms.

The most effective treatment for asthma is identifying triggers, such as cigarette smoke, pets, or aspirin, and eliminating exposure to them. If trigger avoidance is insufficient, the use of medication is recommended. Pharmaceutical drugs are selected based on, among other things, the severity of illness and the frequency of symptoms. Specific medications for asthma are broadly classified into fast-acting and long-acting categories.

Bronchodilators are recommended for short-term relief of symptoms. In those with occasional attacks, no other medication is needed. If mild persistent disease is present (more than two attacks a week), low-dose inhaled corticosteroids or alternatively, an oral leukotriene antagonist or a mast cell stabilizer is recommended. For those who have daily attacks, a higher dose of inhaled corticosteroids is used. In a moderate or severe exacerbation, oral corticosteroids are added to these treatments.

Lifestyle Modification

Avoidance of triggers is a key component of improving control and preventing attacks. The most common triggers include allergens, smoke (tobacco and other), air pollution, non selective beta-blockers, and sulfite-containing foods. Cigarette smoking and second-hand smoke (passive smoke) may reduce the effectiveness of medications such as corticosteroids. Laws that limit smoking decrease the number of people hospitalized for asthma. Dust mite control measures, including air filtration, chemicals to kill mites, vacuuming, mattress covers and others methods had no effect on asthma symptoms. Overall, exercise is beneficial in people with stable asthma. Yoga could provide small improvements in quality of life and symptoms in people with asthma.

Medications

Medications used to treat asthma are divided into two general classes: quick-relief medications used to treat acute symptoms; and long-term control medications used to prevent further exacerbation. Antibiotics are generally not needed for sudden worsening of symptoms.

Fast–Acting

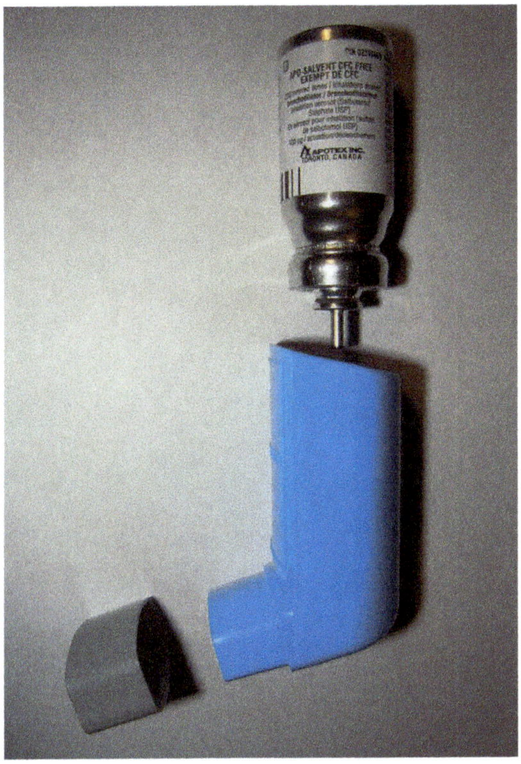

Salbutamol metered dose inhaler commonly used to treat asthma attacks.

- Short-acting beta$_2$-adrenoceptor agonists (SABA), such as salbutamol (*albuterol* USAN) are the first line treatment for asthma symptoms.They are recommended before exercise in those with exercise induced symptoms.
- Anticholinergic medications, such as ipratropium bromide, provide additional benefit when used in combination with SABA in those with moderate or severe symptoms. Anticholinergic bronchodilators can also be used if a person cannot tolerate a SABA. If a child requires admission to hospital additional ipratropium does not appear to help over a SABA.
- Older, less selective adrenergic agonists, such as inhaled epinephrine, have similar efficacy to SABAs. They are however not recommended due to concerns regarding excessive cardiac stimulation.

Long–term Control

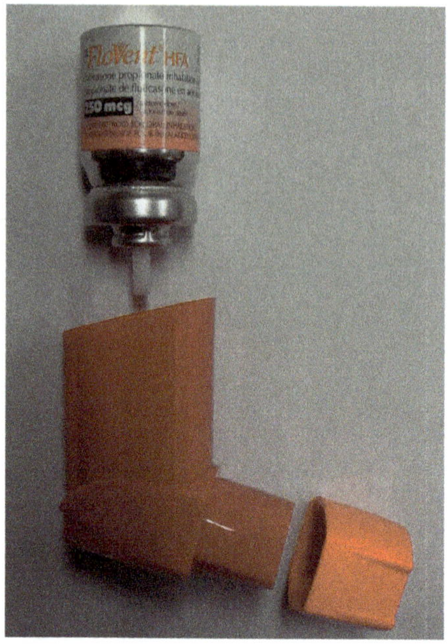

Fluticasone propionate metered dose inhaler commonly used for long-term control.

- Corticosteroids are generally considered the most effective treatment available for long-term control. Inhaled forms such as beclomethasone are usually used except in the case of severe persistent disease, in which oral corticosteroids may be needed. It is usually recommended that inhaled formulations be used once or twice daily, depending on the severity of symptoms.
- Long-acting beta-adrenoceptor agonists (LABA) such as salmeterol and formoterol can improve asthma control, at least in adults, when given in combination with inhaled corticosteroids. In children this benefit is uncertain. When used without steroids they increase the risk of severe side-effects and even with corticosteroids they may slightly increase the risk.
- Leukotriene receptor antagonists (such as montelukast and zafirlukast) may be used in addition to inhaled corticosteroids, typically also in conjunction with a LABA. Evidence is insufficient to support use in acute exacerbations. In children they appear to be of little benefit when added to inhaled steroids, and the same applies in adolescents and adults. They are useful by themselves. In those under five years of age, they were the preferred add-on therapy after inhaled corticosteroids by the British Thoracic Society in 2009. A similar class of drugs, 5-LOX inhibitors, may be used as an alternative in the chronic treatment of mild to moderate asthma among older children and adults. As of 2013 there is one medication in this family known as zileuton.
- Mast cell stabilizers (such as cromolyn sodium) are another non-preferred alternative to corticosteroids.

Delivery Methods

Medications are typically provided as metered-dose inhalers (MDIs) in combination with an asthma spacer or as a dry powder inhaler. The spacer is a plastic cylinder that mixes the medication with air, making it easier to receive a full dose of the drug. A nebulizer may also be used. Nebulizers and spacers are equally effective in those with mild to moderate symptoms. However, insufficient evidence is available to determine whether a difference exists in those with severe disease.

Adverse Effects

Long-term use of inhaled corticosteroids at conventional doses carries a minor risk of adverse effects. Risks include the development of cataracts and a mild regression in stature.

Others

When asthma is unresponsive to usual medications, other options are available for both emergency management and prevention of flareups. For emergency management other options include:

- Oxygen to alleviate hypoxia if saturations fall below 92%.
- Corticosteroid by mouth are recommended with five days of prednisone being the same 2 days of dexamethasone.
- Magnesium sulfate intravenous treatment increases bronchodilation when used in addition to other treatment in moderate severe acute asthma attacks. In adults it results in a reduction of hospital admissions.
- Heliox, a mixture of helium and oxygen, may also be considered in severe unresponsive cases.
- Intravenous salbutamol is not supported by available evidence and is thus used only in extreme cases.
- Methylxanthines (such as theophylline) were once widely used, but do not add significantly to the effects of inhaled beta-agonists. Their use in acute exacerbations is controversial.
- The dissociative anesthetic ketamine is theoretically useful if intubation and mechanical ventilation is needed in people who are approaching respiratory arrest; however, there is no evidence from clinical trials to support this. It is unclear if non-invasive positive pressure ventilation in children is of use as it has not been sufficiently studied.

For those with severe persistent asthma not controlled by inhaled corticosteroids and LABAs, bronchial thermoplasty may be an option. It involves the delivery of controlled thermal energy to the airway wall during a series of bronchoscopies. While it may increase exacerbation frequency in the first few months it appears to decrease the subsequent rate. Effects beyond one year are unknown. Evidence suggests that sublingual immunotherapy in those with both allergic rhinitis and asthma improve outcomes.

Alternative Medicine

Many people with asthma, like those with other chronic disorders, use alternative treatments; surveys show that roughly 50% use some form of unconventional therapy. There is little data to support the effectiveness of most of these therapies. Evidence is insufficient to support the usage of Vitamin C. There is tentative support for its use in exercise induced brochospasm. In people with mild to moderate asthma, treatment with vitamin D supplementation is likely to reduce the risk of asthma exacerbations.

Acupuncture is not recommended for the treatment as there is insufficient evidence to support its use. Air ionisers show no evidence that they improve asthma symptoms or benefit lung function; this applied equally to positive and negative ion generators.

Manual therapies, including osteopathic, chiropractic, physiotherapeutic and respiratory therapeutic maneuvers, have insufficient evidence to support their use in treating asthma. The Buteyko breathing technique for controlling hyperventilation may result in a reduction in medication use; however, the technique does not have any effect on lung function. Thus an expert panel felt that evidence was insufficient to support its use.

Prognosis

Globally it causes moderate or severe disability in 19.4 million people as of 2004 (16 million of which are in low and middle income countries). Of asthma diagnosed during childhood, half of cases will no longer carry the diagnosis after a decade. Airway remodeling is observed, but it is unknown whether these represent harmful or beneficial changes. Early treatment with corticosteroids seems to prevent or ameliorates a decline in lung function. Asthma in children also has negative effects on quality of life of their parents.

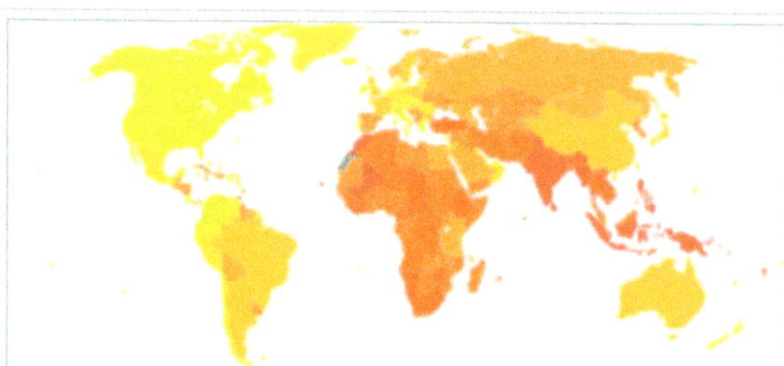

Asthma deaths per million persons in 2012

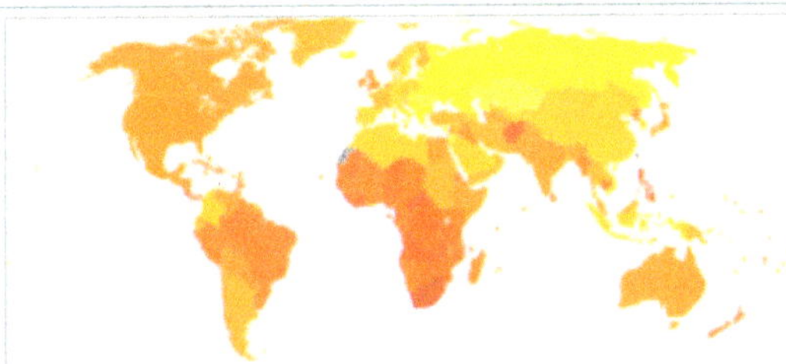

Disability-adjusted life year for asthma per 100,000 inhabitants in 2004.[177]

no data
<100
100–150
150–200
200–250
250–300
300–350
350–400
400–450
450–500
500–550
550–600
>600

The prognosis for asthma is generally good, especially for children with mild disease. Mortality has decreased over the last few decades due to better recognition and improvement in care. In 2010 the death rate was 170 per million for males and 90 per million for females. Rates very between countries by 100 fold.

Epidemiology

As of 2011, 235–330 million people worldwide are affected by asthma, and approximately 250,000–345,000 people die per year from the disease. Rates vary between countries with prevalences between 1 and 18%. It is more common in developed than developing countries. One thus sees lower rates in Asia, Eastern Europe and Africa. Within developed countries it is more common in those who are economically disadvantaged while in contrast in developing countries it is more common in the affluent.The reason for these differences is not well known. Low and middle income countries make up more than 80% of the mortality.

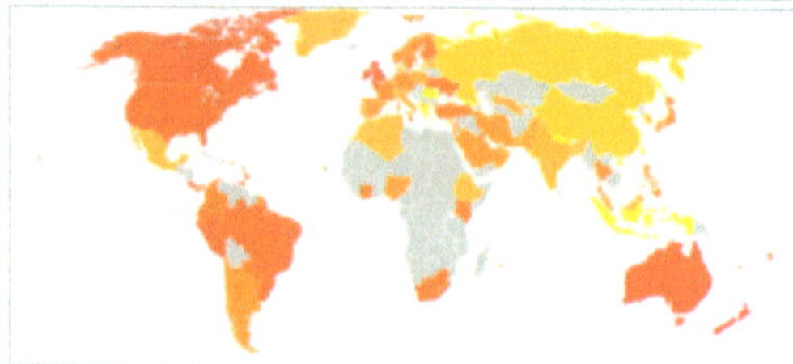

Rates of asthma in different countries of the world as of 2004.

- no data
- <1%
- 1–2%
- 2–3%
- 3–4%
- 4–5%
- 5–6%
- 6–7%
- 7–8%
- 8–10%
- 10–12.5%
- 12.5–15%
- >15%

While asthma is twice as common in boys as girls, severe asthma occurs at equal rates.In contrast adult women have a higher rate of asthma than men and it is more common in the young than the old. In children, asthma was the most common reason for admission to the hospital following an emergency department visit in the US in 2011.

Global rates of asthma have increased significantly between the 1960s and 2008 with it being recognized as a major public health problem since the 1970s. Rates of asthma have plateaued in the developed world since the mid-1990s with recent increases primarily in the developing world. Asthma affects approximately 7% of the population of the United States and 5% of people in the United Kingdom. Canada, Australia and New Zealand have rates of about 14–15%.

Economics

From 2000 to 2010, the average cost per asthma-related hospital stay in the United States for children remained relatively stable at about $3,600, whereas the average cost per asthma-related hospital stay for adults increased from $5,200 to $6,600. In 2010, Medicaid was the most frequent primary payer among children and adults aged 18–44 years in the United States; private insurance was the second most frequent payer. Among both children and adults in the lowest income communities in the United States there is a higher rate of hospital stays for asthma in 2010 than those in the highest income communities.

History

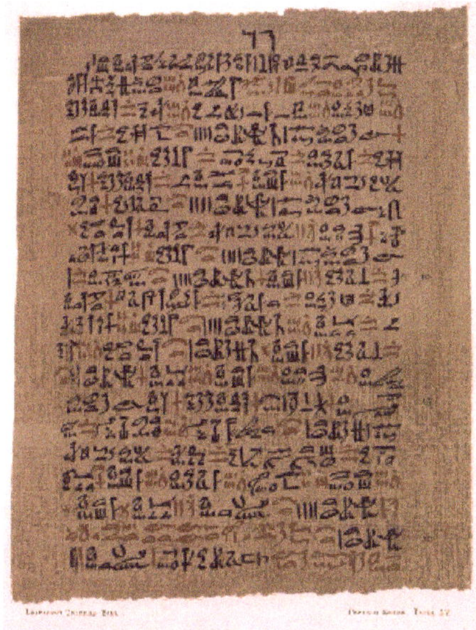

Ebers Papyrus detailing treatment of asthma

Asthma was recognized in Ancient Egypt and was treated by drinking an incense mixture known as kyphi. It was officially named as a specific respiratory problem by Hippocrates circa 450 BC, with the Greek word for "panting" forming the basis of our modern name. In 200 BC it was believed to be at least partly related to the emotions.

In 1873, one of the first papers in modern medicine on the subject tried to explain the pathophysiology of the disease while one in 1872, concluded that asthma can be cured by rubbing the chest with chloroform liniment. Medical treatment in 1880, included the use of intravenous doses of a drug called pilocarpin. In 1886, F.H. Bosworth theorized a connection between asthma and hay fever. Epinephrine was first referred to in the treatment of asthma in 1905. Oral corticosteroids began to be used for this condition in the 1950s while inhaled corticosteroids and selective short acting beta agonist came into wide use in the 1960s.

A notable and well-documented case in the 19th century was that of young Theodore Roosevelt (1858–1919). At that time there was no effective treatment. Roosevelt's youth was in large part shaped by his poor health partly related to his asthma. He experienced recurring nighttime asthma attacks that caused the experience of being smothered to death, terrifying the boy and his parents.

During the 1930s to 1950s, asthma was known as one of the "holy seven" psychosomatic illnesses. Its cause was considered to be psychological, with treatment often based on psychoanalysis and other talking cures. As these psychoanalysts interpreted the asthmatic wheeze as the suppressed cry of the child for its mother, they considered the treatment of depression to be especially important for individuals with asthma.

1907 advertisement for Grimault's Indian Cigarettes, emphasising their alleged efficacy for the relief of asthma.

Beta Thalassemia

Beta thalassemias (β thalassemias) are a group of inherited blood disorders. They are forms of thalassemia caused by reduced or absent synthesis of the beta chains of hemoglobin that result in variable outcomes ranging from severe anemia to clinically asymptomatic individuals. Global annual incidence is estimated at 1 in 100,000. Beta thalassemias are caused by mutations in the HBB gene on chromosome 11, inherited in an autosomal recessive fashion. The severity of the disease depends on the nature of the mutation.

HBB blockage over time leads to decreased beta-chain synthesis. The body's inability to construct new beta-chains leads to the underproduction of HbA. Reductions in HbA available overall to fill the red blood cells in turn leads to microcytic anemia. Microcytic anemia ultimately develops in respect to inadequate HBB for sufficient red blood cell functioning. Due to this factor, the patient may require blood transfusions to make up for the blockage in the beta-chains. Repeated blood transfusions can lead to build-up of iron overload, ultimately resulting in iron toxicity. This iron toxicity can cause various problems, including myocardial siderosis and heart failure leading to the patient's deth.

Signs and Symptoms

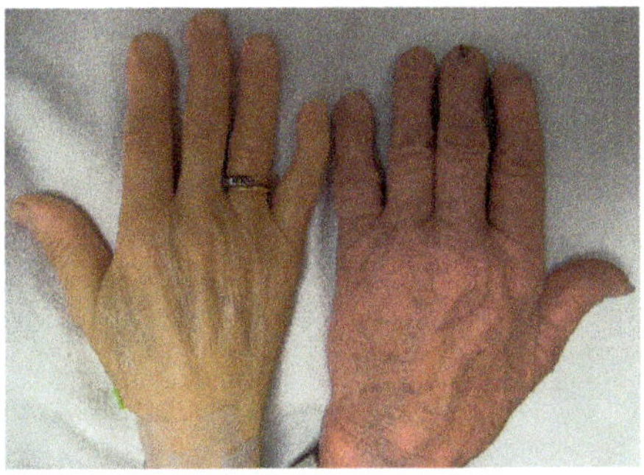

The hand of a person with severe anemia compared to one without

Three main forms have been described: thalassemia major, thalassemia intermedia and thalassemia minor. All people with thalassemia are susceptible to health complications that involve the spleen (which is often enlarged and frequently removed) and gallstones. These complications are mostly found in thalassemia major and intermedia patients. Individuals with beta thalassemia major usually present within the first two years of life with severe anemia, poor growth and skeletal abnormalities during infancy. Untreated thalassemia major eventually leads to death, usually by heart failure; therefore, birth screening is very important.

Excess iron causes serious complications within the liver, heart and endocrine glands. Severe symptoms include liver cirrhosis, liver fibrosis and in extreme cases, liver cancer. Heart failure, growth impairment, diabetes and osteoporosis are life-threatening contributors brought upon by TM.The main cardiac abnormalities seen to have resulted from thalassemia and iron overload include left ventricular systolic and diastolic dysfunction, pulmonary hypertension, valveulopathies, arrhythmias and pericarditis. Increased gastrointestinal iron absorption is seen in all grades of beta thalassemia and increased red blood cell destruction by the spleen due to ineffective erythropoiesis further releases additional iron into the bloodstream.

Cause

Mutations

Two major groups of mutations can be distinguished:

- Nondeletion forms: These defects, in general, involve a single base substitution or small deletion or inserts near or upstream of the β globin gene. Most often, mutations occur in the promoter regions preceding the beta-globin genes. Less often, abnormal splice variants are believed to contribute to the disease.
- Deletion forms: Deletions of different sizes involving the β globin gene produce different syndromes such as (β^{o}) or hereditary persistence of fetal hemoglobin syndromes.

Name	Older synonyms	Description	Alleles
Thalassemia minor		Heterozygous form: Only one of β globin alleles bears a mutation. Individuals will suffer from microcytic anemia. Detection usually involves lower than normal MCV value (<80 fL).	$β^+/β$ $β^o/β$
Thalassemia intermedia		Affected individuals can often manage a normal life but may need occasional transfusions, e.g., at times of illness or pregnancy, depending on the severity of their anemia.	$β^+/β^+$ $β^o/β^+$
Thalassemia major	Mediterranean anemia; Cooley anemia	Homozygous form: Occurs when both alleles have thalassemia mutations. This is a severe microcytic, hypochromic anemia. Untreated, it causes anemia, splenomegaly and severe bone deformities. It progresses to death before age 20. Treatment consists of periodic blood transfusion; splenectomy for splenomegaly and chelation of transfusion-caused iron overload.	$β^o/β^o$

mRNA Assembly

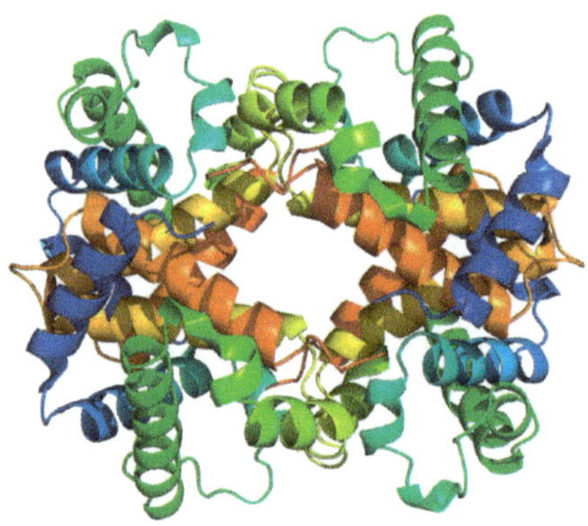

Protein HBB PDB 1a00: this is a healthy Beta Globin Protein

Beta thalassemia is a hereditary disease affecting hemoglobin. As with about half of all hereditary diseases, an inherited mutation damages the assembly of the messenger-type RNA (mRNA) that is transcribed from a chromosome. DNA contains both the instructions (genes) for stringing amino acids together into proteins, as well as stretches of DNA that play important roles in regulating produced protein levels.

In thalassemia, an additional, contiguous length or a discontinuous fragment of non-coding instructions are included in the mRNA. This happens because the mutation obliterates the boundary between the intronic and exonic portions. Because all the coding sections may still be present, normal hemoglobin may be produced and the added material, if it produces pathology, instead disrupts regulatory functions enough to produce anemia. Hemoblogin's normal alpha and beta subunits each have an iron-containing central portion (heme) that allows the protein chain of a subunit to fold around it. Normal adult hemoglobin contains 2 alpha and 2 beta subunits. Thalassemias typically affect only the mRNAs for production of the beta chains (hence the name). Since the mutation may be a change in only a single base (a "Single Nucleotide Polymorphism"), on-going efforts seek gene therapies to make that single correction.

Diagnosis

Abdominal pain due to hypersplenism and splenic infarction and right-upper quadrant pain caused by gallstones are major clinical manifestations. However, diagnosing thalassemiæ from symptoms alone is inadequate. Physicians note these signs as associative due to this disease's complexity. The following associative signs can attest to the severity of the phenotype: pallor, poor growth, inadequate food intake, splenomegaly, jaundice, maxillary hyperplasia, dental malocclusion, cholelithiasis, systolic ejection murmur in the presence of severe anemia and pathologic fractures. Based on symptoms, tests are ordered for a differential diagnosis. These tests include complete blood count; hemoglobin electrophoresis; serum transferrin, ferritin, total iron-binding capacity; urine urobilin and urobilogen; peripheral blood smear, which may show codocytes, or target cells; hematocrit; and serum bilirubin.

DNA Analysis

All beta thalassemias may exhibit abnormal red blood cells, a family history is followed by DNA analysis. This test is used to investigate deletions and mutations in the alpha- and beta-globin-producing genes. Family studies can be done to evaluate carrier status and the types of mutations present in other family members. DNA testing is not routine, but can help diagnose thalassemia and determine carrier status. In most cases the treating physician uses a clinical prediagnosis assessing anemia symptoms: fatigue, breathlessness and poor exercise tolerance. Further genetic analysis may include HPLC should routine electrophoresis prove difficult.

Treatment

Beta Thalassemia Major

Affected children require regular lifelong blood transfusion and can have complications, which may involve the spleen. Bone marrow transplants can be curative for some children. Patients receive frequent blood transfusions that lead to or potentiate iron overload. Iron chelation treatment is necessary to prevent damage to internal organs. Advances in iron chelation treatments allow patients with thalassemia major to live long lives with access to proper treatment. Popular chelators include deferoxamine and deferiprone.

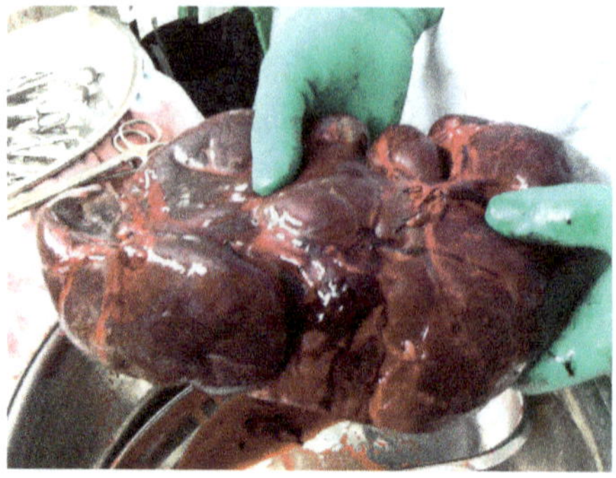

Surgically removed spleen of a Thalassemic child. It is about 15 times larger than normal.

The most common patient deferoxamine complaint is that they are painful and inconvenient. The oral chelator deferasirox was approved for use in 2005 in some countries, it offers some hope with compliance at a higher cost. Bone marrow transplantation is the only cure and is indicated for patients with severe thalassemia major. Transplantation can eliminate a patient's dependence on transfusions. Absent a matching donor, a savior sibling can be conceived by preimplantation genetic diagnosis (PGD) to be free of the disease as well as to match the recipient's human leukocyte antigen (HLA) type.

Scientists at Weill Cornell Medical College have developed a gene therapy strategy that could feasibly treat both beta-thalassemia and sickle cell disease. The technology is based on delivery of a lentiviral vector carrying both the human β-globin gene and an ankyrin insulator to improve gene transcription and translation, and boost levels of β-globin production.

Beta Thalassemia Intermedia

Patients may require episodic blood transfusions. Transfusion-dependent patients develop iron overload and require chelation therapy to remove the excess iron. Transmission is autosomal recessive; however, dominant mutations and compound heterozygotes have been reported. Genetic counseling is recommended and prenatal diagnosis may be offered. Alleles without a mutation that reduces function are characterized as (β). Mutations are characterized as (βo) if they prevent any formation of β chains, mutations are characterized as (β+) if they allow some β chain formation to occur.

Beta Thalassemia Minor

Patients are often monitored without treatment. While many of those with minor status do not require transfusion therapy, they still risk iron overload, particularly in the liver. A serum ferritin test checks iron levels and can point to further treatment. Although not life-threatening on its own, it can affect quality of life due to the anemia. Minor often coexists with other conditions such as asthma and can cause iron overload of the liver and in those with non-alcoholic fatty liver disease, lead to more severe outcomes.

Epidemiology

The beta form of thalassemia is particularly prevalent among the Mediterranean peoples and this geographical association is responsible for its naming: *thalassa* (*θάλασσα*) is the Greek word for sea and *haema* (*αμα*) is the Greek word for blood. In Europe, the highest concentrations of the disease are found in Greece and the Turkish coastal regions. The major Mediterranean islands (except the Balearics) such as Sicily, Sardinia, Corsica, Cyprus, Malta and Crete are heavily affected in particular. Other Mediterranean peoples, as well as those in the vicinity of the Mediterranean, also have high incidence rates, including people from West Asia and North Africa. The data indicate that 15% of the Greek and Turkish Cypriots are carriers of beta-thalassaemia genes, while 10% of the population carry alpha-thalassaemia genes.

Evolutionary Adaptation

The thalassemia trait may confer a degree of protection against malaria, which is or was prevalent in the regions where the trait is common, thus conferring a selective survival advantage on carriers (known as heterozygous advantage), thus perpetuating the mutation. In that respect, the various thalassemias resemble another genetic disorder affecting hemoglobin, sickle-cell disease.

REFERENCES

1. Weismann A. (1889). Essays upon heredity and kindred biological problems. Oxford: Clarendon Press.
2. Yang, Jiang-Nan (2013). "Viscous populations evolve altruistic programmed ageing in ability conflict in a changing environment". Evolutionary Ecology Research. 15: 527–543.
3. Fabian, Daniel; Flatt, Thomas (2011). "The Evolution of Aging". Scitable. Nature Publishing Group. Retrieved May 20, 2014.
4. Medawar, P.B. (1952). An Unsolved Problem of Biology (PDF). London: H.K. Lewis. Edney, E.B. and Gill, R.W. 1968. Delineates the theory of mutation accumulation.
5. Edney EB, Gill RW (October 1968). "Evolution of senescence and specific longevity". Nature. 220 *(5164): 281–2. Bibcode*:1968Natur.220..281E. doi:10.1038/220281a0. PMID 5684860. Further describes theory of mutation accumulation.
6. Guarente L, Kenyon C (November 2000). "Genetic pathways that regulate ageing in model organisms". Nature. 408 *(6809): 255–62.* doi:10.1038/35041700. PMID 11089983. Shows similarities between ageing genes in model organisms.
7. Williams, G.C. (1957). "Pleiotropy, natural selection and the evolution of senescence" (PDF). Evolution. 11 *(4): 398–411.* doi:10.2307/2406060. JSTOR 2406060. Paper in which Williams describes his theory of antagonistic pleiotropy.

8. Leroi, A.M.; Chippindale, A.K.; Rose, M.R. (1994). "Long-term laboratory evolution of a genetic life-history tradeoff in Drosophila melanogaster. 1. The role of genotype-by-environment interaction". Evolution. 48 *(4): 1244–57.* doi:10.2307/2410382. JSTOR 2410383.

9. Kirkwood TB (November 1977). "Evolution of ageing". Nature. 270 *(5635): 301–4. Bibcode*:1977Natur.270..301K. doi:10.1038/270301a0. PMID 593350. Origin of the disposable soma theory.

10. Lorenzini, A, Stamato, T. Sell, C. (2011). "The disposable soma theory revisited: Time as a resource in the theories of aging". 15. doi:10.4161/cc.10.22.18302.

11. Weindruch, R.; Walford, R.L. (1986). The Retardation of Aging and Disease by Dietary Restriction. Springfield, IL: Thomas.

12. Weindruch R (1996). "The Retardation of Aging by Caloric Restriction: Studies in Rodents and Primates". Toxicologic Pathology. 24 *(6): 742–5.* doi:10.1177/019262339602400618. PMID 8994305.

13. Masoro EJ (September 2005). "Overview of caloric restriction and ageing". Mech. Ageing Dev. 126 *(9): 913–22. doi:10.1016/j.mad.2005.03.012. PMID 15885745.* Overview of caloric restriction and aging.

14. Kriete, A. (2013). "Robustness and aging-a systems-level perspective". Biosystems. 112: 37–48. doi:10.1016/j.biosystems.2013.03.014. PMID 23562399.

15. Gourlay CW, Du W, Ayscough KR (December 2006). "Apoptosis in yeast--mechanisms and benefits to a unicellular organism". Mol. Microbiol. 62 *(6): 1515–21. doi:10.1111/j.1365-2958.2006.05486.x. PMID 17087770.*

16. Blasco, M.; Pelengaris, S. (2006-02-28). The molecular biology of cancer?. Blackwell. p. 285. ISBN 978-1-4051-1814-9. OCLC 263712202. This has resulted in speculation that cellular senescence evolved as a cancer suppression mechanism at a time when the life expectancy for humans was far shorter than it is today.

17. Stewart, S. A.; Weinberg, R. A. (2002). "Does senescence function as an anti-neoplastic mechanism in vivo?" (PDF). Oncogene. 21 *(4): 627–630. doi:10.1038/* sj/onc/1205062.

18. Clark, W.R. (1999). A Means to an End: The biological basis of aging and death. New York: Oxford University Press.

19. Mitteldorf, J. (2004). "Ageing selected for its own sake". Evol. Ecol. Res. 6: 937–53.

20. Bredesen DE (October 2004). "The non-existent aging program: how does it work?". *Aging Cell.* 3 *(5): 255–9. doi:10.1111/j.1474-9728.2004.00121.x. PMID 15379848.*

21. Wodinsky, J. (1977). "Hormonal Inhibition of Feeding and Death in Octopus: Control by Optic Gland Secretion". Science. 148 *(4320): 948–51.* doi:10.1126/ science.198.4320.948. PMID 17787564.

22. Wilson, D.S.; Pollock, G.B.; Dugatkin, L.A (1992). "Can altruism evolve in purely viscous populations?". Evol. Ecol. 6: 331–341. doi:10.1007/bf02270969.

23. Taylor, P.D. (1992). "Altruism in viscous populations – an inclusive fitness model.". Evol. Ecol. 6: 352–356. doi:10.1007/bf02270971.

24. Mitteldorf, Joshua; Wilson, D.S. (2000). "Population viscosity and the evolution of altruism" (PDF). J. Theor. Biol. 204: 481–496. doi:10.1006/jtbi.2000.2007.

25. Mitteldorf, J. (2006). "Chaotic population dynamics and the evolution of ageing: proposing a demographic theory of senescence". Evol. Ecol. Res. 8: 561–74.

26. Libertini, G. (2008). "Empirical evidence for various evolutionary hypotheses on species demonstrating increasing mortality with increasing chronological age in the wild". Scientific World Journal. 19 *(8): 182–93. doi:10.1100/tsw.2008.36. PMID 18301820.*

27. Goldsmith TC (June 2008). "Aging, evolvability, and the individual benefit requirement; medical implications of aging theory controversies". J. Theor. Biol. 252 *(4): 764–8. doi:10.1016/j.jtbi.2008.02.035. PMID 18396295.*

28. Skulachev VP (November 1997). "Aging is a specific biological function rather than the result of a disorder in complex living systems: biochemical evidence in support of Weismann's hypothesis". Biochemistry Mosc. 62 *(11): 1191–5. PMID 9467841.*

29. Olshansky, SJ; Hayflick, L; Carnes, BA (2002). "No truth to the fountain of youth". Scientific American. 286 *(6): 92–5.* doi:10.1038/scientificamerican0602-92. PMID 12030096.

30. Holliday R (May 2006). "Aging is no longer an unsolved problem in biology". Ann. N. Y. Acad. Sci. 1067: 1–9. doi:10.1196/annals.1354.002. PMID 16803964.

31. Goldsmith, T. (2009). "Mammal aging: active and passive mechanisms". Journal of Bioscience Hypotheses. 2 *(2): 59–64. doi:10.1016/j.bihy.2008.12.002.*

32. Skulachev, Vladimir P. (2001). "The programmed death phenomena, ageing, and the Samurai law of biology". Experimental Gerontology. 36: 995–1024. doi:10.1016/s0531-5565(01)00109-7.

33. Gavrilov, Leonid A; Gavrilova, Natalia S. (7 February 2002). "Evolutionary Theories of Aging and Longevity". TheScientificWorldJOURNAL. 2 *(2): 339–356.* CiteSeerX *10.1.1.3.4950. doi:10.1100/tsw.2002.96.*

34. Chen D, Guarente L (February 2007). "SIR2: a potential target for calorie restriction mimetics". Trends Mol Med. 13 *(2): 64–71. doi:10.1016/j.molmed.2006.12.004. PMID 17207661.*

Index